ELSEVIER OCEAN ENGINEERING BOOK SERIES

VOLUME 7

TECHNOLOGY AND SAFETY OF MARINE SYSTEMS

Elsevier Internet Homepage:

http://www.elsevier.com

Consult the Elsevier homepage for full catalogue information on all books, journals and electronic products and services.

Elsevier Titles of Related Interest

Published

WATSON
Practical Ship Design
ISBN: 008-042999-8

YOUNG
Wind Generated Ocean Waves
ISBN: 008-043317-0

BAI
Pipelines and Risers
ISBN: 008-043712-5

JENSEN
Local and Global Response of Ships
ISBN: 008-043953-5

TUCKER & PITT
Waves in Ocean Engineering
ISBN: 008-043566-1

Forthcoming titles

BELENKY *et al*
Safety and Stability of Ships
ISBN: 008-043001-5

BOSE & BROOKE
Wave Energy Conversion
ISBN: 008-044212-9

McCORMICK & KRAEMER
Ocean Wave Energy Utilization
ISBN: 008-043932-2

MUNCHMEYER
Boat and Ship Geometry
ISBN: 008-043998-5

OCHI
Hurricane Generated Seas
ISBN: 008-044312-5

Related Journals

Free specimen copy gladly sent on request. Elsevier Ltd, The Boulevard, Langford Lane, Kidlington, Oxford, OX5 1GB, UK

Applied Ocean Research
Advances in Engineering Software
CAD
Coastal Engineering
Composite Structures
Computers and Structures
Construction and Building Materials
Engineering Failure Analysis
Engineering Fracture Mechanics
Engineering Structures
Finite Elements in Analysis and Design
International Journal of Solids and Structures
Journal of Constructional Steel Research
Marine Structures
NDT & E International
Ocean Engineering
Structural Safety
Thin-Walled Structures

To Contact the Publisher

Elsevier welcomes enquiries concerning publishing proposals: books, journal special issues, conference proceedings, etc. All formats and media can be considered. Should you have a publishing proposal you wish to discuss, please contact, without obligation, the publisher responsible for Elsevier's civil and structural engineering publishing programme:

James Sullivan
Publishing Editor
Elsevier Ltd
The Boulevard, Langford Lane
Kidlington, Oxford
OX5 1GB, UK

Phone: +44 1865 843178
Fax: +44 1865 84392
E.mail: j.sullivan@elsevier.com

General enquiries, including placing orders, should be directed to Elsevier's Regional Sales Offices – please access the Elsevier homepage for full contact details (homepage details at the top of this page).

ELSEVIER OCEAN ENGINEERING BOOK SERIES

VOLUME 7

TECHNOLOGY AND SAFETY OF MARINE SYSTEMS

ANAND PILLAY

Lloyds Register

JIN WANG

Liverpool John Moores University, Liverpool, UK

2003

ELSEVIER

Amsterdam - Boston - Heidelberg - London - New York - Oxford - Paris

San Diego - San Francisco - Singapore - Sydney - Tokyo

ELSEVIER SCIENCE Ltd
The Boulevard, Langford Lane
Kidlington, Oxford OX5 1GB, UK

First edition 2003

Library of Congress Cataloging in Publication Data
A catalog record from the Library of Congress has been applied for.

British Library Cataloguing in Publication Data
A catalogue record from the British Library has been applied for.

ISBN: 0 08 044148 3

Transferred to Digital Printing 2008

SERIES PREFACE

In this day and age, humankind has come to the realization that the Earth's resources are limited. In the 19th and 20th Centuries, these resources have been exploited to such an extent that their availability to future generations is now in question. In an attempt to reverse this march towards self-destruction, we have turned our attention to the oceans, realizing that these bodies of water are both sources for potable water, food and minerals and are relied upon for World commerce. In order to help engineers more knowledgeably and constructively exploit the oceans, the **Elsevier Ocean Engineering Book Series** has been created.

The **Elsevier Ocean Engineering Book Series** gives experts in various areas of ocean technology the opportunity to relate to others their knowledge and expertise. In a continual process, we are assembling world-class technologists who have both the desire and the ability to write books. These individuals select the subjects for their books based on their educational backgrounds and professional experiences.

The series differs from other ocean engineering book series in that the books are directed more towards technology than science, with a few exceptions. Those exceptions we judge to have immediate applications to many of the ocean technology fields. Our goal is to cover the broad areas of naval architecture, coastal engineering, ocean engineering acoustics, marine systems engineering, applied oceanography, ocean energy conversion, design of offshore structures, reliability of ocean structures and systems and many others. The books are written so that readers entering the topic fields can acquire a working level of expertise from their readings.

We hope that the books in the series are well-received by the ocean engineering community.

Rameswar Bhattacharyya
Michael E. McCormick

Series Editors

PREFACE

Traditionally, society has regulated hazardous industries by references to engineering codes and standards, and by detailed regulations specifying hardware requirements. Now the trend is to adopt a risk based approach to meeting health, safety and environmental criteria. In such an approach, risk analysis plays a key role as it identifies hazards, categorises the risk and thus provides decision support concerning the choice of arrangements and measures. These are the risk reduction arrangements and measures needed to meet the "safe" yet economical operating level, where the fatality and/or economic loss is minimised to As Low As Reasonably Practicable (ALARP).

Over the years, risk analysis techniques have been developed and perfected. This has led to an abundance of hybrid techniques to either qualitatively or quantitatively express risk levels. This book further adds to the available techniques but more importantly identifies and isolates instances where traditional techniques fall short. The book is based on the extensive research work conducted over the recent years by Dr. Pillay and Professor Wang at Liverpool John Moores University in the United Kingdom. The main driver for this research has been the lack of reliable and useful safety data suffered by the shipping industry. The area of research has predominantly been risk analysis techniques; however, influencing elements have been addressed and captured in this book. These elements include reliability, maintenance, decision-making and human error. Uncertainties that manifest within risk analyses are highlighted and alternative solutions are presented. A considerable body of high quality reference materials Dr. Pillay and Professor Wang have produced, supports the text presented in this book. Many of these references have been peer reviewed and published in world-class journals as well as presented at international conferences.

This book caters to a wide range of readers, among them being industrial safety/design/operation engineers in general and marine/offshore specialists in particular. It also serves as a useful reference material to undergraduate and postgraduate students studying in the fields of Marine Technology, Safety/Reliability Engineering and General Engineering. Just like any other book, which has been written based on in-depth research of a specialist area, it would be beneficial for academics and industrial researchers.

The book serves as a good starting point for young safety engineers and apprentices who are new to risk analysis methods and oblivious to some of the setbacks of the current techniques used in the industry. There are several useful sections that briefly introduce the reader to the shipping scene and set the mood by defining the rules and regulations that govern merchant and fishing vessels. Once these safety requirements are established, the various safety assessment and risk analysis tools available to the reader are highlighted. For the more advanced reader, the following chapters plunge deep into the root of problem of coping with uncertainty in the available data. The methods presented are based on the Formal Ship Safety Assessment framework and are demonstrated using practical examples.

I would also like to acknowledge Dr H. S. Sii, Dr J. B. Yang and Dr D. L. Xu who made significant contribution to Chapter 10.

The views and opinions expressed in this book are strictly those of the authors and do not necessarily reflect those of Lloyds Register and Liverpool John Moores University.

TABLE OF CONTENTS

Chapter 6 - Risk Assessment Using Fuzzy Set Approach

Chapter 7 Modified Failure Mode and Effects Analysis

Chapter 8 Maintenance Modelling

Chapter 9 Human Error Assessment and Decision Making Using Analytical Hierarchy Processing

Chapter 10 Three Novel Risk Modelling and Decision Making Techniques

Chapter 1

Introduction

Summary

This Chapter briefly reviews the historical development of safety and reliability assessments within the maritime industry and outlines the application of such assessments. This is followed by a review of the current status of safety and reliability assessments in the United Kingdom. The different databases available in the maritime industry are described, highlighting the information that each of these databases carries. The contents in the book are finally outlined.

Keywords: Maritime industry, reliability, risk assessment, safety, ships.

1.1 Introduction

Safety was not considered to be a matter of public concern in ancient times, when accidents were regarded as inevitable or as the will of the gods. Modern notions of safety were developed only in the 19th century as an outgrowth of the industrial revolution, when a terrible toll of factory accidents aroused humanitarian concern for their prevention. Today the concern for safety is worldwide and is the province of numerous governmental and private agencies at the local, national and international levels.

The frequency and severity rates of accidents vary from country to country and from industry to industry. A number of accidents in the chemical, oil and gas, marine and nuclear industries over the years have increased the public and political pressure to improve the safety which protects people and the environment. In the evolution of the approach to safety, there has been an increasing move towards risk management in conjunction with more technical solutions. Hazardous industries have developed approaches for dealing with safety and loss prevention, from design standards to plant inspections and technical safety, through to safety auditing and human factors (Trbojevic and Soares (2000)).

As far as the marine industry is concerned, tragic accidents such as the *Herald of Free Enterprise* and *Derbyshire*, together with environmental disasters such as *Exxon Valdez* and *Amoco Cadiz*, have focused world opinion on ship safety and operation (Wang (2002)). This demand for improved safety requires comprehensive safety analyses to be developed. Such safety analyses will ensure efficient, economic and safe ship design and operation.

1.2 Safety and Reliability Development in the Maritime Industry

Reliability and safety methods saw a rapid development after the Second World War. These methods were mainly concerned with military use for electronics and rocketry studies. The first predictive reliability models appeared in Germany on the V1 missile project where a reliability level was successfully defined from reliability requirements and experimentally verified on components during their development stages (Bazovsky (1961)).

The first formal approach to shipboard reliability was the Buships specification, MIL-R-22732 of July 31, 1960, prepared by the United States of America's Department of Defence and addressed ground and shipboard electronic equipment (MIL (1960)). Subsequently in 1961 the Bureau of Weapons issued the MIL standards concerning reliability models for avionics equipment and procedures for the prediction and reporting of the reliability of weapon systems. This was due to the fact that the growing complexities of electronic systems were responsible for the failure rates leading to a significantly reduced availability on demand of the equipment.

In February 1963 the first symposium on advanced marine engineering concepts for increased reliability was held at the office of Naval Research at the University of Michigan. In December 1963 a paper entitled "*Reliability Engineering Applied to the Marine Industry*" (Harrington and Riddick (1963)) was presented at the Society of Naval Architects and Marine Engineers (SNAME) and in June the following year another paper, entitled "*Reliability in Shipbuilding*" (Dunn (1964)), was presented. Following the presentation of these two papers, SNAME in 1965 established Panel M-22 to investigate the new discipline as applied to marine machinery and make it of use to the commercial marine industry.

In the last three decades, stimulated by public reaction and health and safety legislation, the use of risk and reliability assessment methods has spread from the higher risk industries to an even wider range of applications. The Reactor Safety Study undertaken by the U.S.A (U.S Nuclear Regulatory Commission (1975)) and the Canvey studies performed by the UK Health & Safety Executive (HSE (1978, 1981a,b)) resulted from a desire to demonstrate safety to a doubtful public. Both these studies made considerable use of quantitative methods, for assessing the likelihood of failures and for determining consequence models.

1.3 Present Status

There is a long history in the United Kingdom (UK) of research, development and successful practical application of safety and reliability technology. There is a continuing programme of fundamental research in areas such as software reliability and human error in addition to further development of the general methodology. Much of the development work was carried out by the nuclear industry.

Based on the considerable expertise gained in the assessment of nuclear plants, a National Centre for System Reliability (NCSR) was established by the UK Atomic Energy Authority (UKAEA) to promote the use of reliability technology. This organisation plays a leading role in research, training, consultancy and data collection. The NCSR is part of the safety and reliability directorate of the UKAEA, which has played a major role in formulating legislation on major hazards, and has carried out major safety studies on industrial plants. It is noted that some of the major hazard studies commissioned at the national level in the UK have included the evaluation of the risks involved as a result of marine transportation of hazardous materials such as liquefied gases and radioactive substances. It is expected that the recent legislation in relation to the control of major hazards will result in a wider use of quantitative safety assessment methods and this will inevitably involve the marine industry.

Most chemical and petrochemical companies in the UK have made use of safety and reliability assessment techniques for plant evaluation and planning. Similar methods are regularly employed in relation to offshore production and exploration installations.

The Royal Navy has introduced reliability and maintainability engineering concepts in order to ensure that modern warships are capable of a high combat availability at optimum cost (Gosden and Galpin (1999)). The application of these methods has been progressively extended from consideration of the operational phase and maintenance planning to the design phase.

To date, comparatively little use of safety and reliability assessment methods has been made in connection with merchant shipping. Lloyd's Register of Shipping has for a long period, collected information relating to failures and has carried out development work to investigate the application of such methods to the classification of ships. Apart from this, some consultancy work has also been carried out on behalf of ship owners. One example is the *P&O Grand Princess,* for which a comprehensive safety and availability assurance study was carried out at the concept design stage of this cruise ship. Established risk assessment techniques were used including Failure Mode and Effects Analysis (FMEA), flooding risk analysis and fire risk analysis. The resultant ship was believed to be better and safer than it would have been otherwise (Best and Davies (1999)). P&O has now developed an in-house safety management system which is designed to capture any operational feedback, so as to improve the safety and efficiency of its cruise fleet operation and to use it for better design in the future.

The merchant ship-building yards in the UK, having seen the success of the warship yards in applying Availability, Reliability and Maintainability (ARM) studies at the design stage, are actively seeking benefits from adopting a similar approach. Some joint industry-university research projects are being undertaken to explore this area.

1.4 Databases

The early reliability studies, particularly on electronics, made use of failure data obtained by testing a large number of components. As the techniques found more widespread applications, the methods for statistically analysing data from real life experience became more advanced and large communal databases of reliability data were created.

In the 1980's, the maritime classification societies, commercial institutions and other authorities realised the importance of statistical data collection on failure or repair data and eventually, data on general accident statistics were provided (HSE (1992a, b)). These data give general trends and are not directly useable in quantitative assessments. By far the most useful sets of statistics on marine accidents are presented in the publications of the UK Protection and Indemnity (P&I) Club of insurers (P & I Club (1992)).

Accident investigation is a common method used by many organisations in attempt to enhance safety. Discovering the causes of casualties may allow steps to be taken to preclude similar accidents in the future. Since 1981 the United States Coast Guard (USCG) has maintained a computer database summarising the causes of investigated marine casualties. In 1992 the USCG implemented a new computer casualty database, the Marine Investigation Module (MINMOD), which changed the way marine casualty investigations were reported (Hill et al. (1994)). The new system implemented several improvements that were expected to enhance the validity and completeness of the casualty data reported. One of the most important changes made was the adoption of a chain-of-events analysis of accident causes, enabling a more complete description of all accident-related events and their associated causes.

In the past, accident statistics were not gathered systematically and the data type was not consistent. This led to the analyst not knowing if the set of data is applicable to the analysis under consideration. Some commercial institutions have focused on developing databases of maritime accidents. The accident information is presented systematically and in some cases correlation is available. Typical examples include:

- OREDA (Offshore Reliability Data) - A database of offshore accidents which was first published in 1982 and has been updated annually ever since (OREDA (1982)).
- Marine Incident Database System (MIDS) - A database maintained by the Marine Accident Investigation Branch (MAIB).
- World Casualty Statistics - A collection of data published annually by Lloyds Register of Shipping.
- The Institute of London Underwriters.
- CASMAIN - A database maintained by the United States Coast Guard.
- SEAREM - A British Isle database developed and refined under the stewardship of the Royal National Lifeboat Institution (RNLI).

Over the last several years, progressive maritime organisations around the world have been cooperating to form a worldwide information network, called RAM/SHIPNET, to support the optimisation of safety, reliability, and cost effectiveness in vessel operations. The mission of RAM/SHIPNET is to form an efficient information network for vessel operators and other industry participants to collect and share sanitised performance information on vessel equipment. It consists of distributed and partially shared Reliability, Availability, and Maintainability (RAM) databases. RAM/SHIPNET was established to collect equipment performance data and to share this data at different levels by linking chief engineers, ship operators/managers, regulatory agencies, equipment manufacturers and shipyards/designers. First generation stand-alone data collection and processing tools were developed and the system became ready for implementation. The roll-out period is in progress for full validation, demonstration, and implementation of RAM/SHIPNET (Inozu and Radovic (1999)).

The databases that are described in this section, are still lacking specific information of equipment and component failures. Novel methods have to be developed to handle this shortcoming. These novel techniques should integrate expert judgement with available data in a formal manner to ensure the accuracy and the applicability of the safety assessment carried out.

1.5 Description of the Book

The aims of the book are to:

1. Review the current practices employed in the marine industry with particular reference to fishing vessels.
2. Describe the typical safety assessment techniques and their application in order to integrate them into formal safety assessment.

3. Describe several novel safety modelling and decision making techniques that can be applied in situations where traditional methods cannot be applied with confidence.
4. Demonstrate how the adoption of safety analysis methods can facilitate the application of formal safety assessment of ships.
5. Address the issues in formal ship safety assessment.

This book may be used as a reference by marine and offshore safety analysts within industry; by marine and offshore safety researchers; and by undergraduates and postgraduates in marine and offshore technology.

Fishing vessels are chosen as a major test case while other types of ships are also used. Fishing vessels are generally smaller with a unique operating nature and the accidents concerning these vessels have been overlooked in the past. Most fishing vessels are owner operated and lack the organisational structure of merchant vessel owners and operators. This leads to the difficulty in gathering accident/failure information for a safety analysis. Since the fishing industry is starved of safety and reliability data, conventional safety and risk assessment techniques may not be readily applied. The available quantitative techniques require a certain amount of failure data in order to make a reasonable safety prediction. The novel methods described in this book will address this setback of the traditional methods by integrating within their models the ability to handle vague and uncertain data in an effective manner to produce a reasonably accurate safety assessment. These novel methods will integrate hazard identification, risk quantification and ranking with formal decision making techniques so that safety improvements made to new as well as existing vessels are effective and justified.

The body of this book is divided into eleven Chapters. Each Chapter is summarised here, highlighting the salient points delivered.

Chapter 2 highlights the international conventions that govern fishing vessel safety and some of the safety programmes that have been implemented by the International Maritime Organization (IMO) member states. The data that were collected and analysed from various sources including the Department of the Environment, Transport and the Regions (DETR) and the Marine Accident and Investigation Branch (MAIB), are presented. The findings of the accident data analysis are described. A statistical analysis of containership accidents is also briefly conducted.

Chapter 3 gives an introduction to typical safety analysis techniques. The advantages and disadvantages of each method are reviewed. This is followed by a proposed approach to identifying hazards on board ships.

Chapter 4 describes both the offshore safety case approach and formal safety assessment of ships. The current practices and the latest development in safety assessment in both the marine and offshore industries are described. The relationship between the offshore safety case approach and formal ship safety assessment is described and discussed. The study of risk criteria in marine and offshore safety assessment is carried out. The recommendations on further work required are finally given.

Chapter 5 discusses the inception of Formal Safety Assessment (FSA), originally proposed by the UK Maritime & Coastguard Agency, in the maritime applications. The FSA is applied to fishing vessels with an illustrative example. The application of the FSA framework to containerships is also described. Detailed discussions on several aspects of FSA's application to ship design and operation are given.

Chapter 6 describes a new approach for modelling the occurrence probability of a hazard and its severity using Fuzzy Set Theory (FST) with Fault Tree Analysis (FTA). The literature survey indicates that the common problem in quantifying these parameters (of a failure event) is often the small sample size and the statistical uncertainties which are correspondingly high. The approach described utilises FST and expert judgements to deal with this high level of uncertainty. It involves the generation of a fault tree of known events and its synthesis with fuzzy arithmetic to produce a fuzzy probability for the top event (an undesirable event). Linguistic terms such as *Very High, High, Moderate, Low* and *Remote* are used to obtain such an estimate. Mathematical formulas used for calculations in the fault tree are derived from the theory of probability and integrated with fuzzy arithmetic on α-cut sets. The risks associated with failure events are determined by combining the occurrence likelihood and possible consequences to produce a risk ranking. A trial application of the approach is carried out.

Chapter 7 describes a new modified approach to FMEA which incorporates the use of fuzzy rule base and grey theory. The traditional approach utilises the Risk Priority Number (RPN) ranking system. This method determines the RPN by finding the multiplication of factor scores. The three factors considered are the probability of failure, severity and detectability. Traditional FMEA has been criticised to have several weaknesses. These weaknesses are reviewed and are addressed in the approach (modified FMEA) described. The purpose of the new approach is to utilise expert judgement in a formal manner to produce a more logical ranking of the failure events identified during the hazard identification phase. It also allows for the analyst to assign weighting factors to the decision criteria in order to determine where improvements can be made to the system. A test case is presented using the modified FMEA described. The potential of integrating the modified FMEA into the FSA process is discussed.

Chapter 8 presents a maintenance model for reducing machinery failures for ship operations. The results obtained from the data analysis in Chapter 2 show that the failures could have been avoided if a proper maintenance regime had been in place. The current maintenance strategies on vessels are critically reviewed. Upon analysing the present situation of the industry, it is proposed that an inspection regime be implemented to arrest failures before they develop into catastrophic ones. An approach employing delay-time analysis is described to determine the optimal inspection time. Three criteria are modelled, namely, downtime, cost and safety criticality. Based on the criterion selected, an optimum inspection time can be obtained. A best compromise is also discussed where all three criteria are simultaneously minimised to acceptable levels. The described approach is demonstrated on a winch operating system of a fishing vessel. The effect of the integration of an inspection regime within the current maintenance practice is studied and its advantages are highlighted.

Chapter 9 describes a framework for the identification and quantification of human error in fishing vessel operation, following a brief review of human error assessment techniques. This framework ranks the impact of human error and further integrates the available risk control options into the analysis. The approach uses Analytical Hierarchy Processing (AHP) theory to rank the preference of each control option. The advantages of employing the AHP technique are discussed and the integration of such a technique within the FSA framework is described.

Chapter 10 presents three novel safety assessment and decision making approaches. They are (1) a safety based decision support system using artificial neural network techniques, (2) a safety optimisation framework using Taguchi concepts, and (3) A multiple criteria decision making approach applied to safety and cost synthesis. Such approaches provide the safety analyst with more flexibility to facilitate risk modelling and decision making.

Chapter 11 concludes the book by summarising the results and outlining the contributions to formal ship safety assessment.

The safety analysis techniques described in this book will facilitate ship safety assessment in various situations. They can be tailored for safety analysis of any maritime and offshore engineering product with domain-specific knowledge. As some of these approaches described are subjective in nature, they may be more applicable for many engineering applications that lack reliable failure data.

1.6 References (Chapter 1)

1 Bazovsky I., (1961) "Reliability Theory and Practice", Prentice Hall, Englewood Cliffs New Jersey.

2 Best P.J. and Davies W.B., (1999) "The Assessment of Safety for Vessels in Service: Practical Examples of the Application of FSA Techniques from Inland Vessels to Ocean Going Ferries", Transactions of IMarE, Vol. 111, No. 2, pp. 51-58.

3 Dunn T.W., (1964) "Reliability in Shipbuilding", SNAME, Spring Meeting, June 4-5, Boston, Massachusetts, USA.

4 Gosden S.R. and Galpin L.K., (1999) "Breaking the Chains of Preventive Maintenance: The Royal Navy Experience", Maintenance & Asset Management Journal, Vol. 14, No 5, pp. 11-15.

5 Harrington R.L. and Riddick R.P., (1963) "Reliability Engineering Applied to the Marine Industry", SNAME, Hampton Road Section, December 1963.

6 Hill S.G., Byres J.C., Rothblum A.M. and Booth R.L., (1994) "Gathering and Recording Human Related Casual Data in Marine and Other Accident Investigation", Proceedings of the Human Factors and Ergonomics Society 38th Annual Meeting, Nashville, Tennessee, USA, October 24-28, pp. 863-867.

7 HSE, (1978) "Canvey: An Investigation of Potential Hazards from Operations in the Canvey Island/Thurrock Area", Health & Safety Executive, HMSO, United Kingdom.

8 HSE, (1981a) "Canvey: A Second Report. A Review of Potential Hazards from Operations in the Canvey Island/Thurrock Area", Health & Safety Executive, HMSO, United Kingdom.

9 HSE, (1981b) "Canvey: A Second Report. A Summary of a Review of Potential Hazards from Operations in the Canvey Island/Thurrock Area", Health & Safety Executive, HMSO, United Kingdom.

10 HSE, (1992a) "The Offshore Installation (Safety Case) Regulation 1992", Health & Safety Executive, ISBN 00118820559.

11 HSE, (1992b) "The Tolerability of Risk from Nuclear Power Stations", 1992 Edition, Health & Safety Executive, ISBN 00118863681

12 Inozu B. and Radovic I., (1999) "Practical Implementations of Shared Reliability and Maintainability Databases on Ship Machinery: Challenges and Rewards", Transactions of IMarE, Vol. 111, No. 3, pp. 121-133.

13 MIL, (1960) "Buships Specification", MIL-R-22732, Department of Defence, United States of America.

14 OREDA, (1982) "Offshore Reliability Data Handbook (OREDA)", PennWell Books, 1982 onwards.

15 P & I Club, (1992) "Analysis of Major Claims 1992", UK Mutual Steam Ship Association (Bermuda) Ltd Publication.

16 Trbojevic V.M. and Soares C.G., (2000) "Risk Based Methodology for a Vessel Safety Management System", Proceeding ESREL 2000 and SRA-Europe Annual Conference, ISBN: 90 5809 141 4, Scotland, UK, 15-17 May, Vol. 1, pp. 483-488.

17 U.S Nuclear Regulatory Commission, (1975) "Reactor Safety Study: An Assessment of Accident Risks in U.S. Commercial Nuclear Power Plants", October 1975, WASH-1400 (NUREG 75/014).

18 Wang J., (2002) "A Brief Review of Marine and Offshore Safety Assessment", Marine Technology, SNAME, Vol. 39, No. 2, April, pp. 77-85.

Chapter 2

Ship Safety and Accident Statistics

Summary

This Chapter highlights the international conventions that govern fishing vessel safety and some of the safety programmes that have been implemented by the International Maritime Organization (IMO) member states. The data that were collected and analysed from various sources including the Department of the Environment, Transport and the Regions (DETR) and the Marine Accident and Investigation Branch (MAIB), are presented. The findings of the accident data analysis are described. A statistical analysis of containership accidents is also briefly conducted.

Keywords: Accident data, containerships, data analysis, fishing vessels.

2.1 Introduction

Recognising the need for attention to safety of commercial fishing vessels, the IMO organised an international conference, which culminated in the Torremolinos International Convention for the safety of fishing vessels in 1977 (IMO (1977)). It established uniform principles and rules regarding design, construction and equipment for fishing vessels 24m (79 feet) in length and over. This Convention was a major milestone. It provided benchmarks for improving safety, and many fishing nations have adopted its measures into their marine safety programmes.

The IMO convention on Standard of Training, Certification and Watch keeping for seafarers (STCW) 1978 is another important influence on fishing vessel safety. Although the STCW 1978 specifically exempts fishing vessels, it has inspired efforts to develop personnel qualification standards (STCW 95 also exempts fishing vessels). Notable among these efforts is the Document for Guidance on Fishermen's Training and Certification (IMO (1988)) and the Code of Safety for Fishermen and Fishing Vessels (IMO (1975a)). Other IMO codes and guidelines include the Voluntary Guidelines for the Design, Construction and Equipment of Small Fishing Vessels (IMO (1980)) and the Code of Safety for Fishermen and Vessel Design and Construction (IMO (1975b)). These standards are jointly prepared by the IMO and two other United Nations subsidiaries, the Food and Agricultural Organisation (FAO) and the International Labour Organisation (ILO). They provide guidance on training and education and detailed curriculum development.

There are strong safety programmes among the IMO member states that include equipment standards, inspection requirements and certification or licensing of vessel operators and crew. These programmes vary in each country. For example, Canada, Norway and the UK have extensive requirements, while other countries are less stringent. Generally fishing vessels in length 15m or longer are addressed; however some countries address vessels as small as 9m, such as New Zealand and 12m as in the UK.

In the UK, comprehensive regulations have come into force since 1975. Surveys and certification of fishing vessels with the length of 12m or longer are required; they apply to about 2000 vessels. For vessels with the length of over 16.5m, deck officers and engineers have comprehensive entry level professional training, certification, manning and watch keeping requirements.

Studies on the effect of compulsory programmes have been conducted in Norway, The Netherlands, UK and Spain, but they have tended to focus on training, statistics and causes of accidents rather than performance of technical systems in relation to compulsory programmes. It appears that fatalities have generally been reduced, while the rates of incidence for injuries related to vessel casualties and workplace accidents appear unchanged. The lack of apparent change in injury rates may be related to working conditions and methods, vessel design, training deficiencies and changes in the number of fishing vessels and fishermen (Carbajosa (1989), Dahle and Weerasekara (1989), Hoefnagal and Bouwman (1989), Stoop (1989)). The number of vessel casualties over the years has changed. For example, in the UK, since safety rules were applied to all vessels over 12m during the mid 1980's, the number of losses of these vessels has significantly reduced. However, losses of vessels under 12m have more than doubled, perhaps partly because of a large increase in the number of vessels under 12m, to which only life saving and fire safety government regulations apply (Hopper and Dean (1992)).

2.2 The Code of Practice for the Safety of Small Fishing Vessels

The development of a Code of Practice for small fishing vessels marked the beginning of the first major review of fishing vessel safety regulations since 1975. The principal aim in developing the Code was to update the safety equipment requirements for small fishing vessels. Its secondary aim was to build on the concept of hazard identification and risk assessment, and to introduce an assessment by owners of the fitness of their vessels (House of Commons (2000)).

The Code of Practice for the safety of small fishing vessels has been effective since the 1st of April 2001. The aim of this Code of Practice is to improve safety in the under 12 meter sector of the fishing industry and to raise the safety awareness of all those involved with the construction, operation and maintenance of fishing vessels with a registered length of less than 12 meters.

2.2.1 Development

In 1992 the National Audit Office, in its report entitled "Department of Transport: Ship Safety", noted an increase in the fishing vessel accident rate from 1978 to 1989, due in part to an increase in the numbers of smaller vessels (National Audit Office (1992)). It observed the absence, until 1990, of any programme of inspection of fishing vessels with a registered length of less than 12 meters. At about the same time, the House of Lords Select Committee on Science and Technology recommended that fishing vessels down to 7m in length should be brought within the licensing, crew certification and structural safety regimes.

In response, the Surveyor General's Organisation of the Department of Transport (now the Maritime & Coastguard Agency (MCA)), in consultation with industry members of the

Fishing Industry Safety Group (FISG), decided to develop a Code of Practice for fishing vessels with a registered length of less than 12 meters. The content of the Code has been the subject of extensive discussion with representatives of the under 12 meter sector of the fishing industry within a Steering Committee set up by FISG to oversee the Code's development. The Code has been applied from the 1st of April 2001 to all United Kingdom registered fishing vessels with a registered length of less than 12 meters.

2.2.2 Code Requirements

To comply with the Code of Practice, a vessel owner is required:

- To carry safety equipment on the vessel appropriate to its length and construction (i.e. decked or open). The equipment checklist is given in Appendix 1.
- To complete or arrange completion of an assessment of the health and safety risks arising in the normal course of work activities or duties on the vessel in accordance with the provisions of the Merchant Shipping and Fishing Vessels (Health and Safety at Work) Regulations 1997 and MGN (Marine Guidance Note) 20 (M+F) (MSA (1998)).
- To certify annually that the vessel complies with the Code, by declaring that the safety equipment has been properly maintained and serviced in accordance with manufacturers' recommendations and that an appropriate, up-to-date health and safety risk assessment has been completed.
- To present the vessel for inspection either voluntarily or as requested by the MCA.

Appendix 1 gives the checklist of requirements for the Code of Practice for the safety of small fishing vessels in 4 categories. The vessels addressed in this Code of Practice include:

1. Decked vessels 10m and above registered length to less than 12m registered length.
2. All decked vessels up to 10m registered length.
3. Open vessels 7m and above to less than 12m registered length.
4. Open vessels less than 7m registered length.

2.3 The Fishing Vessels (Safety Provisions) Safety Rules 1975

In 1968, three vessels were tragically lost off the coast of Iceland. The investigation of these three vessel accidents determines the loss as 'capsizing due to ice accumulation'. Following the official inquiry into these losses, a rule regime was investigated which eventually arrived on the statute as "The Fishing Vessels (Safety Provisions) Safety Rules 1975". Unfortunately the formulation of the rules did not result in an analysis of the organisational or human failing, present in many safety tragedies within the fishing community. The rules are primarily concerned with vessels of over 12 meters registered length. Smaller vessels are addressed, but only life saving appliances and firefighting measures are included. Appendix 2 gives a list of all equipment addressed in The Fishing Vessel (Safety Provisions) Safety Rules 1975. These rules do not concern themselves with the whole vessel, but may be noted to consider the vessel from the deck and accommodation line downwards. The winches, wires and fishing equipment are not covered by the rules.

Following the introduction of the 1975 Rule, the European Common Fisheries policy brought in a licensing scheme for vessels over 10 meters. This coupled with a de-commissioning scheme for larger vessels, resulted in a huge increase in the number of under 10 meter vessels. These vessels did not need licenses to fish and need not comply with the majority of the 1975 Rules. However, in 1996, the Ministry of Agriculture Fisheries and Food introduced fishing licenses for vessels of under 10 meters overall length. The introduction of this law has reduced the size of the fleet. The greatest incidence of risk has now moved to vessels in the 7 to 20 meter range, with particular safety concern for those vessels under 12 meters. Following concern emanating from the Parliament, inspections on these under 12 meter vessels have been requested. Since 1993, under 12 meter vessels have been subjected to safety inspection.

2.4 Accident Data for Fishing Vessels

Comparisons of the safety record of the fishing industry with other industries indicate that the industry continues to be the most dangerous by a significant margin. In 1995/96 there were 77 fatal injuries per 100,000 fishermen as opposed to 23.2 per 100,000 employees in the mining and quarrying industry (the next highest category in that year) (MAIB (1995)). In 1992 there were 494 reported fishing vessel accidents from a fleet of 10,953 vessels. In 1997, figures indicate 485 reported fishing vessel accidents from a significantly reduced fleet of 7,779 vessels. These statistics do not include personal accidents to fishermen while at sea; it is believed that these are under-reported (MAIB (1997)).

The accident data presented in this section are predominantly gathered from the Marine Accident Investigation Branch (MAIB). The MAIB is a totally independent unit within the Department of the Environment Transport and the Regions (DETR) and reports directly to the Secretary of State. The MAIB received 1,418 accident and incident reports in 1999. Accidents to ships accounted for 641 of those reports.

The data presented here is collected from 1992 to 1999 and reflects all the reported incidents and accidents relating to fishing vessels. It is thought that the actual accident and incident figures are higher than what is presented here, as many accidents are not reported to the coastguard authorities.

Figure 2.1 shows the total number of vessels lost (primary y-axis) and total number of vessels registered (secondary y-axis) from 1992 to 1999. These figures include all vessel sizes ranging from under 12 meter to over 24 meter. From this graph, it is evident that the percentage of vessels lost increased from 1992 to 1994 and then reduced from 1994 to 1998. From 1998 onwards, it is noted that there was a sharp increase in the percentage of vessels lost. Overall, the percentage of vessels lost was between 0.27% (minimum in 1997/98) and 0.45% (maximum in 1999) of the total registered vessels, as seen in Figure 2.2

There were approximately 7,460 UK-registered fishing vessels in 1999 (end December 1999 figure). During the year 370 accidents and incidents involving these vessels were reported to the MAIB. 33 fishing vessels were lost which at 0.45% of the total fleet represent the highest rate since 1994. Machinery damage is noted as the main contributor to the high number of accidents as seen in the pie chart of Figure 2.3.

An analysis of the data from previous years shows that machinery damage has contributed to over 50% of all accidents. This could be attributed to several factors including poorly maintained equipment, incorrect operation, age, lack of automation, etc. The graph in Figure 2.4 shows the number of accidents caused by machinery damage from 1994 to 1999. Although the figures indicate a decreasing trend, the number of accidents related to this category is still high and certainly unacceptable from a safety perspective.

The next highest contributor to accidents is found to be flooding and foundering followed by grounding and then collision and contact. A comparison of all accident types is made as seen in Figure 2.5. Flooding and foundering is estimated to cause almost 15% to 20% of accidents on fishing vessels.

These data are cumulated and presented as a pie chart in Figure 2.6 to reveal the contribution of each accident type for the sampling period. As revealed earlier, machinery damage is found to be the most common cause of accidents on fishing vessels, contributing 64.4% of all accidents. Foundering and flooding (14.2%), grounding (10.2%), collision and contacts (5.7%), and fires and explosions (2.9%) follow.

To determine the severity of the accidents on fishing vessels, data reflecting the accidents to vessel crew together with the number of deaths are gathered and presented in Figures 2.7 and 2.8. These bar charts show that almost 30% of accidents to crew on vessels that are under 12 meter result in deaths and for vessels that are 12-24 meters and more than 24 meters in length, these figures are calculated to be 13% and 15%, respectively. The results indicate that vessels under 12 meters have the highest casualty rates and suffer severe consequences when an accident happens. This could be attributed to the size and stability of these vessels when sailing in bad weather conditions. The number of under 12 meter vessels that were lost is much higher than the other vessels as seen in Figure 2.9. The trend in the number of vessels lost is difficult to determine, as it does not follow any specific mathematical rule. However, by comparing the graphs in Figures 2.1, 2.2 and 2.9, it can be concluded that from 1997, the number of vessels lost increased as the percentage of registered vessels decreased.

Table 2.1 gives the detailed breakdown of accidents by vessel length and accident cause for 1999 (MAIB (1999a)). From this table, it is noted that a great proportion of fishing vessel accidents (20%) is caused by negligence/carelessness of the crew. This could be summarised as human error attributed by several factors including competency of the crew, fatigue, poor manning of vessel and difficult operating conditions. A method assessing human error and means to reduce these errors will be described in Chapter 9. Accidents caused by the lifting gear (15%) and other fishing gear equipment (12%) are also high compared to the other accident causes.

2.5 Data Analysis

In many cases of fishing vessel accidents, information is incomplete or totally lacking. This makes it difficult to analyse the events that lead to the accident. Accurate historical and current data on vessels, fishermen, professional experience, hours and nature of exposure and safety performance of personnel and equipment are fundamental to assessing safety problems, monitoring results of safety programmes and measuring the effectiveness of safety improvement strategies (Loughran et al. (2002)). Very few data are regularly collected or published on these parameters. The limited data make it difficult to quantify safety problems, determine casual relations and assess safety improvement strategies. However, the data that are available indicate that significant safety problems exist and that human error, vessels and equipment inadequacies and environmental conditions all contribute to them.

Marine accidents that have occurred could have been prevented with greater attention to safety. This is particularly true for fishing vessels. Recent inquiries into the losses of fishing vessels "Pescado" (MAIB (1998)) and "Magaretha Maria" (MAIB (1999b)) have raised concerns as to how similar accidents may be prevented in the future. The data analysis in Section 2.4 shows that there is a rise of fishing vessel accidents and the trend seems to be continuing in an upward fashion. From the literature survey, it is found that safety assessment of fishing vessels has been limited to stability consideration and very little work has been carried out on the operational and equipment safety assessment. From the data given in Section 2.4, it can be deduced that fishing vessel safety needs to be addressed and the number of accidents and incidents related to the operation and equipment is to be reduced. In order to direct the attention of the safety assessment on fishing vessels, the probable causes of each accident category have been investigated and are summarised as follows:

2.5.1 Machinery Damage

The highest number of incidents reported in the official statistics relates to machinery damage. Although most machinery failures do not threaten the vessel or lives of the crew, given other factors such as bad weather or being in a tideway, the consequences could be disastrous. Upon investigation of several fishing vessels in the UK, it was found that maintenance activities on board these vessels were almost non-existent. This is thought to lead to the high number of machinery failures. The present situation concerning maintenance on fishing vessels is discussed in detailed in Chapter 8 where a method for improving the current status is described.

2.5.2 Foundering/Flooding

Typically these incidents are caused by burst pipes, fittings working loose, leaking glands and sprung planks. Flooding is a particular problem with smaller wooden vessels. Smaller vessels are often of clinker construction where the strakes are lapped against each other and clenched. They are reliant upon the swelling nature of the wood when soaked for making a good seal. This method of construction is particularly vulnerable in heavy sea conditions. These types of accidents can also happen on vessels that are of metal construction. Sometimes incompatible metals become rapidly corroded in a seawater environment; examples are copper piping adjacent to steel or aluminium structures, which resulted in a relatively new vessel suffering a major flooding incident (Hopper and Dean (1992)).

2.5.3 Grounding

These incidents are associated with all classes of fishing vessels and can be due to various causes. Engine or gearbox failures and propellers fouled by ropes or fishing nets are common causes. However, many cases have been associated with navigational error. This may be a failure to plot a proper course, failure to keep a check on vessel position with wind and tidal drift, reliance on auto-pilots and electronic plotters or a failure to keep a proper lookout. There are no requirements to carry on board a certified navigator (especially for vessels under 12 meters registered length), hence the navigators on these vessels rely heavily upon experience and "gut feeling", which in turn could increase the level of navigator error.

2.5.4 Collisions and Contacts

Almost all collision and contact incidents involve a fishing vessel and a merchant vessel and almost without exception they are due to human error. Large merchant vessels may have a poor line of sight from the wheelhouse and small fishing vessels are not easily seen under the bow. Apart from that, skippers on fishing vessels are too involved in the fishing operation. The fishing operation itself requires sudden stopping or course changing which could lead to unavoidable collisions. Collisions and contacts could also occur involving two or more fishing vessels. This is especially true when pair trawling is in progress. However, the consequences are less severe and the incident normally occurs due to errors of judgement by one or both parties involved.

2.5.5 Fires and Explosions

The investigation of these accidents has shown that in most cases the fire originated from the engine room and was caused by oil or fuel coming into contact with hot exhausts. Other causes are heating and cooking stoves and electrical faults. There have been several cases where the fire had started in the accommodation area due to the crew smoking cigarettes in the sleeping bunk. The number of accidents caused by fire has been relatively low compared to other categories. However, due to the limited fire fighting resources on board fishing vessels, it has the potential to cause severe damage and even loss of life.

2.5.6 Capsizing

From the MAIB reports, it is evident that the majority of capsizing incidents occurred during the fishing and recovery of gear operations. This shows that for the vessels that did capsize, there was an insufficient factor of safety in the present stability criteria. This insufficient factor is introduced by the act of fishing and the associated moment lever introduced by the gear along with the wind lever in the dynamic situation at sea (Loughran et al. (2002)). This is perhaps the most lethal type of incident in terms of loss of life. The capsizing of small fishing vessels happens in a matter of minutes and this leaves little chance for the crew to escape. Extreme sea conditions are one of the many factors that lead to a capsize. As most skippers and crew depend on the catch for their daily income, skippers have been known to put their vessel through extreme sea conditions to get to a fishing ground and sometimes drift within the fishing grounds waiting for the sea to calm in order to resume fishing operations. However, the most common cause of capsizing is when the fishing gear becomes snagged. Trawl gear fouled on some sea bed obstruction is common for a fishing skipper. Attempts to free badly fouled gear by heaving on the winch can result in forces that are large enough to roll the vessel over. Heaving on both warps at the same time will produce a balanced situation but if one side suddenly becomes free, the force on the opposite side may be sufficient to capsize the vessel.

2.5.7 Heavy Weather Damage

The number of vessels suffering weather damage is comparatively low as seen in the graph in Figure 2.5. Small vessels are particularly vulnerable to these accidents, especially when they go out further away from the coastline for their fishing operation (due to the reduced fishing opportunities in British waters). These small vessels will be working far offshore where they cannot withstand the severe weather and wave conditions that can occur unexpectedly. Heavy weather can weaken the hull structure of the vessel and at the same time, cause deck fittings to come loose and lead to an accident.

2.6 Containership Accident Statistics

2.6.1 Introduction to Containerships

Due to a rapidly expanding world trade, the traditional multi-purpose general-cargo liner became increasingly labour and cost intensive. A system was required to accommodate the needs of physical distribution, a system that would offer convenience, speed and above all low cost. By this system, goods should be able to be moved from manufacturer to final distribution using a common carrying unit, compatible with both sea and land legs of transportation. The result was expected to be that all costly and complicated transhipment operations at seaports would be eliminated. The whole process resulted in the development and introduction of the "freight container", a standard box, filled with commodities, detachable from its carrying vehicle, and as easy to carry by sea as by air, road and rail. The beginning of the container era was marked with the sailing of the "container tanker" "MAXTON" on 26th April 1956 from Newark N.J. to Houston, loaded with 58 containers (Chadwin et al. (1999), Stopford (1997)).

During the first years of containerisation, transportation was carried out with modified tankers or dry cargo vessels, broadly accepted as the 1st generation of containerships (Containerisation International (1996), Stopford (1997)). It was not until 1965 that the first orders for purpose built cellular vessels were placed, forming the 2nd generation of container vessels. These were the "Bay Class" ships of 1,600 TEUs (twenty-foot-long equivalent units) capacity. In the late 1970's the 3rd generation appeared increasing the sizes up to Panamax and capacities up to 3,000 TEUs. Following the increasing demand for tonnage but without being prepared to lose the Panama Canal flexibility the industry moved to the development of the 4th generation of container vessels, keeping the Panamax dimensions and increasing the capacity up to 4,200 TEUs represented by the "Econ Class" ships (Containerisation International (1996), Stopford (1997)).

Further development in the shipbuilding industry and the need for the creation of "economies of scale" resulted in the appearance of the 5th generation of container ships, the Post-Panamax in the 1980's (Stopford (1997)). A recent research in the container sector of the shipping industry indicates that the world fully cellular containership fleet increased to more than 3,500 vessels with a total carrying capacity exceeding 4.6 million TEUs in 1999 and with an average annual growth rate up to 11.1% as shown in Table 2.2 (Nippon Yusen Kaisha Research Group (1999)). It is also noteworthy that the growth rate of post-Panamax containerships is the largest of all the containership sizes, amounting up to 26.3%.

Although there were not many major casualties in terms of loss of lives, resulting from accidents involving containerships, this particular ship type has more of its fair share of losses due to incidents involving cargo damage, personal injury, collision, ship structural failure and pollution. Major accidents in the last decade include the total loss of the "C/V Pioneer Container" in 1994 due to a collision in the South China Sea, the loss of the "C/V River Gurara" in 1996, the extensive damages suffered by the "C/V Toyama Maersk" in 1997 due to a collision with a Gas Carrier in the Singapore Strait, the loss of the "C/V MSC Carla" in 1998 which broke into two in bad weather conditions, and the extensive damages suffered by the "M/V APL China" in 1999 due to severe bad weather conditions. Statistics indicate that incidents involving containerships account up to about 7% of the total (Wang and Foinikis (2001)).

In terms of incident categories containerships differ from most other ship types in that shore error accounts for a high percentage of all major incidents. The result is an equally high percentage of cargo damage. Although containerships follow the same pattern as the majority of cargo vessels, as far as the types of damages are concerned, they do differentiate in various aspects. The relative statistics available show that the percentage of incidents is higher in newer containerships, decreasing as they age, while in other cargo ship types, higher incident rates occur at their middle age. The same statistics show that a high percentage of all incidents caused by human error, was due to shore based personnel error, which is far higher than other cargo ship types. As far as ship size is concerned the smaller ships of this type are better placed with fewer incidents.

Other operational characteristics of containerships, such as the fact that they very rarely travel in ballast condition and that there are few opportunities for overnight stay at ports, contribute to the overall performance of these vessels and their operators. It should be stressed that although a relatively large amount of detailed data exists, organisations such as classification societies, as well as private shipping companies are reluctant to release it. This is mainly attributable to the high level of competition in the market. On the other hand, government agencies are either not ready yet to dedicate the necessary resources for data collection, or the time period for which relevant government projects are run is not sufficient to produce reliable data.

2.6.2 Containership Accident Statistics

Classification societies and P&I Clubs can be a very useful source of failure data mainly because of the large amount of vessels each one represents. However, data from these organisations should be critically evaluated before used or combined with others. Classification societies tend to look into safety, mainly from the viewpoint of compliance with the various sets of rules in force (Wang and Foinikis (2001)). On the other hand, P&I Clubs tend to deal with the matter from the viewpoint of financial losses due to lack of safety and are not immediately interested in the regulatory aspect of loss prevention. A recent research carried out by one of the world's leading P&I Clubs, the UK P&I Club shows that for the ten-year period from 1989 to 1999 incidents involving containerships account up to 7% of the total as shown in Figure 2.10 (UK P& I Club (1999)).

In terms of incident categories, containerships differ from most other ship types in that shore error accounts for up to 21% of all major incidents. The result is a fairly high percentage of cargo damage, 54%. All the values of incident categories are shown in Figure 2.11 while the total number of incidents is 273 for the period 1989-1999 (UK P&I Club (1999)).

In terms of ship size and age, the 10-year study shows that the smaller ships of this type are better placed. 87% of the major incidents have occurred on containerships above 10,000 grt as shown in Figure 2.12. Equally interesting is the fact that 44% of incidents involving containerships have occurred on ships of less than 10 years of age as shown in Figure 2.13 (UK P&I Club (1999)). The human error factor in incidents involving containerships is shown to be in decline, following two peak periods in 1988 and 1991 as shown in Figure 2.14 (UK P&I Club (1999)).

Administrations tend to look into marine casualties from the viewpoint of "reportable incidents" within their jurisdiction which results to a differentiation in the relevant numbers, as the sample of vessels considered is smaller than that of P&I Clubs and classification societies. Furthermore, due to their orientation towards ship safety and environmental protection, areas such as cargo damage and third party liability (i.e. fines) may not be considered. Nevertheless, results of such data are equally useful for the identification of major problematic areas of the various ship types.

2.7 Conclusion

A review has been performed on available incident data relevant to fishing vessels. It was found that the amount of data relating to this type of vessel is limited. The only data source that compiles fishing vessel accident/incident data has been identified to be the MAIB. Over the years, the database maintained by the MAIB has considerably improved in terms of its format. However, the database still lacks information about the casual relationship between the causes and effects of the accidents/incidents.

Data interpretation should be carried out with caution, as it is highly likely that there is some degree of underreporting of incidents. This would entail that the actual number of deaths, accidents and vessel losses, would be much higher than the figures presented here. However, the data gathered and analysed in this Chapter show that there is a real problem in the fishing vessel industry. The likelihood of accidents and the associated severity are still high for maritime standards, and the number of accidents/incidents has to be reduced.

A statistical analysis on containership accidents is also carried out. The result indicates that containership accident categories differ from other types of ships. Like fishing vessels, there is also a lack of proper reporting of accidents/incidents for containerships.

2.8 References (Chapter 2)

1. Carbajosa J.M., (1989) "Report on Work Accident in the Sea Fishing Sector", International Symposium of Safety and Working Conditions Aboard Fishing Vessels, L'Université du Quēbec á Rimouski, Quebec, Canada, Aug 22-24.
2. Chadwin M. L., Pope J. A. and Talley W. K., (1999) "Ocean Container Transportation, an Operational Perspective", Taylor & Francis, New York.
3. Containerisation International, (1996) "Year Book on the 40 Years of Containerisation", 1996, pp. v-x.
4. Dahle E.A. and Weerasekara J.C.D., (1989) "Safe Design Construction of Fishing Vessel", International Symposium of Safety and Working Conditions Aboard Fishing Vessels, L'Université du Quēbec á Rimouski, Quebec, Canada, Aug 22-24.
5. Hoefnagal W.A.M. and Bouwman K., (1989) "Safety Aboard Dutch Fishing Vessels", International Symposium of Safety and Working Conditions Aboard Fishing Vessels, L'Université du Quēbec á Rimouski, Quebec, Canada, Aug 22-24.
6. Hopper A.G. and Dean A.J., (1992) "Safety in Fishing-Learning from Experience", Safety Science, Vol. 15, pp. 249-271.

7. House of Commons, (2000) "Fishing Safety, Hansard Debates for 16 Nov 2000 (Part - 1)", Westminster Hall, United Kingdom.

8. IMO, (1975a) "Code of Safety for Fishermen and Fishing Vessels, Part A - Safety & Health Practices for Skippers and Crew", International Maritime Organization, London.

9. IMO, (1975b) "Code of Safety for Fishermen and Fishing Vessels, Part B - Safety & Health Requirements for the Construction and Equipment of Fishing Vessels", International Maritime Organization, London.

10. IMO, (1977) "Final Act of the Conference with Attachments Including the Torremolinos International Convention for Safety of Fishing Vessels", International Conference On Safety Of Fishing Vessels, International Maritime Organization, London.

11. IMO, (1980) "Voluntary Guidelines for the Design, Construction and Equipment of Small Fishing Vessels, An International Maritime Training Guide", International Maritime Organization, London.

12. IMO, (1988) "Document for Guidance on Fishermen's Training and Certification, An International Maritime Training Guide", International Maritime Organization, London.

13. Loughran C.G., Pillay A., Wang J., Wall A.D. and Ruxton T., (2002) "Formal Fishing Vessel Safety Assessment", Journal of Risk Research, Vol. 5, No. 1, pp. 3-21.

14. MAIB, (1995) "Marine Accident Investigation Branch - Safety Digest Summary", No. 2/95, United Kingdom.

15. MAIB, (1997) "MAIB Annual Report", Department of the Environment, Transport and the Regions, United Kingdom.

16. MAIB, (1998) "Marine Accident Investigation Branch: Marine Accident Report 1/98", Department of the Environment, Transport and the Regions, United Kingdom, September, ISBN 1 85 112101 3.

17. MAIB, (1999a) "Marine Accident Investigation Branch: Annual Report 1999", Department of the Environment, Transport and the Regions, United Kingdom.

18. MAIB, (1999b) "Marine Accident Investigation Branch: Marine Accident Report 3/99", Department of the Environment, Transport and the Regions, United Kingdom, July, ISBN 1 85112 109 9.

19. MSA, (1998) "Implementation of EC Directive 89/391 – Merchant Shipping and Fishing Vessels (Health and Safety at Work) Regulations 1997", Marine Safety Agency, Department of the Environment, Transport and the Regions, Southampton, UK.

20. National Audit Office, (1992) "Department of Transport: Ship Safety, HC 186, Parliamentary Session 1991-92", United Kingdom, ISBN 010 218692 8.

21. Nippon Yusen Kaisha Research Group, (1999) "World Containership Fleet and Its Operations 1999", Internal Report, Tokyo, 11th June 1999.

22. Stoop J., (1989) "Safety and Bridge Design", International Symposium of Safety and Working Conditions Aboard Fishing Vessels, L'Université du Quēbec á Rimouski, Quebec, Canada, Aug 22-24.

23. Stopford M., (1997) "Maritime Economics", 2nd edition, Routledge, London, 1997.

24. UK P&I CLUB, (1999) "Analysis of Major Claims – Ten Years Trends in Maritime Risk", London, 1999.

25. Wang J. and Foinikis P., (2001) "Formal Safety Assessment of Containerships", Marine Policy, Vol. 21, pp. 143-157.

Table 2.1 Accidents by Vessel Length and Accident Cause

Accidents by Vessel Length and Accident Cause (more than one cause may be applicable to a particular accident)				
Accident	*Under 12 metres*	*12-24 metres*	*Over 24 metres*	*Total*
Negligence/carelessness of injured person	4	10	9	23
Ship movement	1	3	3	7
Lifting gear	2	8	7	17
Miscellaneous fishing gear and equipment	3	5	6	14
Failure of deck machinery and equipment	-	2	5	7
Sea washing inboard	3	1	4	8
No known cause	2	-	2	4
Trawl boards	-	1	4	5
Door or hatch not secured	1	1	-	2
Failure to comply with warnings/orders	-	1	-	1
Unsecured non-fishing gear on deck	-	-	-	-
Unfenced opening	1	-	-	1
Fatigue	-	-	-	-
Failure to use protective clothing or equipment	3	1	1	5
Slippery surface	-	-	3	3
Lifting/carrying by hand incorrectly	-	-	3	3
Failure of engine room and workshop equipment	-	-	-	-
Others	2	5	4	11

Table 2.2 World Fully Cellular Containerships in TEUs

	Under 1000 TEU		1000-1999 TEU		2000-2999 TEU	
	VSL	TEU	VSL	TEU	VSL	TEU
1999	1,836	765,922	851	1,177,368	426	1,060,460
(1998)	1,751	714,155	807	1,117,310	381	956,349

	3000-3999 TEU		4000 + TEU		Total	
	VSL	TEU	VSL	TEU	VSL	TEU
1999	205	711,498	188	889,982	3,506	4,605,230
(1998)	189	653,444	152	704,559	3280	4,145,817

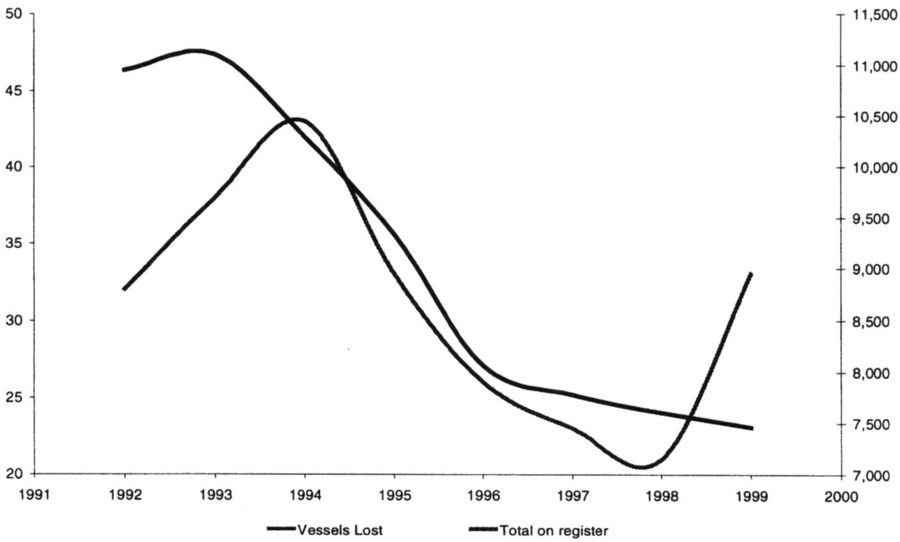

Figure 2.1 Vessels registered and lost (1992-1999)

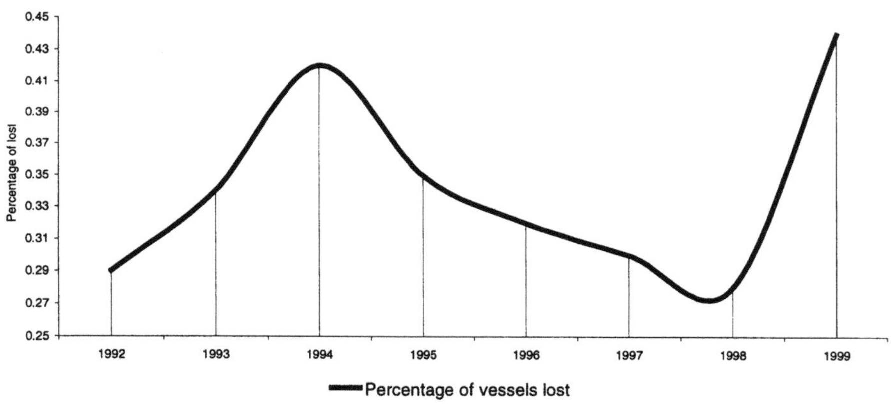

Figure 2.2 Proportion of vessels lost

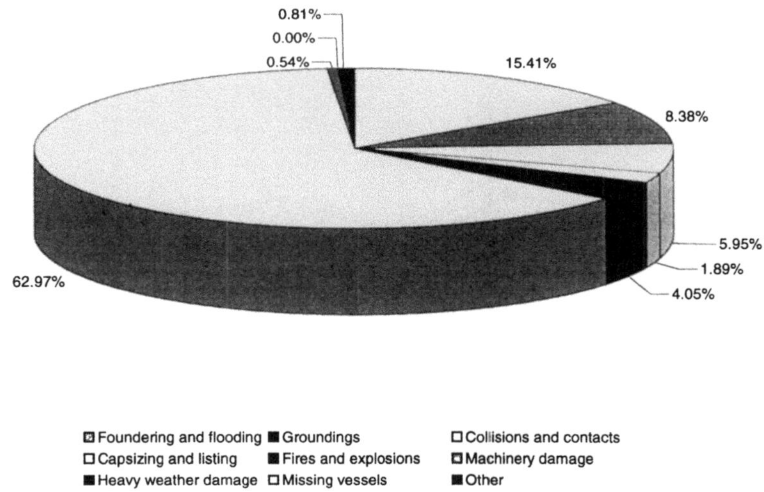

Figure 2.3 Accidents to vessels by accident type in 1999

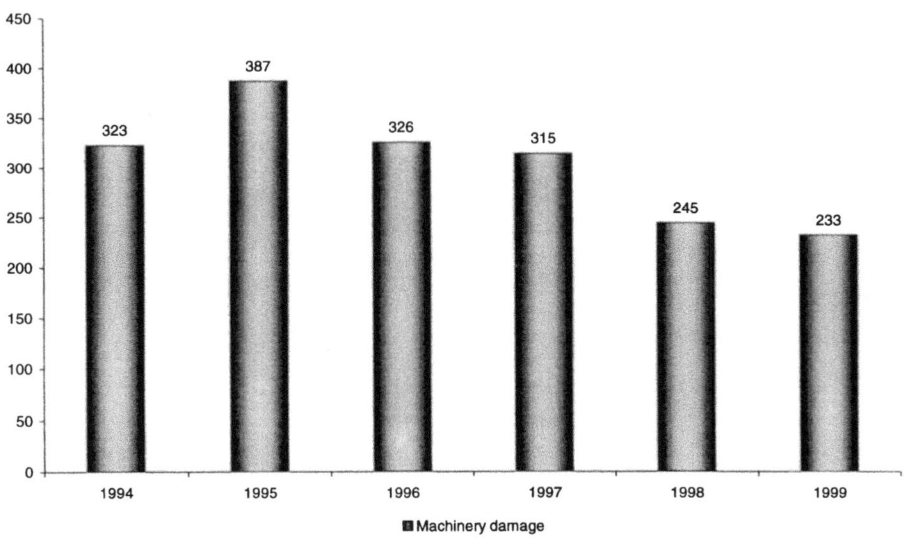

Figure 2.4 Accidents caused by machinery damage

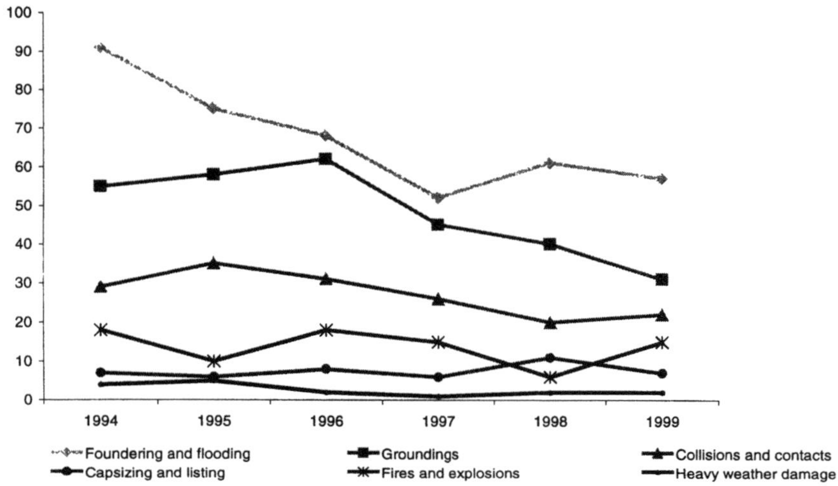

Figure 2.5 Accidents to fishing vessels by accident type

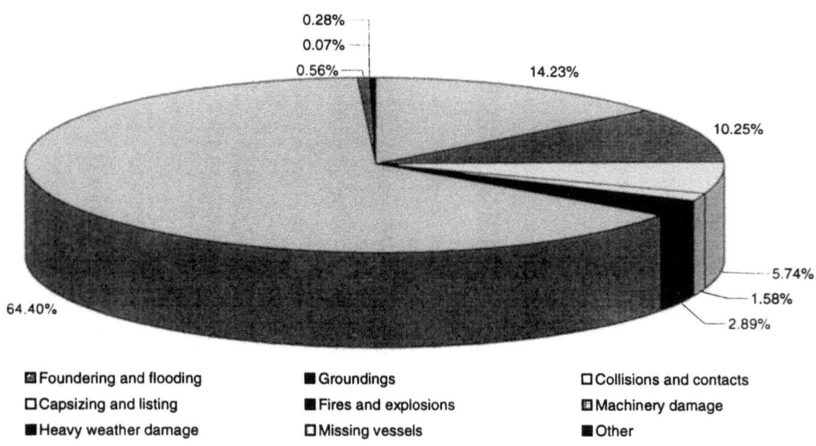

Figure 2.6 Accidents by nature (1994 - 1999)

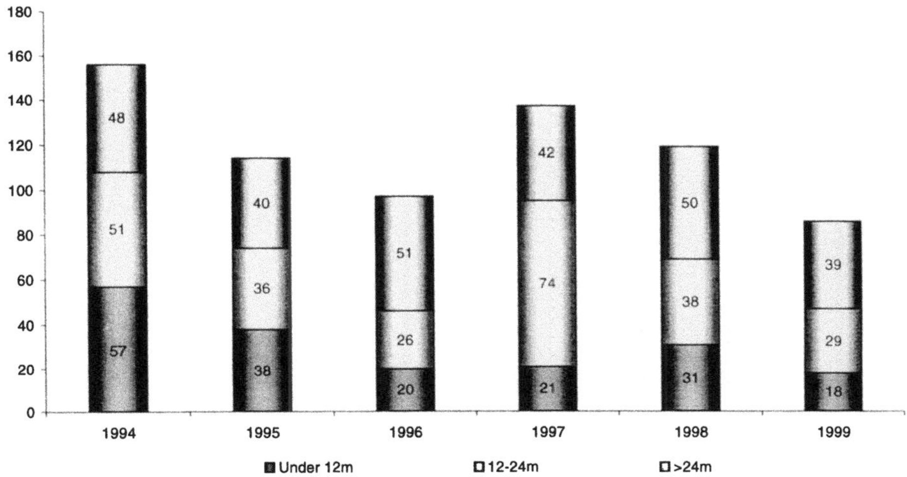

Figure 2.7 Accidents to crew

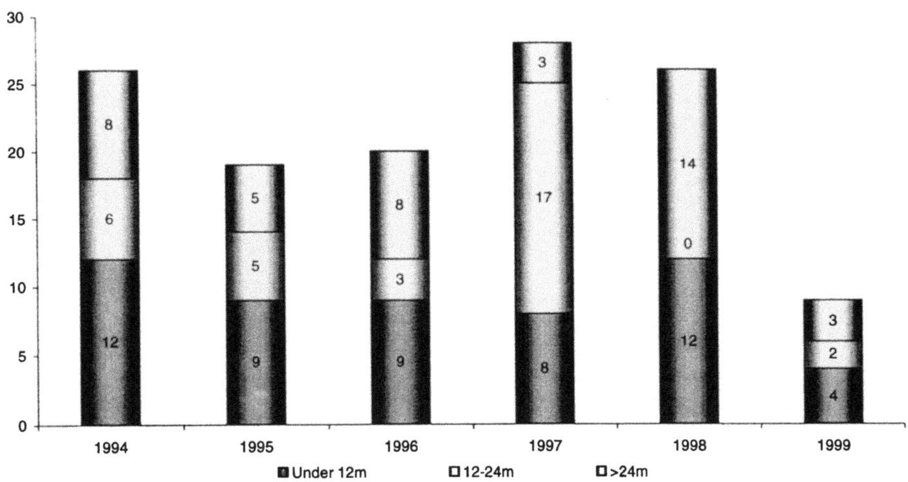

Figure 2.8 Deaths to crew

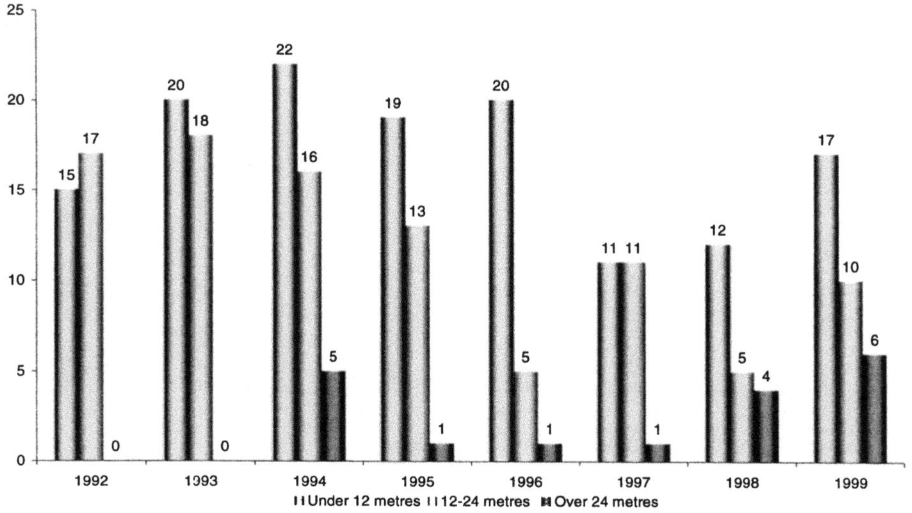

Figure 2.9 Vessels lost

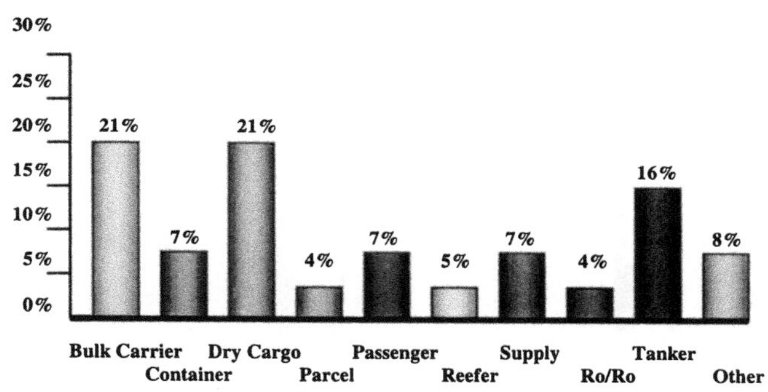

Figure 2.10 Distribution of incidents per ship type

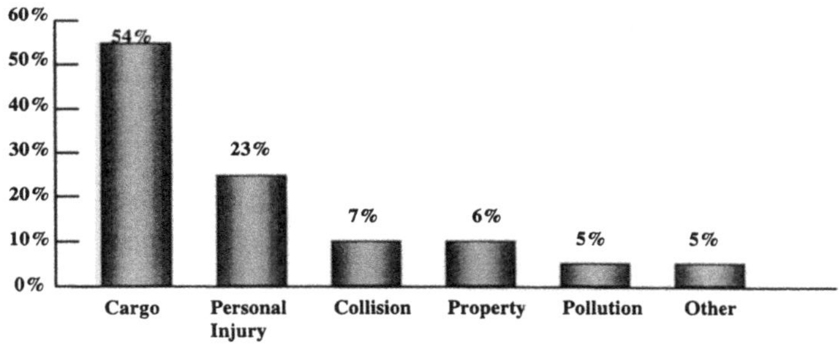

Figure 2.11 Incident categories involving containerships

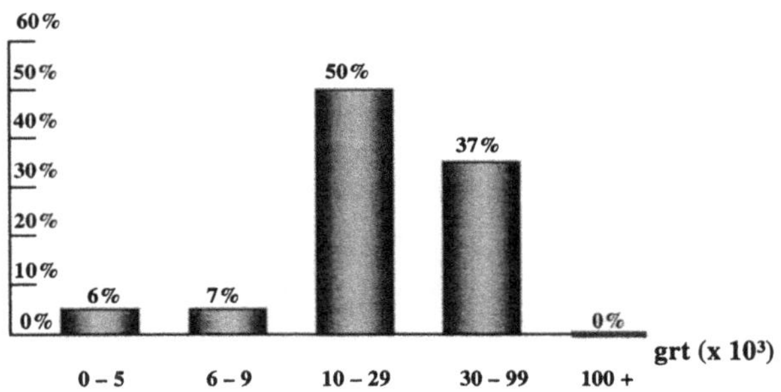

Figure 2.12 Distribution of incidents as per ship size (in grt)

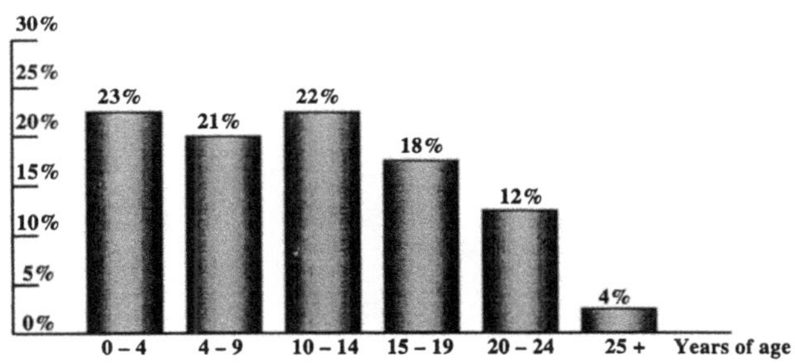

Figure 2.13 Distribution of incidents as per ships age

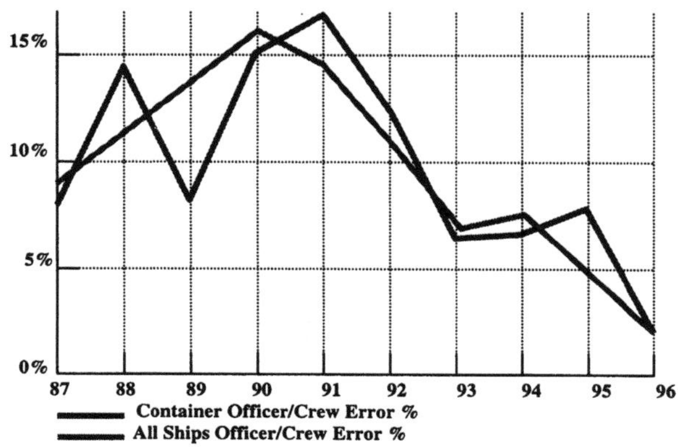

Figure 2.14 Containership- officer/crew error- frequency trend

Chapter 3

Safety Analysis Techniques

Summary

This Chapter gives an introduction to some typical safety analysis techniques. A detailed discussion is carried out on HAZard and OPerability studies (HAZOP) and this is followed by a proposed approach using HAZOP to identify hazards on board ships. Advantages and disadvantages of the safety analysis techniques described are discussed.

Keywords: Qualitative analysis, quantitative analysis, safety analysis techniques.

3.1 Introduction

Reliability and safety analyses are different concepts that have a certain amount of overlapping between them. Reliability analysis of an item involves studying its characteristics expressed by the probability that it will perform a required function under stated conditions for a stated period of time. If such an analysis is extended to involve the study of the consequences of the failures in terms of possible damage to property and the environment or injuries/deaths of people, the study is referred to as safety analysis.

Risk is a combination of the probability and the degree of the possible injury or damage to health in a hazardous situation (British Standard (1991)). Safety is the ability of an entity not to cause, under given conditions, critical or catastrophic consequences. It is generally measured by the probability that an entity, under given conditions, will not cause critical or catastrophic consequences (Villemuer (1992)).

Safety assessment is a logical and systematic way to seek answers to a number of questions about the system under consideration. The assessment of the risk associated with an engineering system or a product may be summarised to answer the following three questions:

1. What can go wrong?
2. What are the effects and consequences?
3. How often will they happen?

The answer obtained from these questions will provide the information about the safety of the system. Such information is interesting but is of no practical significance unless there is a method for controlling and managing the risks associated with specific hazards to tolerable levels. Hence, a complete safety assessment will require a fourth question to be answered:

4. What measures need to be undertaken to reduce the risks and how can this be achieved?

Safety analysis can be generally divided into two broad categories, namely, quantitative and qualitative analysis (Wang and Ruxton (1997)). Depending on the safety data available to the analyst, either a quantitative or a qualitative safety analysis can be carried out to study the risk of a system in terms of the occurrence probability of each hazard and its possible consequences.

3.2 Qualitative Safety Analysis

Qualitative safety analysis is used to locate possible hazards and to identify proper precautions that will reduce the frequencies or consequences of such hazards. Generally this technique aims to generate a list of potential failures of the system under consideration. Since this method does not require failure data as an input to the analysis, it relies heavily on engineering judgement and past experience.

A common method employed in qualitative safety analysis is the use of a risk matrix method (Halebsky (1989), Tummala and Leung (1995)). The two parameters that are considered are the occurrence likelihood of the failure event and the severity of its possible consequences. Upon identifying all the hazards within the system under consideration, each hazard is evaluated in terms of these two parameters. The severity of all the failure events could be assessed in terms of the four categories (i.e. Negligible, Marginal, Critical and Catastrophic) as shown in Table 3.1.

The occurrence likelihood of an event is assessed qualitatively as frequent, probable, occasional, remote or improbable as depicted in Table 3.2 (Military Standard (1993)). Each of these categories can be represented quantitatively by a range of probabilities. For example, such a range of probabilities can be seen in column three of Table 3.2. This is to provide a rough guideline for the experts or analysts who are providing the information or carrying out the analysis.

It is reasonable to assign a high priority if the hazard has a catastrophic consequence and a frequent probability. On the other hand, it is also reasonable to assign a low priority if the hazard has a negligible consequence and an improbable probability. Based on this logic, certain acceptable criteria can be developed. All identified hazards can be prioritised corresponding to safety and reliability objectives by appropriate hazard indexes using the hazard severity and the corresponding hazard probabilities as shown in Table 3.3 (Military Standard (1980)). The hazard probabilities shown in this table are used to carry out qualitative analysis for a military defence system. These probabilities can be assigned appropriately when different systems are considered. If an identified hazard is assigned with a hazard index of 4C. 3D, 4D, 2E, 3E or 4E, it needs an immediate corrective action. A hazard with an index 3B, 4B, 2C, 2D or 3C would require a possible corrective action. Similarly, a hazard with index 3A, 4A, 2B, 1D or 1E would be tracked for a corrective action with low priority; or it may not warrant any corrective action. On the other hand, a hazard with index 1A, 2A, 1B or 1C might not even require a review for action.

All the identified hazards within the system under study can be evaluated using this method to produce a risk ranking based on the highest priority down to the lowest priority. A variation of this qualitative risk matrix approach will be presented in Chapter 5 with its application to the safety analysis of a ship.

3.3 Quantitative Safety Analysis

Quantitative safety analysis utilises what is known and assumed about the failure characteristics of each individual component to build a mathematical model that is associated with some or all of the following information:

- Failure rates.
- Repair rates.
- Mission time.
- System logic.
- Maintenance schedules.
- Human error.

Similar to the qualitative analysis, the occurrence probability of each system failure event and the magnitude of possible consequences are to be obtained. However, these parameters are to be quantified.

3.3.1 Event Probabilities

There are predominantly three methods that could be used to determine the occurrence probability of an event, namely (Preyssl (1995)):

1. Statistical method.
2. Extrapolation method.
3. Expert judgement method.

The statistical method involves the treatment of directly relevant test of experience data and the calculation of the probabilities. The extrapolation method involves the use of model prediction, similarity considerations and Bayesian concepts. Limited use of expert judgement is made to estimate unknown values as input to the extrapolation method. The expert judgement method involves direct estimation of probabilities by specialists.

These methods can be used together in an effective way to produce a reasonable estimate of the probability of an event occurring. The flowchart in Figure 3.1 shows the type of event probability produced depending on the available data.

3.3.2 Failure Probability Distributions

There are a number of probability distributions to model failures. The distribution types can be found in various sources (Henley and Kumamoto (1992), Hoover (1989), Law and Kelton (1982), Rubinstein (1981), Savic (1989)). The typical ones are listed as follows:

- Beta.
- Exponential.
- Gamma.

- Lognormal.
- Normal.
- Triangular.
- Uniform.
- Weibull.

In this Chapter, only two particular types of distributions (i.e. Exponential and Normal distributions) are briefly described.

For many items, the relationship of failure rate versus time can be commonly referred to as the "bathtub" curve. The idealised "bathtub" curve shown in Figure 3.2 has the following three stages:

1. Initial period

   The item failure rate is relatively high. Such failure is usually due to factors such as defective manufacture, incorrect installation, learning curve of equipment user, etc. Design should also aim at having a short "initial period".

2. Useful life.

   In this period of an item, the failure rate is constant. Failures appear to occur purely by chance. This period is known as the "useful life" of the item.

3. Wear-out period

   In this period of an item, the item failure rate rises again. Failures are often described as wear-out failures.

3.3.2.1 Exponential Distribution

A risk assessment mainly concentrates on the useful life in the "bathtub" curve in Figure 3.2. In the useful life region, the failure rate is constant over the period of time. In other words, a failure could occur randomly regardless of when a previous failure occurred. This results in a negative exponential distribution for the failure frequency. The failure density function of an exponential distribution is as follows:

$$f(t) = \lambda e^{-\lambda t}$$

where failure rate $\lambda = 1/MTBF$ and t = time of interest.

(*MTBF*: Mean Time Between Failure)

Failure probability of an item at time t is:

$$P(t) = 1 - e^{-\lambda t}$$

Example

Given that the Mean Time Between Failure for an item is 10,000 hours, calculate the failure probabilities of the item at t = 0, 10,000 and 100,000 hours if failures follow an exponential distribution.

Solution

$\lambda = 1/MTBF = 0.00001$ per hour

When $t = 0$, $P(0) = 1 - e^{-\lambda t} = 1 - e^{0} = 0$

When $t = 10{,}000$, $P(10{,}000) = 1 - e^{-\lambda t} = 1 - e^{-0.00001 \times 10{,}000} = 0.632$

When $t = 100{,}000$, $P(100{,}000) = 1 - e^{-\lambda t} = 1 - e^{-0.00001 \times 100{,}000} = 1$

From the above, it can be seen that at $t = 0$ the item does not fail and after a considerable time it fails.

3.3.2.2 Normal Distribution

Normal distributions are widely used in modelling repair activities. The failure density function of a normal distribution is:

$$f(t) = \frac{1}{\sqrt{2\pi}\sigma} e^{-(t-\mu)^2/2\sigma^2}$$

where μ = mean and σ^2 = standard deviation of t.

An application of this type of distribution can be seen in Chapter 8.

3.3.3 Event Consequences

The possible consequences of a system failure event can be quantified in terms of the possible loss of lives and property damage, and the degradation of the environment caused by the occurrence of the failure event (Smith (1985, 1992)). Experts of the particular operating situation normally quantify these elements in monetary terms. Quantifying human life in monetary terms could be difficult as it involves several moral issues that are constantly debated. Hence, it is normally expressed in terms of the number of fatalities (Henley and Kumamoto (1992)).

The process of risk assessment is initially performed qualitatively and later extended quantitatively to include data when it becomes available. The interactions and outcomes of both these methods are seen in Figure 3.3. Using the quantified method, risk evaluation can be carried out to determine the major risk contributors and the analysis can be attenuated to include cost benefit assessment of the risk control options.

3.4 Cause and Effect Relationship

As discussed in the previous two sections, safety analysis techniques can be initially categorised either as qualitative or quantitative methods. However, the way each analysis explores the relationship between causes and effects can be categorised further into four different categories, namely,

1. Deductive techniques.
2. Inductive techniques.
3. Exploratory techniques.
4. Descriptive techniques.

Deductive techniques start from known effects to seek unknown causes, whereas inductive techniques start from known causes to forecast unknown effects. Exploratory techniques establish a link between unknown causes to unknown effects while descriptive techniques link known causes to known effects. These four ways to investigate the relationship between causes and effects are illustrated in Table 3.4 (Pillay (2001)).

3.5 Preliminary Hazard Analysis (PHA)

Preliminary Hazard Analysis (PHA) was introduced in 1966 after the Department of Defence of the United States of America requested safety studies to be performed at all stages of product development. The Department of Defence issued the guidelines that came into force in 1969 (Military Standard (1969, 1999)).

Preliminary Hazard Analysis is performed to identify areas of the system, which will have an effect on safety by evaluating the major hazards associated with the system. It provides an initial assessment of the identified hazards. PHA typically involves:

1. Determining hazards that might exist and possible effects.
2. Determining a clear set of guidelines and objectives to be used during a design.
3. Creating plans to deal with critical hazards.
4. Assigning responsibility for hazard control (management and technical).
5. Allocating time and resources to deal with hazards.

"Brainstorming" techniques are used during which the design or operation of the system is discussed on the basis of the experience of the people involved in the brainstorming activity. Checklists are commonly used to assist in identifying hazards.

The results of the PHA are often presented in tabular form, which would typically include information such as but not limited to (Henley and Kumamoto (1992), Smith (1992), Villemuer (1992)):

1. A brief description of the system and its domain.
2. A brief description of any sub-systems identified at this phase and the boundaries between them.
3. A list of identified hazards applicable to the system, including a description and unique reference.
4. A list of identified accidents applicable to the system including a description, a unique reference and a description of the associated hazards and accident sequences.
5. The accident risk classification.
6. Preliminary probability targets for each accident.
7. Preliminary predicted probabilities for each accident sequence.
8. Preliminary probability targets for each hazard.
9. A description of the system functions and safety features.
10. A description of human error which could create or contribute to accidents.

The advantages of using the PHA method include:

1. It identifies the potential for major hazards at a very early stage of project development.

2. It provides basis for design decisions.
3. It helps to ensure plant to plant and plant to environment compatibility.
4. It facilitates a full hazard analysis later.

The disadvantage of PHA is that it is not comprehensive and must be followed by a full HAZard and OPerability (HAZOP) study.

3.5.1 Subsystem Hazard Analysis/System Hazard Analysis

Subsystem Hazard Analysis (SSHA) or System Hazard Analysis (SHA) is one requiring detailed studies of hazards, identified in the PHA, at the subsystem and system levels, including the interface between subsystems and the environment, or by the system operating as a whole. Results of this analysis include design recommendations, changes or controls when required, and evaluation of design compliance to contracted requirements. Often subsystem and system hazards are easily recognised and remedied by design and procedural measures or controls. These hazards are often handled by updating and expanding the PHA, with timing of the SSHA/SHA normally determined by the availability of subsystem and system design data (usually begins after the preliminary design review and completed before the critical design review).

3.5.2 Operating and Support Hazard Analysis

Operating and Support Hazard Analysis (OSHA) is an analysis performed to identify those operating functions that may be inherently dangerous to test, maintenance, handling, transportation or operating personnel or in which human error could be hazardous to equipment or people. The information for this analysis is normally obtained from the PHA. The OSHA should be performed at the point in system development when sufficient data is available, after procedures have been developed. It documents and evaluates hazards resulting from the implementation of operations performed by personnel. It also considers:

1. The planned system configuration at each phase of activity.
2. The facility interfaces.
3. The planned environments.
4. The support tools or other equipment specified for use.
5. The operation or task sequence.
6. Concurrent task effects and limitations.
7. Regulatory or contractually specified personnel safety and health requirements.
8. The potential for unplanned events including hazards introduced by human error.

OSHA identifies the safety requirements (or alternatives) needed to eliminate identified hazards or to reduce the associated risk to an acceptable level.

3.6 What-If Analysis

What-If analysis uses a creative team brainstorming "what if" questioning approach to the examination of a process to identify potential hazards and their consequences. Hazards are identified, existing safeguards noted, and qualitative severity and likelihood ratings are assigned to aid in risk management decision making. Questions that begin with "what-if" are formulated by engineering personnel experienced in the process or operation preferably in advance.

There are several advantages and disadvantages of using the What-If technique. The advantages include:

1. Team of relevant experts extends knowledge and creativity pool.
2. Easy to use.
3. Ability to focus on specific element (i.e. human error or environmental issues).

The disadvantages include:

1. Quality is dependent on knowledge, thoroughness and experience of team.
2. Loose structure that can let hazards slip through.
3. It does not directly address operability problems.

3.7 HAZard and OPerability (HAZOP) Studies

A HAZard and OPerability (HAZOP) study is an inductive technique, which is an extended Failure Mode, Effects and Criticality Assessment (FMECA). The HAZOP process is based on the principle that a team-approach to hazard analysis will identify more problems than when individuals working separately combine results.

The HAZOP team is made up of individuals with varying backgrounds and expertise. The expertise is brought together during HAZOP sessions and through a collective brainstorming effort that stimulates creativity and new ideas, a thorough review of the process under consideration is made. In short it can be applied by a multidisciplinary team using a checklist to stimulate systematic thinking for identifying potential hazards and operability problems, particularly in the process industries (Bendixen et al. (1984)).

The HAZOP team focuses on specific portions of the process called "nodes". A process parameter (e.g. flow) is identified and an intention is created for the node under consideration. Then a series of guidewords is combined with the parameter "flow" to create a deviation. For example, the guideword "no" is combined with the parameter "flow" to give the deviation "no flow". The team then focuses on listing all the credible causes of a "no flow" deviation beginning with the cause that can result in the worst possible consequences the team can think of at the time. Once the causes are recorded, the team lists the consequences, safeguards and any recommendations deemed appropriate. The process is repeated for the next deviation until completion of the node. The team moves on to the next node and repeats the process.

3.7.1 Guidewords, Selection of Parameters and Deviations

The HAZOP process creates deviations from the process design intent by combining guidewords (no, more, less, etc.) with process parameters resulting in a possible deviation from the design intent. It should be pointed out that not all guideword/parameter combinations would be meaningful. A sample list of guidewords is given below:

- No
- More
- Less
- As Well As
- Reverse
- Other Than

The application of parameters will depend on the type of process being considered, the equipment in the process and the process intent. The most common specific parameters that should be considered are flow, temperature, pressure, and where appropriate, level. In almost all instances, these parameters should be evaluated for every node. The scribe shall document, without exception, the team's comments concerning these parameters. Additionally, the node should be screened for application of the remaining specific parameters and for the list of applicable general parameters. These should be recorded only if there is a hazard or an operability problem associated with the parameter. A sample set of parameters includes the following:

- Flow
- Temperature
- Pressure
- Composition
- Phase
- Level
- Relief
- Instrumentation

3.7.2 HAZOP Process

A HAZOP study can be broken down into the following steps (McKelvey (1988)):

1. Define the scope of the study.
2. Select the correct analysis team.
3. Gather the information necessary to conduct a thorough and detailed study.
4. Review the normal functioning of the process.
5. Subdivide the process into logical, manageable sub-units for efficient study and confirm that the scope of the study has been correctly set.
6. Conduct a systematic review according to the established rules for the procedure being used and ensure that the study is within the special scope.
7. Document the review proceedings.

8. Follow up to ensure that all recommendations from the study are adequately addressed.

The detailed description of the methodology can be found in (Bendixen et al. (1984), McKelvey (1988), Kletz (1992), Wells (1980)),

3.7.3 HAZOP Application to Fishing Vessels

To apply the HAZOP process for the study of a fishing vessel system, the conventional method given in the previous sub-section is modified and can be summarised as follows (Pillay (2001)):

1. Define the system scope and team selection
   - Firstly define the scope of the study and then accordingly select the appropriate team to be involved in the study
2. Describe the system
   - Describe the system in some detail. This description should clarify the intention of the system as a whole from an operational viewpoint.
   - The information generated here will help the analyst understand the system and its criticality to the safe operation of the vessel. The data will later prove to be useful when used to determine the consequences of component failure in Step 6 of the approach.
3. Break it down into smaller operations for consideration and identify each component within the considered system.
   - Having attained the overall picture, break it down into its sub-operations/routines. It is difficult to see all the problems in a complex process but when each individual process is analysed on its own, the chances are that little will be missed out. Ideally, each operation should be singled out, but it is frequently more convenient to consider more than one operation at a time due to its inter-relationship and dependency.
   - The identification of each component can be achieved by first looking at historical failure data that is available and then complementing it with components identified from equipment drawings. Component failure data can be obtained from logbooks, docking reports, Chief engineer's reports and maintenance reports.
4. Determine design intention for each component that is identified.
   - At this stage, the purpose or intention of each component is ascertained. This helps to determine the functional purpose of the specific operation and shows how it relates/interacts to achieve the process intentions.
5. Apply a series of guidewords to see how that intention may be frustrated.
   - This is the heart of HAZOP. Having decided the intention of a process, this stage analyses the ways in which it can go wrong.
   - Examples of guide words are as illustrated in Table 3.5.
6. For meaningful deviations from the intention, look for possible causes and likely consequences.

- At this stage, the root of the problem is identified and the possible consequences are predicted and complemented with any historical data available. The consequences are considered for four major categories (personnel, environment, equipment and operation). At this point, it is determined how the failure of a component will affect the safety and integrity in terms of these four categories.

7. Consider possible action to remove the cause or reduce the consequences.

   - A HAZOP team usually provides ideas to remove a cause or deal with the possible consequences. This could be suggestion of improvements in design, operational procedure, maintenance periods and redundancy arrangements. It would be very unusual for every single one of these actions to be put into practice, but at least a rational choice could be made.

8. Reiteration

   - Consider how the improvements will affect the operation of the system and re-evaluate what can go wrong (with the improvements incorporated).

These steps can be illustrated in the flowchart in Figure 3.4. There are several advantages of using HAZOP to assess the safety of fishing vessels. These include:

1. It is the most systematic and comprehensive PHA methodology.
2. It provides greatest safety assurance.
3. It can be used in conjunction with Human Error Analysis (HEA).
4. It is the only PHA to address both safety/operability problems and environmental hazards.

The HAZOP process can be time consuming and costly if it is not well prepared in advance and can be tedious if it is not well facilitated. A comprehensive HAZOP study will require many experts and a considerable duration.

3.8 Fault Tree Analysis (FTA)

Fault Tree Analysis (FTA) is a formal deductive procedure for determining combinations of component failures and human errors that could result in the occurrence of specified undesired events at the system level (Ang and Tang (1984)). It is a diagrammatic method used to evaluate the probability of an accident resulting from sequences and combinations of faults and failure events. This method can be used to analyse the vast majority of industrial system reliability problems. FTA is based on the idea that:

1. A failure in a system can trigger other consequent failures.
2. A problem might be traced backwards to its root causes.

The identified failures can be arranged in a tree structure in such a way that their relationships can be characterised and evaluated.

3.8.1 Benefits to Be Gained from FTA

There are several benefits of employing FTA for use as a safety assessment tool. These include:

1. The Fault Tree (FT) construction focuses the attention of the analyst on one particular undesired system failure mode, which is usually identified as the most critical with respect to the desired function (Andrews and Moss (2002)).
2. The FT diagram can be used to help communicate the results of the analysis to peers, supervisors and subordinates. It is particularly useful in multi-disciplinary teams with the numerical performance measures.
3. Qualitative analysis often reveals the most important system features.
4. Using component failure data, the FT can be quantified.
5. The qualitative and quantitative results together provide the decision-maker with an objective means of measuring the adequacy of the system design.

An FT describes an accident model, which interprets the relation between malfunction of components and observed symptoms. Thus the FT is useful for understanding logically the mode of occurrence of an accident. Furthermore, given the failure probabilities of the corresponding components, the probability of a top event occurring can be calculated. A typical FTA consists of the following steps:

1. System description.
2. Fault tree construction.
3. Qualitative analysis.
4. Quantitative analysis.

These steps are illustrated in Figure 3.5.

3.8.2 System Definition

FTA begins with the statement of an undesired event, that is, failed state of a system. To perform a meaningful analysis, the following three basic types of system information are usually needed:

1. Component operating characteristics and failure modes: A description of how the output states of each component are influenced by the input states and internal operational modes of the component.
2. System chart: A description of how the components are interconnected. A functional layout diagram of the system must show all functional interconnections of the components.
3. System boundary conditions: These define the situation for which the fault tree is to be drawn.

3.8.3 Fault Tree Construction.

FT construction, which is the first step for a failure analysis of a technical system, is generally a complicated and time-consuming task. An FT is a logical diagram constructed by deductively developing a specific system failure, through branching intermediate fault events until a primary event is reached. Two categories of graphic symbols are used in an FT construction, logic symbols and event symbols.

The logic symbols or logic gates are necessary to interconnect the events. The most frequently used logic gates in the fault tree are **AND** and **OR** gates. The **AND** gate produces an output if all input events occur simultaneously. The **OR** gate yields output events if one or more of the input events are present.

The event symbols are rectangle, circle, diamond and triangle. The rectangle represents a fault output event, which results from combination of basic faults, and/or intermediate events acting through the logic gates. The circle is used to designate a primary or basic fault event. The diamond describes fault inputs that are not a basic event but considered as a basic fault input since the cause of the fault has not been further developed due to lack of information. The triangle is not strictly an event symbol but traditionally classified as such to indicate a transfer from one part of an FT to another. Figure 3.6 gives an example of a fault tree. The fault tree in Figure 3.6 is constructed using Fault Tree+ (Isograph Limited (1995)). In the fault tree in Figure 3.6, it can be seen that the occurrence probabilities of basic events *A*, *B* and *C* are assumed to be 0.1, 0.2 and 0.3 under certain conditions for a given period of time, respectively.

To complete the construction of a fault tree for a complicated system, it is necessary first to understand how the system works. This can be achieved by studying the blue prints of the system (which will reflect the interconnections of components within the system). In practice, all basic events are taken to be statistically independent unless they are common cause failures. Construction of an FT is very susceptible to the subjectivity of the analyst. Some analysts may perceive the logical relationships between the top event and the basic events of a system differently. Therefore, once the construction of the tree has been completed, it should be reviewed for accuracy, completeness and checked for omission and oversight. This validation process is essential to produce a more useful FT by which system weakness and strength can be identified.

3.8.4 Qualitative Fault Tree Evaluation

Qualitative FTA consists of determining the minimal cut sets and common cause failures. The qualitative analysis reduces the FT to a logically equivalent form, by using the Boolean algebra, in terms of the specific combination of basic events sufficient for the undesired top event to occur (Henley and Kumamoto (1992)). In this case each combination would be a critical set for the undesired event. The relevance of these sets must be carefully weighted and major emphasis placed on those of greatest significance.

3.8.5 Quantitative Fault Tree Evaluation

In an FT containing independent basic events, which appear only once in the tree structure, then the top event probability can be obtained by working the basic event probabilities up through the tree. In doing so, the intermediate gate event probabilities are calculated starting at the base of the tree and working upwards until the top event probability is obtained.

When trees with repeated events are to be analysed, this method is not appropriate since intermediate gate events will no longer occur independently. If this method is used, it is entirely dependent upon the tree structure whether an overestimate or an underestimate of the top event probability is obtained. Hence, it is better to use the minimal cut-set method.

In Boolean algebra, binary states 1 and 0 are used to represent the two states of each event (i.e. occurrence and non-occurrence). Any event has an associated Boolean variable. Events A and B can be described as follows using Boolean algebra:

$$A = \begin{cases} 1 & event\ occurs \\ 0 & event\ does\ not\ occur \end{cases}$$

$$B = \begin{cases} 1 & event\ occurs \\ 0 & event\ does\ not\ occur \end{cases}$$

Suppose "+" stands for "*OR*" and "·" for "*AND*". Suppose "$\overline{A}$" stands for "not A". Then the typical Boolean algebra rules are described as follows:

Identity laws

$$A + 0 = A$$

$$A + 1 = 1$$

$$A \cdot 0 = 0$$

$$A \cdot 1 = A$$

Indempotent laws

$$A + A = A$$

$$A \cdot A = A$$

Complementative laws

$$A \cdot \overline{A} = 0$$

$$A + \overline{A} = 1$$

Commutative laws

$$A + B = B + A$$

$$A \cdot B = B \cdot A$$

Associative laws

$$(A + B) + C = A + (B + C)$$

$$(A \cdot B) \cdot C = A \cdot (B \cdot C)$$

Distributive laws

$$A \cdot (B + C) = A \cdot B + A \cdot C$$

$$A + (B \cdot C) = (A + B) \cdot (A + C)$$

Absorption laws

$$A + A \cdot B = A$$

$$A \cdot (A + B) = A$$

De Morgan's laws

$$\overline{A \bullet B} = \overline{A} + \overline{B}$$

$$\overline{A + B} = \overline{A} \bullet \overline{B}$$

The above rules can be used to obtain the minimum cut sets leading to a top event in a fault tree. The occurrence probability of a top event can then be obtained from the associated minimum cut sets. The following two mini-trees are used to demonstrate how the occurrence probability of a top event can be obtained:

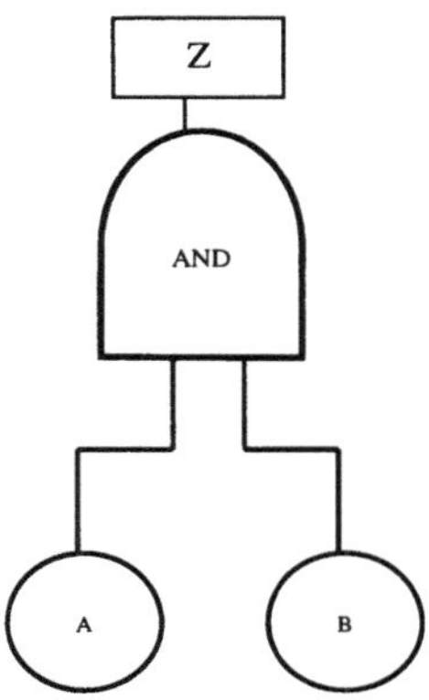

Obviously the minimum cut set for the mini-tree on the left is $A \cdot B$.

If one event is independent from the other, the occurrence probability of top event Z is

$$P(Z) = P(A \cdot B) = P(A) \times P(B)$$

where $P(A)$ and $P(B)$ are the occurrence probabilities of events A and B.

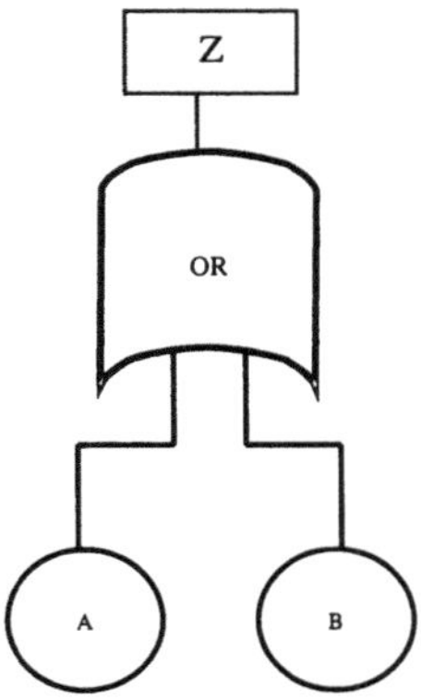

Obviously the minimum cut set for the mini-tree on the left is $A + B$.

If one event is independent from the other, the occurrence probability of top event Z is

$$P(Z) = P(A + B)$$

$$= P(A) + P(B) - P(A \cdot B)$$

$$= P(A) + P(B) - P(A) \times P(B)$$

where $P(A)$ and $P(B)$ are the occurrence probabilities of events A and B.

FTA may be carried out in the hazard identification and risk estimation phases of the safety assessment of ships to identify the causes associated with serious system failure events and to assess the occurrence likelihood of them. It is worth noting that in situations where there is a lack of the data available, the conventional FTA method may not be well suited for such an application. As such, a new modified method incorporating FTA and Fuzzy Set Theory (FST) will be presented and discussed in detail in Chapter 6.

3.8.6 FTA Example

An example

The risk assessment of a marine system is carried out at the early design stages. It has been identified that a serious hazardous event (top event) arises if

events *X1* and *X2* happen; or

event *X3* occurs.

X1 occurs when events *A* and *B* happen.

X2 occurs when

event *B* happens; or

events *B* and *C* occur.

Event *X3* occurs when

events *C* and *D* happen; or

events *A, C* and *D* happen.

Events *A, B, C* and *D* are basic events. It is assumed that events *A, B, C* and *D* follow an exponential distribution. The failure rates (1/hour) for events *A, B, C* and *D* are 0.0001, 0.0002, 0.0003 and 0.0004, respectively.

i. Draw the fault tree for the above problem.

ii. Find the minimum cut sets.

iii. Discuss how the likelihood of occurrence of the top event can be reduced/eliminated.

iv. Calculate the occurrence likelihood of the top event at time t = 10,000 hours assuming that events *A, B, C* and *D* are independent of each other.

Solution

i. The fault tree is built as shown in Figure 3.7.

ii. Top event

$$= X1 \cdot X2 + X3$$

$$= A \cdot B \cdot (B + B \cdot C) + C \cdot D + A \cdot C \cdot D$$

$$= A \cdot B \cdot B + C \cdot D$$

$$= A \cdot B + C \cdot D$$

iii. When events *A* and *B* or events *C* and *D* happen, the top event happens. Therefore, to avoid the occurrence of the top event, it is required to make sure that events *A* and *B* do not happen simultaneously and events *C* and *D* do not happen simultaneously. To reduce the occurrence likelihood of the top event, it is required to reduce the occurrence likelihood of four basic events *A, B, C* and *D*.

iv. At t = 10,000 hours

$$P(A) = 1 - e^{-\lambda t} = 1 - e^{-0.0001 \times 10{,}000} = 0.632$$

$$P(B) = 1 - e^{-\lambda t} = 1 - e^{-0.0002 \times 10{,}000} = 0.865$$

$$P(C) = 1 - e^{-\lambda t} = 1 - e^{-0.0003 \times 10{,}000} = 0.95$$

$$P(D) = 1 - e^{-\lambda t} = 1 - e^{-0.0004 \times 10{,}000} = 0.982$$

$$P(\text{Top event}) = P(A \cdot B + C \cdot D) = P(A \cdot B) + P(C \cdot D) - P(A \cdot B \cdot C \cdot D)$$

$$= P(A) \times P(B) + P(C) \times P(D) - P(A) \times P(B) \times P(C) \times P(D)$$

$$= 0.97$$

The likelihood of occurrence of the top event at time t = 10,000 hours is 0.97.

It should be noted that when calculating the failure probability of the top event, the application of the simplification rules may be required, This is demonstrated by the following example:

Example

Given that $P(A) = P(B) = P(C) = P(D) = 0.5$ and also that basic events *A, B, C* and *D* are independent, calculate $P(A \cdot B + B \cdot C + A \cdot C)$.

Solution

$$P(A \cdot B + B \cdot C + A \cdot C)$$

$$= P(A \cdot B) + P(B \cdot C + A \cdot C) - P(A \cdot B \cdot (B \cdot C + A \cdot C))$$

$$= P(A) \times P(B) + P(B \cdot C) + P(A \cdot C) - P(B \cdot C \cdot A \cdot C) - P(A \cdot B \cdot B \cdot C + A \cdot B \cdot A \cdot C)$$

$$= P(A) \times P(B) + P(B) \times P(C) + P(A) \times P(C) - P(A \cdot B \cdot C) - P(A \cdot B \cdot C + A \cdot B \cdot C)$$

$$= P(A) \times P(B) + P(B) \times P(C) + P(A) \times P(C) - P(A \cdot B \cdot C) - P(A \cdot B \cdot C)$$

$$= P(A) \times P(B) + P(B) \times P(C) + P(A) \times P(C) - 2 \times P(A \cdot B \cdot C)$$

$$= P(A) \times P(B) + P(B) \times P(C) + P(A) \times P(C) - 2 \times P(A) \times P(B) \times P(C)$$

$$= 0.5$$

The top events of a system to be investigated in FTA may also be identified through a PHA or may correspond to a branch of an event tree or a system Boolean representation table (Wang et al. (1995)). The information produced from FMECA may be used in construction of fault trees. Detailed description of FTA and its applications can be found in various published documents such as (Andrews and Moss (2002), Ang and Tang (1984), Halebsky (1989), Henley and Kumamoto (1992)).

3.9 Event Tree Analysis

In the case of standby systems and in particular, safety and mission-oriented systems, the Event Tree Analysis (ETA) is used to identify the various possible outcomes of the system following a given initiating event which is generally an unsatisfactory operating event or situation. In the case of continuously operated systems, these events can occur (i.e. components can fail) in any arbitrary order. In the ETA, the components can be considered in any order since they do not operate chronologically with respect to each other. ETA provides a systematic and logical approach to identify possible consequences and to assess the occurrence probability of each possible resulting sequence caused by the initiating failure event (Henley and Kumamoto (1992), Villemuer (1992)).

3.9.1 Event Tree Example

A simple example of an event tree is shown in Figure 3.8. In the event tree, the initiating event is "major overheats" in an engine room of a ship. It can be seen that when the initiating event "major overheats" takes place and if there is no fuel present, the consequences will be negligible in terms of fire risks. If there is fuel present, then it is required to look at if the detection fails. If the answer is no, then the consequences are minor damage, otherwise it is required to investigate if the sprinkler fails. If the sprinkler works, then the consequences will be smoke, otherwise it is required to see if the alarm system works. If the alarm system works, then the consequences will be major damage, otherwise injuries/deaths will be caused.

ETA has proved to be a useful tool for major accident risk assessments. Such an analysis can be effectively integrated into the hazard identification and estimation phases of a safety assessment programme. However, an event tree grows in width exponentially and as a result it can only be applied effectively to small sets of components.

3.10 Markov Chains

Markov methods are useful for evaluating components with multiple states, for example, normal, degraded and critical states (Norris (1998)). Consider the system in Figure 3.9 with three possible states, 0, 1 and 2 with failure rate λ and repair rate μ. In the Markovian model, each transition between states is characterised by a transition rate, which could be expressed as failure rate, repair rate, etc. If it is defined that:

$P_i(t)$ = probability that the system is in state i at time t.

$\rho_{ij}(t)$ = the transition rate from state i to state j.

and if it is assumed that $P_i(t)$ is differentiable, it can be shown that:

$$\frac{dP_i(t)}{dt} = \left(\sum_j \rho_{ij}(t) \right) \bullet P_i(t) + \sum_j \left(\rho_{ij}(t) \bullet P_j(t) \right)$$

If a differential equation is written for each state and the resulting set of differential equation is solved, the time dependent probability of the system being in each state is obtained (Modarres (1993)). Markov chains are mainly a quantitative technique, however, using the state and transition diagrams, qualitative information about the system can be gathered.

3.11 Failure Mode, Effects and Critical Analysis (FMECA)

The process of conducting a Failure Mode, Effects and Critical Analysis (FMECA) can be examined in two levels of detail. Failure Mode and Effects Analysis (FMEA) is the first level of analysis, which consists of the identification of potential failure modes of the constituent items (components or sub-systems) and the effects on system performance by identifying the potential severity of the effect. The second level of analysis is Criticality Analysis for criticality ranking of the items under investigation. Both of these methods are intended to provide information for making risk management decisions.

FMEA is an inductive process that examines the effect of a single point failure on the overall performance of a system through a "bottom-up approach" (Andrews and Moss (2002)). This analysis should be performed iteratively in all stages of design and operation of a system.

The first step in performing an FMEA is to organise as much information as possible about the system concept, design and operational requirements. By organising the system model, a rational, repeatable, and systematic means to analyse the system can be achieved. One method of system modelling is the system breakdown structure model - a top down division of a system (e.g. ship, submarine, propulsion control) into functions, subsystems and components. Block diagrams and fault-tree diagrams provide additional modelling techniques for describing the component/function relationships.

A failure mode is a manner that a failure is observed in a function, subsystem, or component (Henley and Kumamoto (1992), Villemuer (1992)). Failure modes of concern depend on the specific system, component, and operating environment. Failure modes are sometimes described as categories of failure. A potential failure mode describes the way in which a product or process could fail to perform its desired function (design intent or performance requirements) as described by the needs, wants, and expectations of the internal and external customers/users. Examples of failure modes are: fatigue, collapse, cracked, performance deterioration, deformed, stripped, worn (prematurely), corroded, binding, seized, buckled, sag, loose, misalign, leaking, falls off, vibrating, burnt, etc. The past history of a component/system is used in addition to understanding the functional requirements to determine relevant failure modes. For example, several common failure modes include complete loss of function, uncontrolled output, and premature/late operation (IMO (1995)).

The causes of a failure mode (potential causes of failure) are the physical or chemical processes, design defects, quality defects, part misapplication, or others, which are the reasons for failure (Military Standard (1980)). The causes listed should be concise and as complete as possible. Typical causes of failure are: incorrect material used, poor weld, corrosion, assembly error, error in dimension, over stressing, too hot, too cold, bad maintenance, damage, error in heat treat, material impure, forming of cracks, out of balance, tooling marks, eccentric, etc. It is important to note that more than one failure cause is possible for a failure mode; all potential causes of failure modes should be identified, including human error.

The possible effects are generally classified into three levels of propagation: local, next higher level, and end effect. An effect is an adverse consequence that the customer/user might experience. The customer/user could be the next operation, subsequent operations, or the end user. The effects should be examined at different levels in order to determine possible corrective measures for the failure (Military Standard (1980)). The consequences of the failure mode can be assessed by a severity index indicating the relative importance of the effects due to a failure mode. Some common severity classifications include (1) Negligible, (2) Marginal, (3) Critical and (4) Catastrophic.

Criticality analysis allows a qualitative or a quantitative ranking of the criticality of the failure modes of items as a function of the severity classification and a measure of the frequency of occurrence. If the occurrence probability of each failure mode of an item can be obtained from a reliable source, the criticality number of the item under a particular severity class may be quantitatively calculated as follows:

$$C = \sum_{i=1}^{N} E_i L_i t$$

where:

E_i = failure consequence probability of failure mode i (the probability that the possible effects will occur, given that failure mode i has taken place.

L_i = occurrence likelihood of failure mode i.

N = number of the failure modes of the item, which fall under a particular severity classification.

t = duration of applicable mission phase.

Once all criticality numbers of the item under all severity classes have been obtained, a criticality matrix can be constructed which provides a means of comparing the item to all others. Such a matrix display shows the distributions of criticality of the failure modes of the item and provides a tool for assigning priority for corrective action. Criticality analysis can be performed at different indenture levels. Information produced at low indenture levels may be used for criticality analysis at a higher indenture level. Failure modes can also be prioritised for possible corrective action. This can be achieved by calculating the Risk Priority Number (RPN) associated with each failure mode. This will be studied in detail in Chapter 7.

Part of the risk management portion of the FMEA is the determination of failure detection sensing methods and possible corrective actions (Modarres (1993)). There are many possible sensing device alternatives such as alarms, gauges and inspections. An attempt should be made to correct a failure or provide a backup system (redundancy) to reduce the effects propagation to rest of system. If this is not possible, procedures should be developed for reducing the effect of the failure mode through operator actions, maintenance, and/or inspection.

FMEA/FMECA is an effective approach for risk assessment, risk management, and risk communication concerns. This analysis provides information that can be used in risk management decisions for system safety. FMEA has been used successfully within many different industries and has recently been applied in maritime regulations to address safety concerns with relatively new designs. While FMEA/FMECA is a useful tool for risk management, it also has qualities that limit its application as a complete system safety approach. This technique provides risk analysis for comparison of single component failures only.

3.11.1 FMECA Example

Example

Table 3.6 shows an FMEA for a control system of a marine crane hoisting system (Wang (1994), Wang et al. (1995)). It can be seen that for the control system there are five failure modes. Failure mode rate is the ratio of the failure rate of the failure mode to the failure rate of the item. From Table 3.6 it can be seen that the sum of the five failure mode rates is equal to 1.

Suppose the failure consequence probabilities for the failure modes in Table 3.6 are 20%, 100%, 20%, 10% and 30%, respectively. The duration of interest is 10,000 hours. Formulate the criticality matrix of the above system.

Solution

From Table 3.6, it can be seen that failure mode 2 is classified as severity class 1, failure mode 3 as severity class 2 and failure mode 1 as severity class 2 while failure modes 4 and 5 are classified as severity class 4.

Severity class 1: Criticality number

$= E_2 \times L_2 \times t$

$= 1 \times 0.31 \times 0.000036 \times 10000$

$= 0.1116$

Severity class 2: Criticality number

$= E_3 \times L_3 \times t$

$= 0.2 \times 0.365 \times 0.000036 \times 10000$

$= 0.02628$

Severity class 3: Criticality number

$= E_2 \times L_2 \times t$

$= 0.2 \times 0.015 \times 0.000036 \times 10000$

$= 0.00108$

Severity class 4: Criticality number

$= E_4 \times L_4 \times t + E_5 \times L_5 \times t$

$= 0.1 \times 0.155 \times 0.000036 \times 10000 + 0.3 \times 0.155 \times 0.000036 \times 10000$

$=0.02233$

The criticality matrix can be formulated as follows:

Severity class	Criticality number
1	0.1116
2	0.02628
3	0.00108
4	0.2232

If the criticality matrices for other systems are produced, comparisons can be made to determine which system needs more attention in the design stages.

3.12 Other Analysis Methods

Apart from the methods described above, several other methods have gained popularity in the industry. Many of these methods have been developed to a very advanced stage and have been integrated with other analysis tools to enhance their applicability.

3.12.1 Diagraph-based Analysis (DA)

Diagraph-based Analysis (DA) is a bottom up, event-based, qualitative technique. It is commonly used in the process industry, because relatively little information is needed to set up the diagraph (Kramer and Palowitch (1987)). In a DA, the nodes correspond to the state variables, alarm conditions or failure origins and the edges represent the casual influences between the nodes. From the constructed diagraph, the causes of a state change and the manner of the associated propagation can be found out (Umeda (1980)). Diagraph representation provides explicit casual relationships among variables and events of a system with feedback loops. The DA method is effective when used together with HAZOP (Vaidhyanathan and Venkatasubramanian (1996)).

3.12.2 Decision Table Method

Decision table analysis uses a logical approach that reduces the possibility of omission, which could easily occur in a fault tree construction (Dixon (1964)). A decision table can be regarded as a Boolean representation model, where an engineering system is described in terms of components and their interactions (Wang et al. (1995)). Given sufficient information about the system to be analysed, this approach can allow rapid and systematic construction of the Boolean representation models. The final system Boolean representation table contains all the possible system top events and the associated cut sets. This method is extremely useful for analysing systems with a comparatively high degree of innovation since their associated top events are usually difficult to obtain by experience, from previous accident and incident reports of similar products, or by other means. A more detailed discussion on the use of this method for safety assessment can be found in (Wang (1994), Wang et al. (1995)).

3.12.3 Limit State Analysis

Limit state analysis is readily applicable to failure conditions, which occur when the demand imposed on the component, or system exceeds its capability. The probability of failure is the probability that the limit state functions are violated. These probabilities are estimated by the statistical analysis of the uncertainty or variability associated with the functions' variables. In most cases, the analytical solution of the probability of failure is very difficult and sometimes almost practically impossible. However, by incorporating the Monte Carlo simulation method this setback can be addressed. This method is normally used in structural reliability predictions and represents only half of a safety assessment (as it does not consider the severity of the failure) (Bangash (1983), Damkilde and Krenk (1997)).

3.13 Conclusion

In this Chapter, typical safety analysis methods are outlined in terms of their requirements, advantages and limitations. Some of these techniques have been successfully used in the industry and still continue to be used. However, the application of these conventional techniques to ship safety assessment may not be as straightforward as it may seem. Certain modifications are needed to enhance the application of such methods to the maritime industry. These modifications include the ability of the analysis methods to handle data that is associated with a high degree of uncertainty and the integration of expert opinion in a formal manner, where there is no bias of opinion.

The conventional methods can be used together within the framework of a formal safety assessment process. The formal safety assessment process will be described and discussed in Chapters 4 and 5, detailing how the analysis methods identified here can be used effectively together with some of the novel techniques described in the following Chapters of this book.

3.14 References (Chapter 3)

1. Andrews J.D. and Moss T.R., (2002) "Reliability and Risk Assessment", Professional Engineering Publishing Ltd, London and Bury St Edmunds, UK.
2. Ang A.H.S. and Tang W.H., (1984) "Probability Concepts in Engineering Planning and Design", John Wiley & Sons, UK.
3. Bangash Y., (1983) "Containment Vessel Design and Practice", Progress-in-Nuclear-Energy, Vol. 11, No. 2, pp. 107-181.
4. Bendixen L.M., O'Neil J.K. and Little A.D., (1984) "Chemical Plant Risk Assessment Using HAZOP and Fault Tree Methods", Plant/Operations Progress, Vol. 3, No. 3, pp. 179-184.
5. British Standard, (1991) "BS EN 292: Safety of Machinery - Basic Concepts, General Principles for Design, Part 1: Basic Terminology Methodology; Part 2: Technical Principles and Specification".
6. Damkilde L. and Krenk S., (1997) "Limits - a System for Limit State Analysis and Optimal Material Layout", Computers & Structures, Vol. 64, No. 1, pp. 709-718.
7. Dixon P., (1964) "Decision Tables and Their Applications", Computer and Automation, Vol. 13, No. 4, pp. 376-386.
8. Halebsky M., (1989) "System Safety Analysis Techniques as Applied to Ship Design", Marine Technology, Vol. 26, No. 3, pp. 245-251.
9. Henley E.J. and Kumamoto H., (1992) "Probabilistic Risk Assessment", IEEE Press, NY, USA.
10. Hoover S. and Perry R., (1989) "Simulation, a Problem-solving Approach", Addison-Wesley Publishing Company, USA.
11. IEE, (1999) "Health and Safety Information, Quantified Risks Assessment Techniques (Part 2) Event Tree Analysis – ETA", No. 26(b), Institution of Electrical Engineers, UK.
12. IMO, (1995) "International Code of Safety for High-Speed Craft", International Maritime Organisation, London, pp. 175-185.
13. Isograph Limited, (1995) "Fault Tree+ Version 6.0", Manchester, United Kingdom.
14. Kletz T.A., (1992) "HAZOP and HAZAN: Identifying and Assessing Process Industry Hazards", 3rd Edition, AIChE.

15. Kramer M.A. and Palowitch B.L., (1987) "A Rule Based Approach to Fault Diagnosis Using the Signed Directed Graph", AIChE Journal, Vol. 33, No. 7, pp. 1067-1077.

16. Law A.M. and Kelton W.D., (1982) "Simulation Modelling and Analysis", McGraw-Hill Book Company,

17. McKelvey T.C., (1988) "How to Improve the Effectiveness of Hazard and Operability Analysis", IEEE Transaction on Reliability, Vol. 37, No. 1, pp. 167-170.

18. Military Standard, (1969) "Department of Defence; System Safety Program Requirements", MIL-STD-882, USA.

19. Military Standard, (1980) "Procedures for Performing a Failure Mode, Effects and Criticality Analysis", MIL-STD-1629, November, AMSC Number N3074.

20. Military Standard, (1993) "System Safety Program Requirements", MIL-STD-882c, January, AMSC Number F6861.

21. Military Standard, (1999) "Department of Defence, Military Standards; System Safety Program Requirements", MIL-STD-882D, USA America.

22. Modarres M., (1993) "What Every Engineer Should Know about Reliability and Risk Analysis", Marcel Dekker Inc., NY, USA.

23. Norris J.R., (1998) "Markov Chains, Statistical and Probabilistic Mathematics: Series 2", Cambridge University, ISBN. 0521633966253.

24. Pillay A., (2001) "Formal Safety Assessment of Fishing Vessels", PhD Thesis, School of Engineering, Liverpool John Moores University, UK.

25. Preyssl C., (1995) "Safety Risk Assessment and Management- the ESA Approach", Reliability Engineering and System Safety, Vol. 49, No. 3, pp. 303-309.

26. Rubinstein R., (1981) "Simulation and the Monte Carlo Method", John Wiley & Sons, NY, USA.

27. Savic D., (1989) "Basic Technical Systems Simulation", Butterworths Basic Series.

28. Smith D.J., (1985) "Reliability and Maintainability and Perspective", 2nd Edition, Macmillan Publishers Ltd, London, UK.

29. Smith D.J., (1992) "Reliability, Maintainability and Risk", 4th Edition, Butterworths-Heinemann Ltd, Oxford, UK.

30. Tummala V.M.R. and Leung Y.H., (1995) "A Risk Management Model to Assess Safety and Reliability Risks", International Journal of Quality and Reliability Management, Vol. 13, No. 8, pp. 53-62.

31. Umeda T., Kuryama T.E., O'Shima E. and Matsuyama H., (1980) "A Graphical Approach to Cause and Effect Analysis of Chemical Processing Systems", Chem. Eng. Sci., Vol. 35, pp. 2379-2386.

32. Vaidhyanathan R. and Venkatasubramanian V., (1996) "Digraph-based Models for Automated HAZOP Analysis", Computer Integrated Process Operations Consortium (CIPAC), School of Chemical Engineering, Prudue University, USA.

33. Villemeur A., (1992) "Reliability, Availability, Maintainability and Safety Assessment, Vol.1: Methods and Techniques", John Wiley & Sons, Chichester, UK.

34. Wang J., (1994) "Formal Safety Analysis Methods and Their Application to the Design Process", PhD thesis, University of Newcastle upon Tyne, UK.

35. Wang J., Ruxton T. and Labrie C.R., (1995) "Design for Safety of Marine Engineering Systems with Multiple Failure State Variables", Reliability Engineering & System Safety, Vol. 50, No. 3, pp. 271-284.

36. Wang J. and Ruxton T., (1997) "A Review of Safety Analysis Methods Applied to the Design Process of Large Engineering Products", Journal of Engineering Design, Vol. 8, No. 2, pp. 131-152.

37. Wang J., Ruxton T. and Labrie C.R., (1995) "Design for Safety of Marine Engineering Systems with Multiple Failure State Variables", Reliability Engineering & System Safety, Vol. 50, No. 3, pp. 271-284.

38. Wells G.L., (1980) "Safety in Process Plants Design", John Wiley & Sons, NY, USA.

Table 3.1 Assessment of Hazard Severity and Categories

Hazard Consequences	Hazard severity	Category
Less than minor injury or less than minor system or environmental damage, etc	Negligible	1
Minor injury or minor system or environmental damage, etc	Marginal	2
Severe injury or major system or environmental damage, etc	Critical	3
Death, system loss or severe environmental damage, etc	Catastrophic	4

Table 3.2 Assessment of Hazard Probabilities and Levels

Hazard Categories	Qualitative	Quantitative	Level
Improbable	So unlikely, it can be assumed occurrence may not be experienced	The probability is less than 10^{-6}	A
Remote	Unlikely but possible to occur in the lifetime of an item	The probability is between 10^{-6} and 10^{-3}	B
Occasional	Likely to occur sometime in the life of an item	The probability is between 10^{-3} and 10^{-2}	C
Probable	Will occur several times in the life time of an item	The probability is between 10^{-2} and 10^{-1}	D
Frequent	Likely to occur frequently	The probability is greater than 10^{-1}	E

Table 3.3 Priority Matrix Based on Hazard Severity and Hazard Probability

Hazard probability	Hazard Severity			
	(1) Negligible	(2) Marginal	(3) Critical	(4) Catastrophic
(A) Improbable ($x < 10^{-6}$)	1A	2A	3A	4A
(B) Remote ($10^{-3} > x > 10^{-6}$)	1B	2B	3B	4B
(C) Occasional ($10^{-2} > x > 10^{-3}$)	1C	2C	3C	4C
(D) Probable ($10^{-1} > x > 10^{-2}$)	1D	2D	3D	4D
(E) Frequent ($x > 10^{-1}$)	1E	2E	3E	4E

Table 3.4 Ways to Investigate Cause-Effect Relationship

		Effects	
		Known	Unknown
Cause	Known	Descriptive techniques	Inductive techniques
	Unknown	Deductive techniques	Exploratory techniques

Table 3.5 Examples of Guidewords

Guide words	Examples
No	No flow, no signal
Less	Less flow, less cooling
More	Excess temperature, excess pressure
Opposite	Cooling instead of heating
Also	Water as well as lubricating oil
Other	Heating instead of pumping
Early	Opening the drain valve too soon
Late	Opening the drain valve too late
Part of	Incomplete drainage

Table 3.6 An Example of FMEA

Name		Control system			
Function		Controlling the servo hydraulic transmission system			
Failure rate		36 (failures per million hours)			
Failure mode no.	Failure mode rate	Failure mode	Effects on system	Detecting method	Severity
1	0.015	Major leak	Loss of hoisting pressure in lowering motion. Load could fall.	Self-annunciation	Critical (3)
2	0.31	Minor leak	None.	Self-annunciation	Negligible (1)
3	0.365	No output when required.	Loss of production ability.	Self-annunciation & by maintenance	Marginal (2)
4	0.155	Control output for lowering motion cannot be stopped when required.	Possibility of fall or damage of load. Possibility of killing and/or injuring personnel.	Self-annunciation & by maintenance	Catastrophic (4)
5	0.155	Control output for hoisting up motion cannot be stopped when required.	Possibility of fall or damage of load. Possibility of killing and/or injuring personnel.	Self-annunciation & by maintenance	Catastrophic (4)

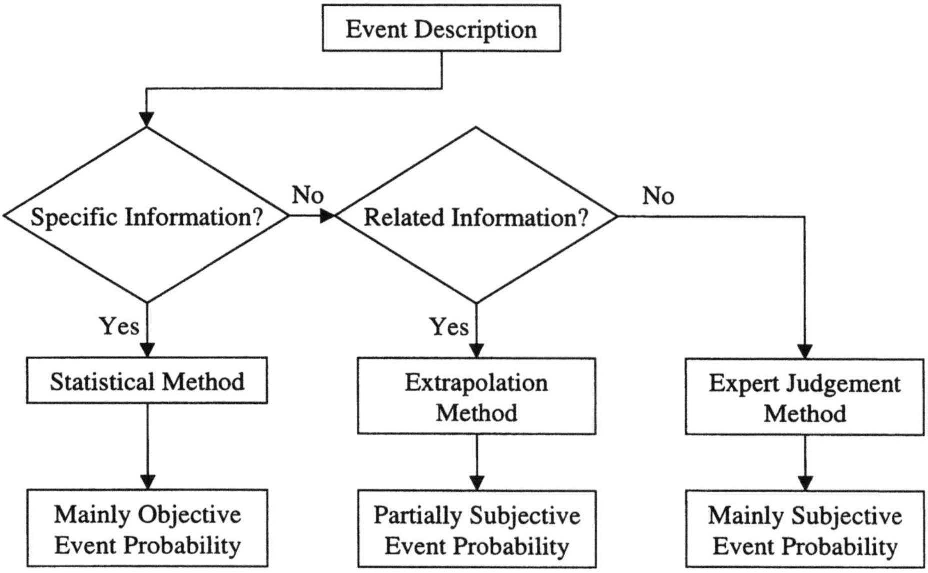

Figure 3.1 Event probability determination

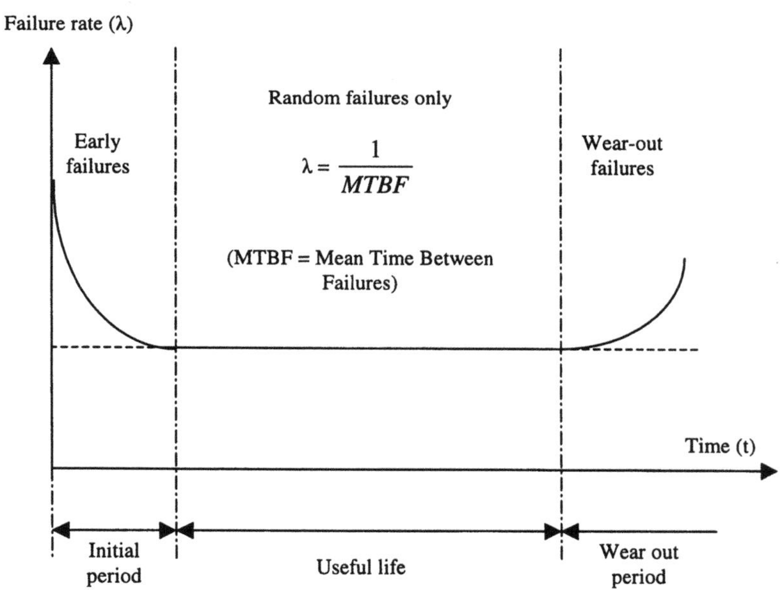

Figure 3.2 The "bathtub" curve

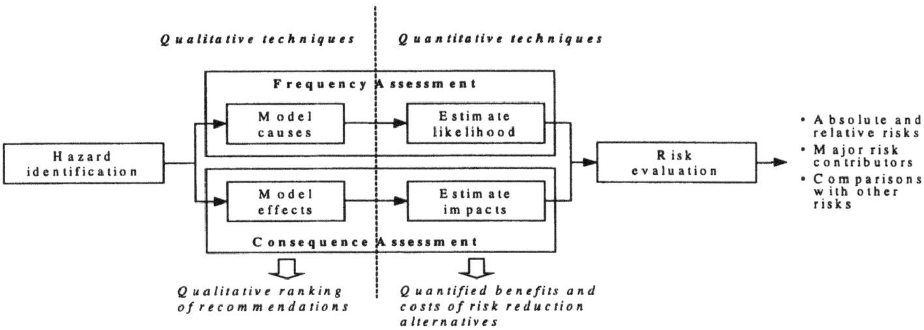

Figure 3.3 Qualitative and quantitative analysis

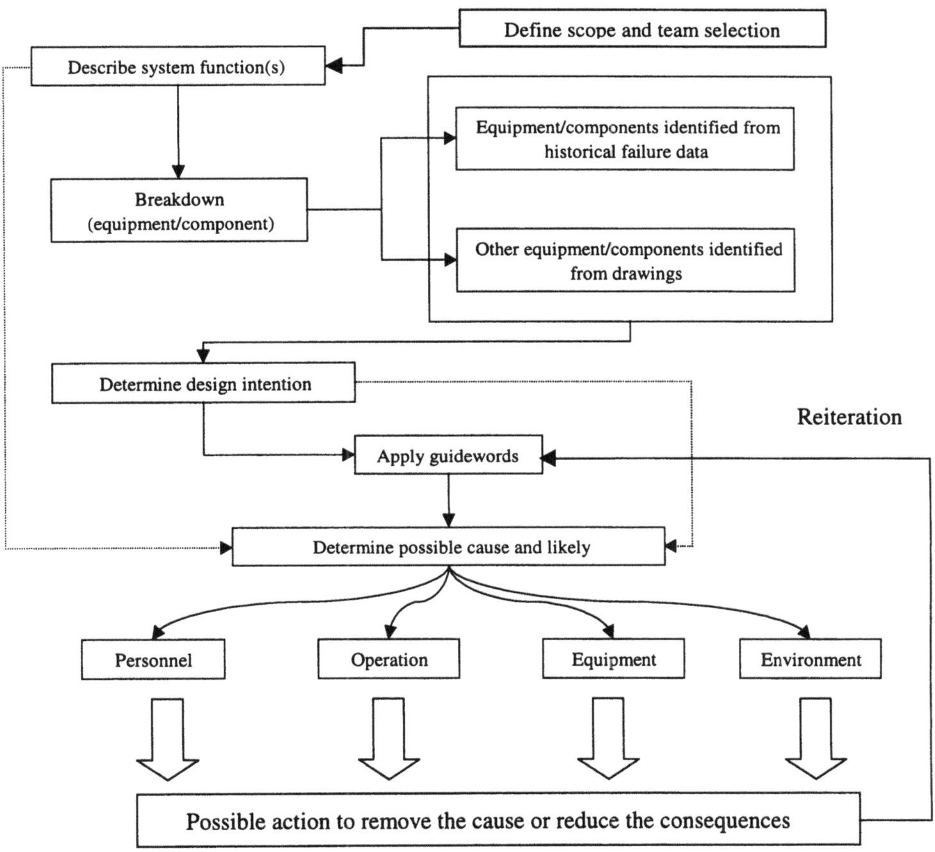

Figure 3.4 Flowchart of HAZOP process applied to fishing vessels

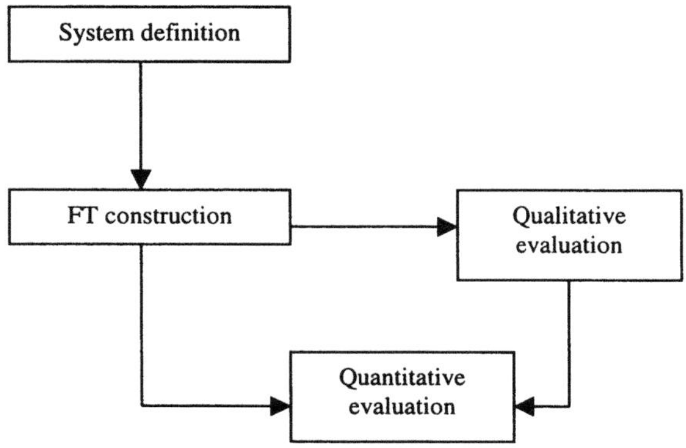

Figure 3.5 FTA method

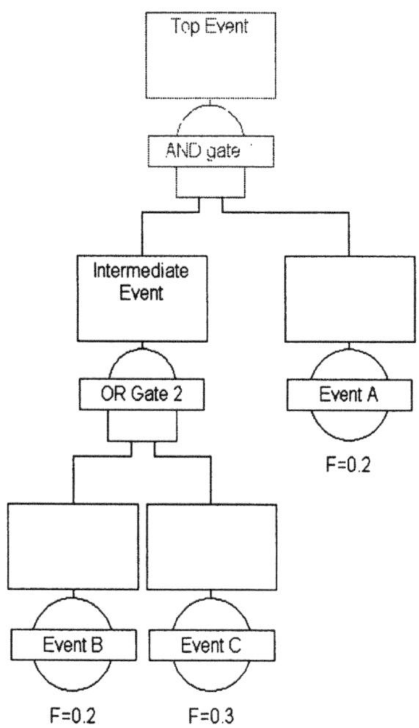

Figure 3.6 Fault tree example

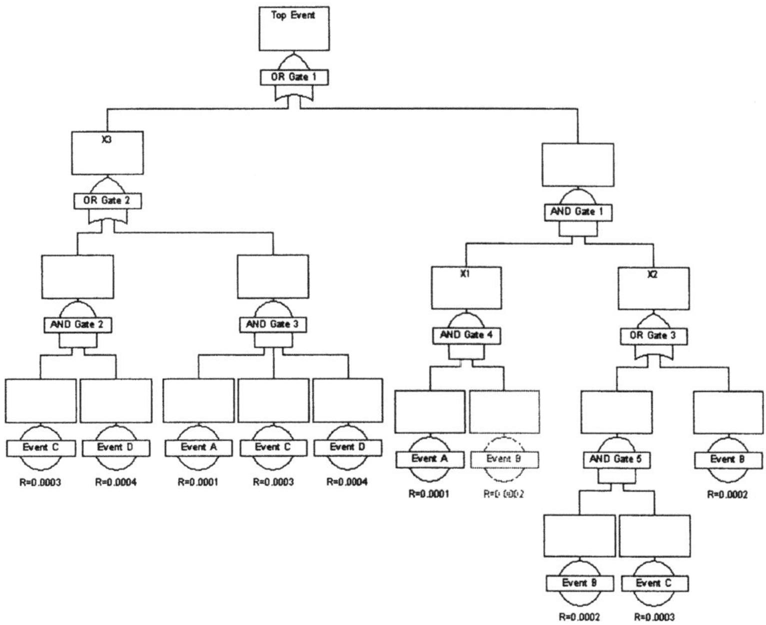

Figure 3.7 A fault tree

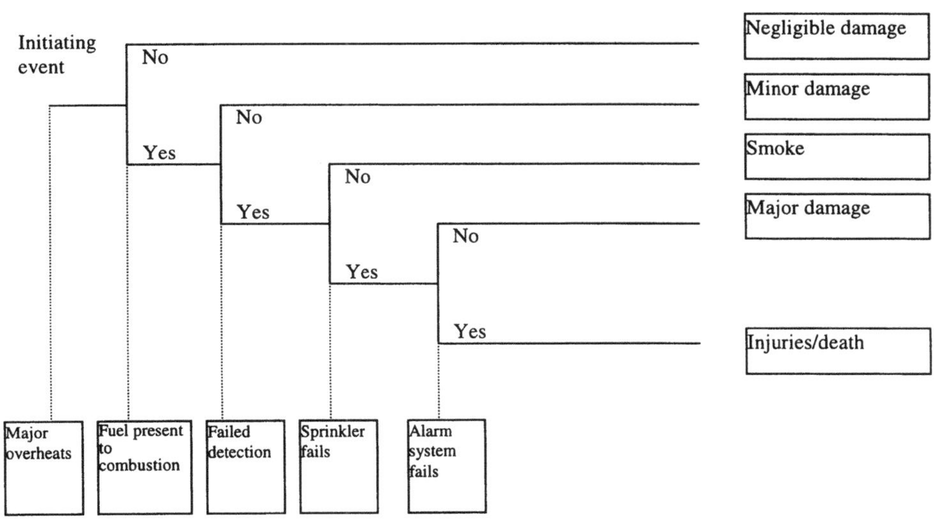

Figure 3.8 Example of an event tree

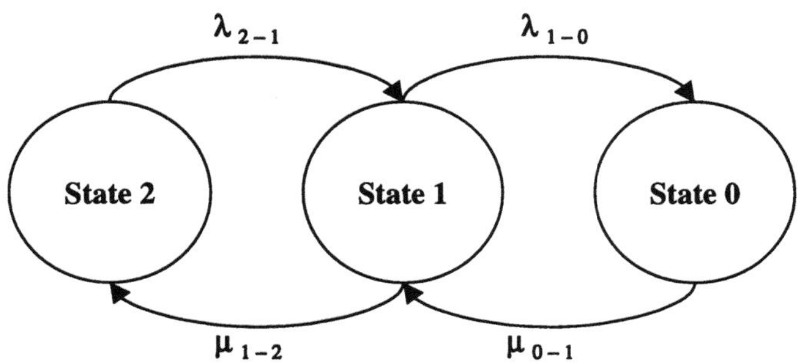

Figure 3.9 Markovian model for a system with three states

Chapter 4

Formal Safety Assessment of Ships and Its Relation to Offshore Safety Case Approach

Summary

This Chapter briefly describes both the offshore safety case approach and formal safety assessment of ships. The current practices and the latest development in safety assessment in both the marine and offshore industries are outlined. The relationship between the offshore safety case approach and formal ship safety assessment is described and discussed. The study of risk criteria in marine and offshore safety assessment is carried out. The recommendations on further work required are finally given.

Keywords: Formal safety assessment, marine safety, offshore safety, risk assessment, safety case.

4.1 Offshore Safety Assessment

Following the public inquiry into the Piper Alpha accident (Department of Energy (1990)), the responsibilities for offshore safety regulations were transferred from the Department of Energy to the Health & Safety Commission (HSC) through the Health & Safety Executive (HSE) as the single regulatory body for offshore safety. In response to the accepted findings of the Piper Alpha enquiry the HSE Offshore Safety Division launched a review of all offshore safety legislation and implemented changes. The changes sought to replace legislation which was seen as prescriptive with a more "goal setting" regime. The mainstay of the regulations is the health and safety at work act. Under that act, a draft of the offshore installations (safety case) regulations (SCR-1992) was produced in 1992 (HSE (1992)). It was then modified, taking into account the comments arising from public consultation. The regulations came into force in two phases - at the end of May 1993 for new installations and November 1993 for existing installations. The regulations require operational safety cases to be prepared for all offshore installations. Both fixed and mobile installations are included. Additionally all new fixed installations require a design safety case. For mobile installations the duty holder is the owner whereas for fixed installations the duty holder is the operator.

A safety case covers all aspects of the safety of the plant or process in question, and determines how the risks involved are to be minimised. It should include sufficient data to demonstrate that (HSE (1992)):

- Hazards with the potential to cause major accidents have been identified.

- Risks have been evaluated and measures have been taken to reduce them to a As Low As Reasonably Practicable level (ALARP).

A safety case should be prepared demonstrating safety by design, describing operational requirements, providing for continuing safety assurance by means of regular review, and setting out the arrangements for emergency response. It should also include identification of a representative sample of major accident scenarios and assessments of the consequences of each scenario together with an assessment in general terms of the likelihood of its happening. The report suggests that innovative safety analysis methods and cost-benefit analysis may be beneficially used for the prediction and control of safety.

The report on the public inquiry into the Piper Alpha accident (the Cullen Report) recommends Quantitative Risk Assessment (QRA) to be used in the process of hazard identification and risk assessment in preparing a safety case. QRA can help to provide a structured objective approach to the assessment of risks, provided that it relies on and is supplemented by good engineering judgement and the limitation of the data used is roughly understood. The significant pathway leading to serious failure conditions can be systematically identified using QRA and hence all reasonably practicable steps can be taken to reduce them.

The HSE framework for decisions on the tolerability of risk is shown in Figure 4.1 where there are three regions: intolerable, ALARP and broadly acceptable. Offshore operators must submit operational safety cases for all existing and new offshore installations to the HSE Offshore Safety Division for acceptance. An installation cannot legally operate without an accepted operational safety case. To be acceptable a safety case must show that hazards with the potential to produce a serious accident have been identified and that associated risks are below a tolerability limit and have been reduced as low as is reasonably practicable. For example, the occurrence likelihood of events causing a loss of integrity of the safety refuge should be less than 10^{-3} per platform year (Spouse (1997)) and associated risks should be reduced to an ALARP level.

Management and administration regulations (MAR-1995) were introduced to cover areas such as notification to the HSE of changes of owner or operator, functions and powers of offshore installation managers (HSE (1995b)). MAR-1995 is applied to both fixed and mobile offshore installations excluding sub-sea offshore installations. The importance of safety of offshore pipelines has also been recognised. As a result, pipeline safety regulations (PSR-1996) were introduced to embody a single integrated, goal setting, risk based approach to regulations covering both onshore and offshore pipelines (HSE, 1996d)). Fires and explosions may be the most significant hazards with potential to cause disastrous consequences in offshore installations. Prevention of fire and explosion, and emergency response regulations (PFEER-1995) were therefore developed in order to manage fire and explosion hazards and emergency response from protecting persons from their effects (HSC (1997)). A risk-based approach is promoted to be used to deal with problems involving fire and explosion, and emergency response. PFEER-1995 supports the general requirements by specifying goals for preventive and protective measures to manage fire and explosive hazards and to secure effective emergency response and ensure compliance with regulations by the duty holder.

After several years' experience of employing the safety case approach in the UK offshore industry, the safety case regulations were amended in 1996 to include verification of safety-critical elements and the offshore installations and wells (design and construction, etc.) regulations 1996 (DCR-1996) were introduced to deal with various stages of the life cycle of the installation (HSE (1996b)). From the earliest stages of the life cycle of the installation the duty holder must ensure that all safety-critical elements be assessed. Safety-critical elements are such parts of an installation and such of its plant (including computer programs), or any part thereof the failure of which could cause or contribute substantially to; or a purpose of which is to prevent, or limit the effect of a major accident (HSE (1996c)). In DCR-1996, a verification scheme is also introduced to ensure that a record is made of the safety-critical elements; comment on the record by an independent and competent person is invited; a verification scheme is drawn up by or in consultation with such person; a note is made of any reservation expressed by such person; and such scheme is put into effect (HSE (1996c)). All such records are subject to the scrutiny of the HSE at any time. More detailed information about the DCR-1996 can be found in (HSE (1996a, b, c)). DCR-1996 allows offshore operators to have more flexibility to tackle their own offshore safety problems. Offshore duty holders may use various safety assessment approaches and safety based decision making tools to study all safety-critical elements of offshore installations and wells to optimise safety. This may encourage offshore safety analysts to develop and employ novel safety assessment and decision making approaches and to make more efforts to deal with offshore safety problems.

The relationships between such typical offshore safety regulations can be seen in Figure 4.2 where the core is the safety case regulations and others are closely related to them.

Compliance with the current offshore safety regulations is achieved by applying an integrated risk-based approach, starting from feasibility studies and extending through the life cycle of the installation. This is achieved through stages of hazard identification for the life cycle of installation from concept design to decommissioning and the use of state-of-the-art risk assessment methods (Janardhanan and Grillo (1998)). In a risk-based approach, early considerations are given to those hazards which are not foreseeable to design out by progressively providing adequate measures for prevention, detection, control and mitigation and further integration of emergency response.

The main feature of the new offshore safety regulations in the UK is the absence of a prescriptive regime, defining specific duties of the operator and definition as regard to what are adequate means. The regulations set forth high level safety objective while leaving the selection of particular arrangements to deal with hazards in the hands of the operator. This is in recognition of the fact that hazards related to an installation are specific to its function and site conditions.

Recently, the industrial guidelines on a framework for risk related decision support have been produced by the UK Offshore Operators Association (UKOOA) (UKOOA (1999)). In general, the framework could be usefully applied to a wide range of situations. Its aim is to support major decisions made during the design, operation and abandonment of offshore installations. In particular, it provides a sound basis for evaluating the various options that need to be considered at the feasibility and concept selection stages of a project, especially with respect to "major accidents hazards" such as fire, explosion, impact, loss of stability, etc. It can also be combined with other formal decision making aids such as Multi-Attribute Utility Analysis (MAUA), Analytical Hierarchy Process (AHP) or decision trees if a more detailed or quantitative analysis of the various decision alternatives is desired.

It should be noted that there can be significant uncertainties in the information and factors that are used in the decision making process. These may include uncertainties in estimates of the costs, time-scales, risks, safety benefits, the assessment of stakeholder views and perceptions, etc. There is a need to apply common sense and ensure that any uncertainties are recognised and addressed.

The format of safety case regulations was advocated by Lord Robens in 1972 when he laid emphasis on the need for self regulation, and at the same time he pointed out the drawbacks of a rule book approach to safety (Sii (2001)). The concept of the safety case has been derived and developed from the application of the principles of system engineering for dealing with the safety of systems or installations for which little or no previous operational experience exists (Kuo (1998)). The five key elements of the safety case concepts are illustrated in Figure 4.3. These elements are discussed as follows:

1. Hazard identification

   This step is to identify all hazards with the potential to cause a major accident.

2. Risk estimation

   Once the hazards have been identified, the next step is to determine the associated risks. Hazards can generally be grouped into three risk regions known as the intolerable, tolerable and negligible risk regions as shown in Figure 4.1.

3. Risk reduction

   Following risk assessment, it is required to reduce the risks associated with significant hazards that deserve attention.

4. Emergency preparedness

   The goal of the emergency preparedness is to be prepared to take the most appropriate action in the event that a hazard becomes a reality so as to minimise its effects and, if necessary, to transfer personnel from a location with a higher risk level to another one with a lower risk level.

5. Safety management system

   The purpose of a safety management system (SMS) is to ensure that the organisation is achieving the goals safely, efficiently and without damaging the environment. One of the most important factors of the safety case is an explanation of how the operator's management system will be adapted to ensure that safety objectives are actually achieved.

A safety case is a written submission prepared by the operation of an offshore installation. It is a standalone document which can be evaluated on its own but has cross-references to other supporting studies and calculations. The amount of detail contained in the document is a matter of agreement between the operator and the regulating authority. In general, the following elements of an offshore installation are common for many safety cases:

1. A comprehensive description of the installation.
2. Details of hazards arising from the operation installation.
3. Demonstrations that risks from these hazards have been properly addressed and reduced to an ALARP level.
4. Description of the safety management system, including plans and procedures in place for normal and emergency operations.
5. Appropriate supporting references.

The following activities characterise the development of a safety case:

1. Establish acceptance criteria for safety, including environment and asset loss.

2. Consider both internal and external hazards, using formal and rigorous hazard identification techniques.
3. Estimate the frequency or probability of occurrence of each hazard.
4. Analyse the consequences of occurrence of each hazard.
5. Estimate the risk and compare with criteria.
6. Demonstrate ALARP.
7. Identify remedial measures for design, modification or procedure to reduce the frequency of occurrence or to mitigate the consequences.
8. Prepare the detailed description of the installation including information on protective systems and measures in place to control and manage risk.
9. Prepare a description of the safety management system and ensure that the procedures, which are appropriate for the hazards, are identified.

The following seven parts drawn from a safety case (Sii (2001), Wang (2002)) are subjects that can be found in a typical safety case for the operations of an offshore installation:

Part I Introduction and management summary

Part I of an operational safety case is an introduction and management summary. It will:

- Describe the scope and structure of the safety case.
- Describe the ownership and operatorship of the installation.
- Provide brief summaries of Part II to VII, highlighting major conclusions.

A summary of all key features contained in the safety case is outlined, including:

- Definition of the safety case.
- Objectives.
- Scope and structure of the seven parts of the safety case.
- Usage of the safety case.
- Custodian of the safety case.
- Review periods and updates.
- Application of the hazard management process to the operation.
- Hazard analysis of the operation.
- Remedial work.
- Conclusions drawn concerning the safety of the operation.

Part II: Operations safety management system

Part II is a concise description of the safety management system at the installation. It summarises both the corporate and installation specific policies, organisational structures, responsibilities, standards, procedures, processes, controls and resources which are in place to manage safety on safety case.

The six main sections of Part II cover the following:

- Policies and objectives.
- Organisation, responsibilities and resources.
- Standards and procedures.
- Performance monitoring.
- Audits and audit compliance.
- Management review and improvement.

Part III: Activities catalogue

Part III contains the activities catalogue which lists all safety activities applicable to the operation in the activity specification sheet. The activity specification sheet describes the activity and the hazard management objectives of that activity, safety related inputs and outputs, methods used to achieve the hazard management objectives along with management controls applied and the accountability for meeting the stated objectives. Any areas of concern arising from these sheets are noted as deficiencies.

Part IV: Description of operations

Part IV describes the essential features of the installation in sufficient detail to allow the effectiveness of safety systems to be appreciated. As such it describes the purpose of the installation and the processes performed there and its relationship to the location, reservoir and other facilities. Operational modes and manning for the installation are described, e.g. normal operation, shut down configurations, maintenance modes, etc.

The essence of Part IV is not to give a detailed physical description but to explain how the various systems relate to the safety of the installation and how their use can affect safety.

Part V: Hazard analysis, hazard register and manual of permitted operations (MOPO)

Part V provides a description of the hazards, their identification, ranking and assessment, the means by which they are to be controlled, and the recovery mechanisms. The design reviews and audits carried out to identify and assess hazards are also described.

Part V contains the following four sections.

- Hazard assessment.
- Hazard register (including the hazard/activity matrix).
- Safety critical operational procedures (SCOP).
- Manual of permitted operations (MOPO).

The sections are constructed as follows:

(a) A summary of all hazard investigations, design reviews and audits carried out, stating the major findings and recommendations from those investigations and the follow up of recommended action items.

(b) The hazard register, which describes each hazard in terms of

- The way it was identified.
- The methods used to assess the possible dangers presented by the hazard.
- The measures in place to control the hazard.

- The methods used to recover from any effects of the hazard.

Part V also contains the hazard/activity matrix, which cross-refers the activities identified in Part III with their effects on the identified hazards.

(c) The MOPO defines the limits of safety operation permitted when the defences are reduced, operating conditions are unusually severe or during accidental activities.

(d) A list of all safety critical operations procedures (SCOP), identifying the key hazard controls and recovery procedures required for the installation.

Part VI: Remedial action plan

Part VI records any deficiencies identified during the studies which lead to Parts II, III, IV and V, that require action to be taken. The record known as the "remedial action plan" includes:

- A statement of each identified deficiency.
- The proposed modifications to address the problem.
- An execution plan to show action parties and planned completion dates.

This remedial action plan will be used as the basis of the improvement plan, and as such the plan will be regularly reviewed and updated annually.

Part VII: Conclusion and statement of fitness

Part VII includes summaries of the major contributors to risk, the acceptance criteria for such risks, deficiencies identified and planned remedial actions.

Part VII ends with a "statement of fitness" which is the asset owner's statement that he appreciates and understands the hazards of the operation and considers that sufficient hazard control mechanisms are in place for the operation to continue. This statement is signed by the asset owner and approved by the signature of the operations directors.

In offshore safety analysis, it is expected to make safety based design/operation decisions at the earliest stages in order to reduce unexpected costs and time delays regarding safety due to late modifications. It should be stressed that a risk reduction measure that is cost-effective at the early design stages may not be ALARP at the late stage. HSE's regulations aim to have risk reduction measures identified and in place as early as possible when the cost of making any necessary changes is low. Traditionally, when making safety based design/operation decisions for offshore systems, the cost of a risk reduction measure is compared with the benefit resulting from reduced risks. If the benefit is larger than the cost, then it is cost-effective, otherwise it is not. This kind of cost benefit analysis based on simple comparisons has been widely used as a general principle in offshore safety analysis.

Conventional safety assessment methods and cost benefit analysis approaches can be used to prepare a safety case. As the safety culture in the offshore industry changes, more flexible and convenient risk assessment methods and decision making approaches can be employed to facilitate the preparation of a safety case. The UKOOA framework for risk related decision support can provide an umbrella under which various risk assessment and decision making tools are employed.

The guidelines in the UKOOA framework set out what is generally regarded in the offshore industry as good practice. These guidelines are a living document. The experience in the application of the framework changes in working practices, the business and social environment and new technology may cause them to need to be reviewed and updated to ensure that they continue to set out good practice. It should be noted that the framework produced by the UKOOA is only applicable to risks falling within the ALARP region shown in Figure 4.1.

Life-cycle approach is required to manage the hazards that affect offshore installations. It should be noted that offshore safety study has to deal with the boundaries of other industries such as marine operations and aviation. In offshore safety study, it is desirable to obtain the optimum risk reduction solution for the total life cycle of the operation or installation, irrespective of the regulatory boundaries (UKOOA (1999)). The basic idea is to minimise/eliminate the source of hazard rather than place too high reliance on control and mitigatory measures. To reduce risks to an ALARP level, the following hierarchical structure of risk control measures should follow:

1. Elimination and minimisation of hazards by "inherently safer" design.
2. Prevention.
3. Detection.
4. Control.
5. Mitigation of consequences.

Decisions evolve around the need to make choices, either to do something or not to do something, or to select one option from a range of options. These can either take the form of rigid criteria, which must be achieved, or take the form of goals or targets which should be aimed for, but which may not be met. The UK offshore oil and gas industry operates in an environment where safety and environmental performances are key aspects of successful business. The harsh marine environment and the remoteness of many of the installations also provide many technical, logistic and operational challenges. Decision making can be particularly challenging during the early stages of design and sanction of new installations where the level of uncertainty is usually high.

In many situations, there may be several options which all satisfy the requirements. It may also be difficult to choose a particular option, which is not obviously the best. If this is the case, then there is a need to consider what is or may be "reasonably practicable" from a variety of perspectives and to identify and assess more than just the basic costs and benefits. The decision making process can be set up to (UKOOA (1999)):

1. Define the issue.
2. Examine the options.
3. Take the decision.
4. Implement, communicate and review the decision.

Making risk based decisions may be very challenging because it may be difficult to:

1. Ensure the choices have been properly selected and defined.
2. Find ways to set out criteria and objectives.
3. Identify risk issues and perceptions.
4. Assess the performance of options against aspects that may not be quantifiable or which may involve judgements and perceptions that vary or are open to interpretation.
5. Establish the relative importance of often widely different types of objectives and factors.

6. Deal with uncertainties in estimates, data and analyses.
7. Deal with conflicting objectives and aspects of performance.
8. Deal with differences in resolution of estimates, data and analyses – these may not be able to provide a fair reflection of the actual differences between the options being considered.
9. Deal with or avoid hidden assumptions or biases.

A narrow view in the decision making process may result in decisions creating problems in other areas later on. For example, in a lifecycle view of the project or installation, decisions made during design to cut engineering and installation costs may lead to higher operating costs, reducing the overall profitability of the venue.

Safety and risk factors in the decision making process include risk transfer, risk quantification, cost benefit analysis, risk levels and gross disproportion, risk aversion, perception, risk communication, stakeholders and uncertainties. In general, decision making can be carried out on a "technological" basis or based on value judgements and perception. It can be difficult to determine which basis is most appropriate to a given decision, especially where the different bases may indicate conflicting best outcomes. It is therefore important to understand the decision context and use this to identify the importance of the various decision bases in any given decision situation (UKOOA (1999)).

The factors that affect offshore safety based decision making include degree of novelty vs. well understood situation or practice; degree of risk trade-offs and uncertainties; strength of stakeholder views and risk perceptions; and degree of business and economic implications (UKOOA (1999)). The study of such factors will determine the basis on which decision making can be best conducted. Different means of calibrating or checking the basis of the decision need to be matched to the type of decision context as shown in Figure 4.4 (UKOOA (1999)). Calibration should be used to ensure that the basis of decision making has been properly assessed and is still appropriate. Decision calibration changes with decision context. As the design context moves from prescription to strong views and perceptions, means of calibration change from codes and standards to external stakeholder consultation through verification, peer review, benchmarking and internal stakeholder consultation.

The framework proposed by the UKOOA is also capable of reflecting the differences between the design for safety approach for fixed offshore installations operating in the UK continental shelf and mobile offshore installation operating in an international market. Fixed offshore installations in the UK continental shelf are usually uniquely designed and specified for the particular duty and environment, and their design basis can be set against very specific hazards and specific processing and operation requirements. Many of the more complex design decisions therefore often fall into the "Type B" context in the detailed framework shown in Figure 4.5. Mobile offshore installations have to operate in very different environments and tackle a wide range of operational activities and reservoir conditions (UKOOA (1999)). The design cannot be based on specific hazards or duties but needs to address more global duties and operating envelopes. It is less likely to make specific risk based approaches to the design. It is usually the case that the installation is designed around a generic operating envelope. Because many of the hazards are generally well understood, common solutions incorporated into code, standards and ship classification society rules have been developed. Therefore, many mobile offshore installation design decisions fall into the "Type A" context. Where neither codes and rules can be effectively applied nor traditional analysis can be carried with confidence, such installations may be categorised as the "Type C" context (UKOOA (1999)).

4.2 Formal Ship Safety Assessment

As serious concern is raised over the safety of ships all over the world, the International Maritime Organization (IMO) has continuously dealt with safety problems in the context of operation, management, survey, ship registration and the role of the administration. The improvement of safety at sea has been highly stressed. The international safety-related marine regulations have been governed by serious marine accidents that have happened. The lessons were first learnt from the accidents and then the regulations and rules were produced to prevent similar accidents to occur. For example, the capsize of the *Herald of Free Enterprise* in 1987 greatly affected the rule developing activities of the IMO (Cowley (1995)), Sekimizu (1997)). The accident certainly raised serious questions on operation requirements and the role of management, and stimulated discussions in those areas at the IMO. This finally resulted in the adoption of the International Management System (ISM) Code. The *Exxon Valdes* accident in 1989 seriously damaged the environment by the large scale oil spill. It facilitated the implementation of the international convention on Oil Pollution Preparedness, Response and Co-operation (OPRC) in 1990. Double hull or mid-deck structural requirements for new and existing oil tankers were subsequently applied (Sekimizu (1997)). The *Scandinavian Star* disaster in 1990 resulted in the loss of 158 lives. Furthermore, the catastrophic disaster of the *Estonia*, which capsized in the Baltic Sea in September 1994, caused more than 900 people to lose their lives. Those accidents highlighted the role of human error in marine casualties, and as a result, the new Standards for Training, Certificates and Watchkeeping (STCW) for seafarers were subsequently introduced.

After Lord Carver's report on the investigation of the capsize of the *Herald of Free Enterprise* was published in 1992, the UK Maritime & Coastguard Agency (MCA, previously named as Marine Safety Agency (MSA)) quickly responded and in 1993 proposed that the IMO should explore the concept of formal safety assessment and introduce it in relation to ship design and operation. This proposal was submitted to the 62nd session of the Maritime Safety Committee (MSC) held from 24-28 May 1993 (MSA (1993)). The IMO reacted favourably to the UK's formal safety assessment submission. At the 65th meeting of the MSC in May 1995, strong support was received from the member countries and a decision was taken to make formal safety assessment a high priority item on the MSC's agenda. Accordingly, the UK decided to embark on a major series of research projects to further develop an appropriate framework and conduct a trial application on the chosen subject of high speed passenger catamaran ferries. The framework produced was delivered to MSC 66 in May 1996, with the trial application programmed for delivery to MSC 68 in May 1997. An international formal safety assessment working group was formulated at MSC 66 and MSC 67 where draft international guidelines were generated, including all key elements of the formal safety assessment framework developed by the UK.

Several applications of formal safety assessment have been attempted by the IMO on various vessels and systems. These include the application to the transportation of dangerous goods on passenger/ro-ro cargo vessels (IMO (1998a)), the effects of introducing Helicopter Landing Areas (HLA) on cruise ships (IMO (1998b)), high speed catamaran passenger vessels (IMO (1997b)), novel emergency propulsion and steering devices for oil tankers (IMO (1998d)) and the trial application is on a bulk carrier (IMO (1998c), MCA (1998)). The IMO has approved the application of formal safety assessment for supporting rule making process (IMO (1997a), Wang (2001, 2002)).

Formal safety assessment in ship design and operation may offer great potential incentives. The application of it may:

1. Improve the performance of the current fleet, be able to measure the performance change and ensure that new ships are good designs.
2. Ensure that experience from the field is used in the current fleet and that any lessons learned are incorporated into new ships.
3. Provide a mechanism for predicting and controlling the most likely scenarios that could result in incidents.

The possible benefits have already been realised by many shipping companies. For example, the P & O Cruises Ltd in the UK has reviewed the implementation of risk assurance methods as a strategic project and proposed short-term/medium-term and long-term objectives (Vie and Stemp (1997)). Its short-term/medium-term objectives are to provide a reference point for all future risk assurance work; to develop a structure chart that completely describes vessel operation; to complete a meaningful hazard identification as the foundation of the data set; to enable identification of realistic options for vessel improvement; to be a justified record of modifications adopted or rejected; and to be capable of incorporating and recording field experience to ensure that the knowledge is not lost. The idea of formal safety assessment may well be fitted to the above objectives in order to improve the company's performance.

Formal safety assessment is a new approach to maritime safety which involves using the techniques of risk and cost-benefit assessment to assist in the decision making process. It should be noted that there is a significant difference between the safety case approach and formal safety assessment in terms of their application to design and operations (Wang (2002)). A safety case approach is applied to a particular ship, whereas formal safety assessment is designed to be applied to safety issues common to a ship type (such as high-speed passenger vessel) or to a particular hazard (such as fire). The philosophy of formal safety assessment is essentially same as the one of the safety case approach. Many ship owners have begun to develop their own ship safety cases. The major difference between such ship specific applications of the approach and its generic application by regulators is that whilst features specific to a particular ship cannot be taken into account in a generic application, the commonalities and common factors which influence risk and its reduction can be identified and reflected in the regulator's approach for all ships of that type (IMarE (1998)). This should result in a more rational and transparent regulatory regime being developed. Use of formal safety assessment by an individual owner for an individual ship on the one hand and by the regulator for deriving the appropriate regulatory requirements on the other hand, are entirely consistent (IMarE (1998)).

It has been noted that many leading classification societies including Lloyds Register of Shipping and American Bureau of Shipping are moving towards a risk based regime. It is believed that the framework of formal safety assessment can facilitate such a move.

A formal ship safety assessment framework proposed by the UK MCA consists of the following five steps:

1. The identification of hazards.
2. The assessment of risks associated with those hazards.
3. Ways of managing the risks identified.
4. Cost benefit assessment of the options.
5. Decisions on which options to select.

Formal safety assessment involves much more scientific aspects than previous conventions. The benefits of adopting formal safety assessment as a regulatory tool include (MSA (1993)):

1. A consistent regulatory regime which addresses all aspects of safety in an integrated way.
2. Cost effectiveness, whereby safety investment is targeted where it will achieve the greatest benefit.
3. A pro-active approach, enabling hazards that have not yet given rise to accidents to be properly considered.
4. Confidence that regulatory requirements are in proportion to the severity of the risks.
5. A rational basis for addressing new risks posed by ever changing marine technology3

4.3 Risk Criteria

Risk criteria are standards which represent a view, usually that of a regulator, of how much risk is acceptable/tolerable (HSE (1995a)). In the decision making process, criteria may be used to determine if risks are acceptable, unacceptable or need to reduce to an ALARP level. When QRA is performed, it is required to use numerical risk criteria. The offshore industry has extensively used QRA and significant experience has been gained. The shipping industry has functioned reasonably well for a long time without consciously making use of risk criteria. Recently QRA has been used extensively for ships carrying hazardous cargoes in port areas and for ships operating in the offshore industry (Spouse (1997)). It is noted that in general there is no quantitative criteria in formal safety assessment even for a particular type of ship although the MCA trial applications have used QRA to a certain extent. As time goes on, it is believed that more QRA will be conducted in marine safety assessment. Therefore, numerical risk criteria in the shipping industry need to be dealt with in more detail.

As described previously in this Chapter, risk assessment involves uncertainties. Therefore it may not be suitable to use risk criteria as inflexible rules. The application of numerical risk criteria may not always be appropriate because of uncertainties in inputs. Accordingly, acceptance of a safety case is unlikely to be based solely on a numerical assessment of risk.

Risk criteria may be different for different individuals. They would also vary between societies and alter with time, accident experience and changing expectation of life. Risk criteria can therefore only assist judgements and be used as guidelines for decision making.

In different industries, risk criteria are also different. For example, in the aviation industry, a failure with catastrophic effects must have a frequency less than 10^{-9} per aircraft flying hour. In the nuclear industry, the basic principles of the safety policy recommended by the International Commission Radiological Protection (ICRP) are that no practice shall be adopted unless it has a positive net benefit; that all exposures shall be kept As Low As Reasonably Achievable (ALARA), taking economic and social factors into account; and that individual radiation doses shall not exceed specific criteria (ICRP (1977)). There are no explicit criteria used by ICRP.

As far as risk criteria for ships are concerned, the general criteria may include: (1) the activity should not impose any risks which can reasonably be avoided; (2) the risks should not be disproportionate to the benefits; (3) the risks should not be unduly concentrated on particular individuals; and (4) the risks of catastrophic accidents should be a small proportion of the total (Spouse (1997)). More specifically, individual risk criteria and social risk criteria need to be defined. For example, the maximum tolerable risk for workers may be 10^{-6} per year according to the HSE industrial risk criteria. In the regions between the maximum tolerable and broadly acceptable levels, risks should be reduced to an ALARP level, taking costs and benefits of any further risk reduction into account (Wang (2001, 2002)).

4.4 Discussion and Conclusion

An offshore installation/a ship is a complex and expensive engineering structure composed of many systems and is usually unique with its own design/operational characteristics (Wang and Ruxton (1997)). Offshore installations/ships need to constantly adopt new approaches, new technology, new hazardous cargoes, etc. and each element brings with it a new hazard in one form or another. Therefore, safety assessment should cover all possible areas including those where it is difficult to apply traditional safety assessment techniques, some of which are described in Chapter 3. Such traditional safety assessment techniques are considered to be mature in many application areas. Depending on the uncertainty level/the availability of failure data, appropriate methods can be applied individually or in combination to deal with the situation. All such techniques can be integrated in a sense that they formulate a general structure to facilitate risk assessment.

Lack of reliable safety data and lack of confidence in safety assessment have been two major problems in safety analysis of various engineering activities. To solve such problems, further development may be required to develop novel and flexible safety assessment techniques for dealing with uncertainty properly and also to use decision making techniques on a rational basis.

In offshore safety assessment, a high level of uncertainty in failure data has been a major concern that is highlighted in the UKOOA's framework for risk related decision support. Different approaches need to be applied with respect to different levels of uncertainty.

Software safety analysis is another area where further study is required. In recent years, advances in computer technology have been increasingly used to fulfil control tasks to reduce human error and to provide operators with a better working environment in ships. This has resulted in the development of more and more software intensive systems. However, the utilisation of software in control system has introduced new failure modes and created problems in the development of safety-critical systems. The DCR-1996 has dealt with this issue in the UK offshore industry. In formal ship safety assessment, every safety-critical system also needs to be investigated to make sure that it is impossible or extremely unlikely that its behaviour will lead to a catastrophic failure of the system and also to provide evidence for both the developers and the assessment authorities that the risk associated with the software is acceptable within the overall system risks (Wang (1997)).

The formal safety assessment philosophy has been approved by the IMO for reviewing the current safety and environmental protection regulations and studying any new element proposal by the IMO; and justifying and demonstrating a new element proposal to the IMO by an individual administration. Further applications may include the use of formal safety assessment for granting exemptions or accepting equivalent solutions for specific ships under the provisions of SOLAS Chapter 1 by an individual administration; for demonstrating the safety of a specific ship and its operation in compliance with mandatory requirements to the acceptance of the Flag Administration by an individual owner; as a management tool to facilitate the identification and control of risks as a part of the Safety Management System in compliance with the ISM Code by an individual owner. Several possible options regarding the application of formal safety assessment have been under investigation at the IMO. Among the possible application options, the individual ship approach may have the great impact on marine safety and change the nature of the safety regulations at sea since it may lead to deviation from traditional prescriptive requirements in the conventions towards performance-based criteria. This may be supported by ship type specific information. However, this would raise concerns due to the difficulty in the safety evaluation process by other administrations particularly when acting as port states although the merits of it may also be very significant. At the moment, unlike in the UK offshore industry, there is no intention to put in place a requirement for individual ship safety cases.

It is also very important to take into account human error problems in formal safety assessment. Factors such as language, education and training, that affect human error, need to be taken into account. The application of formal safety assessment may also encourage the Flag States to collect operation data. Another important aspect that needs to be considered is the data problem. The confidence of formal safety assessment greatly depends on the reliability of failure data. If formal safety assessment is applied, it may facilitate the collection of useful data on operational experience which can be used for effective pro-active safety assessment.

More test case studies also need to be carried out to evaluate and modify formal ship safety assessment and associated techniques and to provide more detailed guidelines for the employment of them. This would enable validation of them and can also direct the further development of flexible risk modelling and decision making techniques and facilitate the technology transfer to industries.

It is clear that it would be possible to reduce marine accidents by good design, training, and operation in an appropriate systematic management system. As the public concern regarding maritime safety increases, more and more attention has been directed to the wide application of formal safety assessment of ships as a regulatory tool. It is believed that the adoption of such a tool in ship design and operation will reduce maritime risks to a minimum level.

4.5 References (Chapter 4)

1. Cowley J., (1995) "The Concept of the ISM Code", Proceeding of Management and Operation of Ships: Practical Techniques for Today and Tomorrow, The Institute of Marine Engineers, 24-25 May, London, Paper 1.
2. Department of Energy, (1990) "The Public Inquiry into the Piper Alpha Disaster", (Cullen Report), ISBN 0 10 113102, London.
3. HSC, (1997), "Prevention of Fire and Explosion, and Emergency Response on Offshore Installations", Health & Safety Executive, HSE Books, Suffolk, UK, ISBN: 0717613860.
4. HSE, (1992) "The Offshore Installations (Safety Case) Regulations 1992", Health & Safety Executive, ISBN 00118820559.
5. HSE, (1995a) "Generic Terms and Concepts in the Assessment and Regulation of Industrial Risks", Health & Safety Executive, Discussion Document DDE2, HSE Books, Suffolk, UK.

6. HSE, (1995b) "A Guide to the Offshore Installations and Pipeline Works (Management and Administration) Regulations 1995", Health & Safety Executive, HSE Books, Suffolk, UK, ISBN: 0717609383.
7. HSE, (1996a) "A Guide to the Integrity, Workplace Environment and Miscellaneous Aspects of the Offshore Installations and Wells (Design and Construction, Etc.) Regulations 1996", Health & Safety Executive, HSE Books, Suffolk, UK, ISBN 0717611647.
8. HSE, (1996b) "A Guide to the Installation Verification and Miscellaneous Aspects of Amendments by the Offshore Installations and Wells (Design and Construction, etc.) Regulations 1996 to the Offshore Installations (Safety Case) Regulations 1992", Health & Safety Executive, HSE Books, Suffolk, UK, ISBN 0717611930.
9. HSE, (1996c) "The Offshore Installations and Wells (Design and Construction, etc.) Regulations 1996, ISBN 0-11-054451-X, No. 913.
10. HSE, (1996d), "A Guide to the Pipelines Safety Regulations 1996", HSE Books, Suffolk, UK, ISBN: 0717611825.
11. ICRP, (1977) "Recommendations of the ICRP", ICRP Publication 26, International Commission on Radiological Protection, Pergamon Press, Oxford.
12. IMarE, (1998) "Proceeding of New Safety Culture", Organised by the Institute of Marine Engineers (IMarE) and the MCA, London, 4th December.
13. IMO, (1997a) "IMO/MSC Circular 829, Interim Guidelines for the Application of Formal Safety Assessment to the IMO Rule-Making Process", International Maritime Organization, London, 17th November.
14. IMO, (1997b) "FSA Trial Application to High Speed Passenger Catamaran Vessel", International Maritime Organization, MSC 68/14/2 & 68/INF, UK.
15. IMO, (1998a) "Trial Application of FSA to the Dangerous Goods on Passenger/Ro-Ro Vessels", MSC 69/INF.24, International Maritime Organization, Submitted by IMO Finland.
16. IMO, (1998b) "FSA Study on the Effects of Introducing Helicopter Landing Areas (HLA) on Cruise Ships", MSC 69/INF.31, International Maritime Organization, Submitted by IMO Italy.
17. IMO, (1998c) "Report of the Working Group on FSA", IMO MSC 66/WP 1.4, International Maritime Organization, United Kingdom.
18. IMO, (1998d) "Novel Emergency Propulsion and Steering Devices for Oil Tanker Analysed with the FSA Method", MSC 69/14/1, International Maritime Organization, Submitted by IMO Germany.
19. Janardhanan K. and Grillo P., (1998) "Latest UK Offshore Safety Legislation", The Journal of Offshore Technology, May, pp.37-39.
20. Kuo C., (1998) "Managing Ship Safety", LLP, London, ISBN 1-85978-841-6.
21. Loughran C., Pillay A., Wang J., Wall A. and Ruxton T., (2002) "A Preliminary Study of Fishing Vessel Safety", Journal of Risk Research, Vol.5, No.1, pp. 3-21.

22. MCA, (1998) "Formal Safety Assessment for Bulk Carriers" (Including Annexes A-I), Informal Paper submitted by UK to IMO/MSC, 70th session, London, 27th November, (IMO/MSC 70/INF PAPER).

23. MSA, (1993) "Formal Safety Assessment MSC66/14", Submitted by the United Kingdom to IMO Maritime Safety Committee, Marine Safety Agency, 1993.

24. Sekimizu K., (1997) "Current Work at IMO on Formal Safety Assessment", Proceeding of Marine Risk Assessment: A Better Way to Manage Your Business, The Institute of Marine Engineers, London, 8-9 April.

25. Sii H.S., (2001) "Marine and Offshore Safety Assessment", PhD Thesis, Staffordshire University/Liverpool John Moores University, UK.

26. Spouse J., (1997) "Risk Criteria for Use in Ship Safety Assessment", Proceeding of Marine Risk Assessment: A Better Way to Manage Your Business, The Institute of Marine Engineers, London, 8-9 April.

27. UKOOA, (1999) "Industry Guidelines on a Framework for Risk Related Decision Making", Published by the UK Offshore Operators Association, April.

28. Vie R.H. and Stemp J.B., (1997) "The Practical Application of Risk Assurance Technology Techniques to Cruise Vessel Design and Operation", Proceeding of Marine Risk Assessment: A Better Way to Manage Your Business, The Institute of Marine Engineers, London, 8-9 April.

29. Wang J., (1997) "A Subjective Methodology for Safety Analysis of Safety Requirements Specifications", IEEE Transactions on Fuzzy Systems, Vol. 5, No. 3, pp. 418-430.

30. Wang J., (2001) "Current Status of Future Aspects of Formal Safety Assessment of Ships", Safety Science, Vol. 38, pp. 19-30.

31. Wang J., (2002) "Offshore Safety Case Approach and Formal Safety Assessment of Ships", Journal of Safety Research, Vol. 33, No. 1, pp. 81-115.

32. Wang J. and Ruxton T., (1997) "A Review of Safety Analysis Methods Applied to the Design Process of Large Engineering Products", Journal of Engineering Design, Vol. 8, No. 2, pp. 131-152.

33. Wang J., Yang J. B., Sen P. and Ruxton T., (1996) "Safety Based Design and Maintenance Optimisation of Large Marine Engineering Systems", Applied Ocean Research, Vol. 18, No. 1, pp. 13-27.

34. Yang J.B. and Sen P., (1994) "A General Multi-level Evaluation Process for Hybrid MADM with Uncertainty", IEEE Transactions on Systems, Man and Cybernetics, Vol. 24, pp. 1458-147

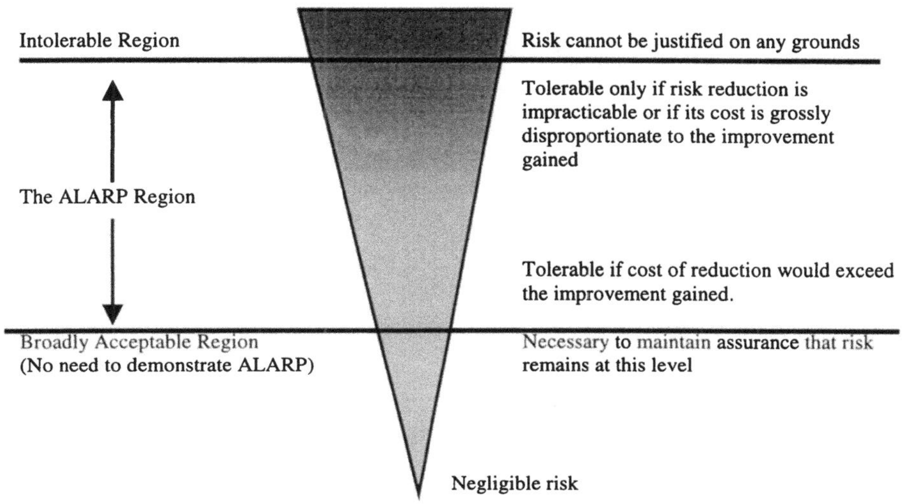

Figure 4.1 The HSE framework for decisions on the tolerability of risk

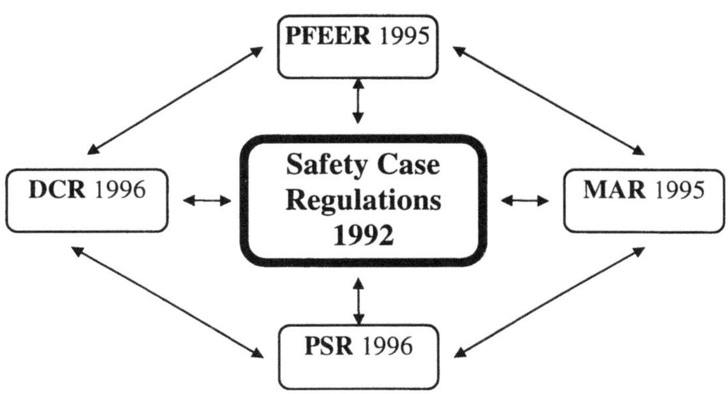

Figure 4.2 Relationships between offshore safety regulations

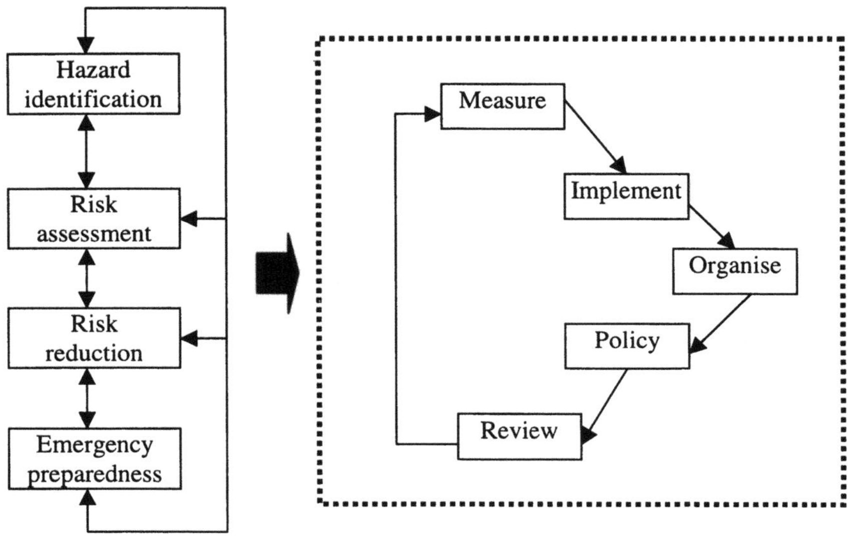

Figure 4.3 The five key elements of the safety case concepts

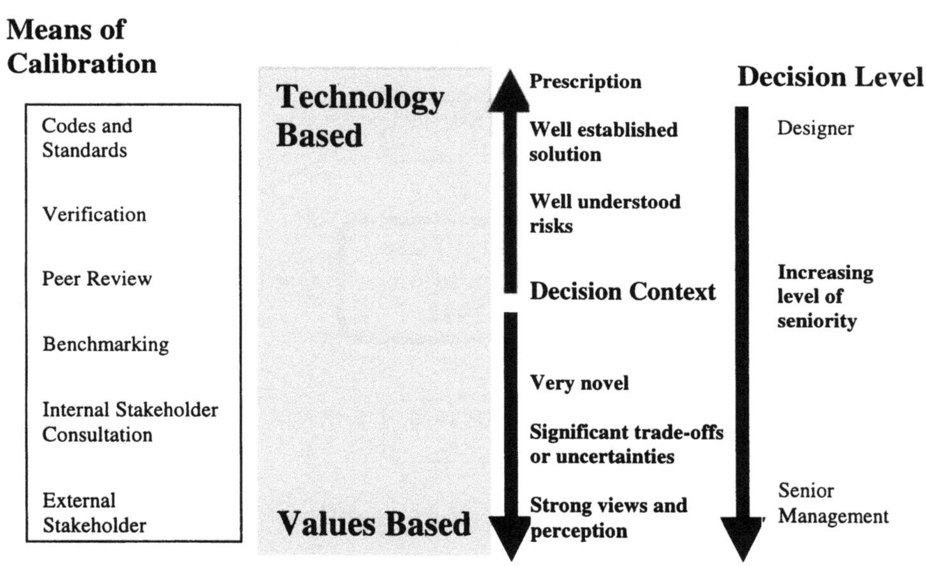

Figure 4.4 Decision levels

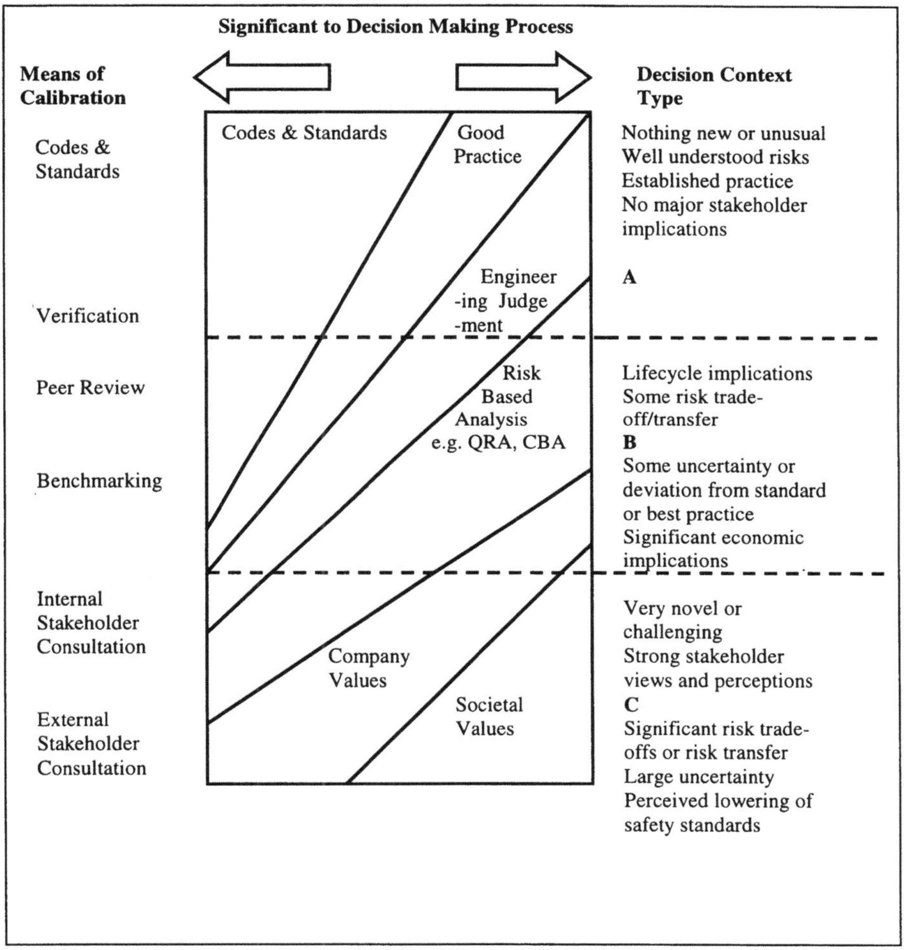

Figure 4.5 The detailed UKOOA framework

Chapter 5

Formal Safety Assessment

Summary

This Chapter discusses the inception of Formal Safety Assessment (FSA), originally proposed by the UK Maritime & Coastguard Agency, in the maritime applications. The FSA is applied to fishing vessels with an illustrative example. The application of the FSA framework to containerships is also described. Detailed discussions on several aspects of FSA's application to ship design and operation are given.

Keywords: Containerships, fishing vessels, formal safety assessment, HAZIN, risk, safety.

5.1 Formal Safety Assessment

Formal Safety Assessment (FSA) has as its objective the development of a framework of safety requirements for shipping in which risks are addressed in a comprehensive and cost effective manner. The adoption of FSA for shipping represents a fundamental cultural change, from a largely reactive approach, to one, which is integrated, proactive and soundly based upon the evaluation of risk.

As described in Chapter 4, the FSA framework consists of five steps. The interaction between the five steps can be illustrated in a process flowchart as shown in Figure 5.1. As it can be seen, there are repeated iterations between the steps, which makes FSA effective as it constantly checks itself for changes within the analysis. Each step within the FSA can be further broken down into individual tasks and is represented in Figure 5.2. The execution and documentation of each task is vital, as it will enable the preceding tasks/steps to be carried out with ease. In order for the assessment to be accurate, the analyst must understand and appreciate the objectives of each step and execute it without any "short-cuts".

Depending on the requirement of the safety analysts and the safety data available, either a qualitative or a quantitative safety analysis can be carried out to study the risks of a system in terms of the occurrence probability of each hazard and possible consequences. As described in Chapter 3, qualitative safety analysis is used to locate possible hazards and to identify proper precautions (design changes, administrative policies, maintenance strategies, operational procedures, etc.) that will reduce the frequencies or/and consequences of such hazards.

5.1.1 Step 1 - Hazard Identification

Various methods may be used individually or in a combination to carry out Step 1 of the FSA approach. Such typical methods include: Preliminary Hazard Analysis (PHA), Fault Tree Analysis (FTA), Event Tree Analysis (ETA), Cause-Consequence Analysis (CCA), Failure Mode, Effects and Criticality Analysis (FMECA), HAZard and OPerability analysis (HAZOP), Boolean Representation Method (BRM) and Simulation analysis (Henley and Kumamoto (1996), Smith (1992), Villemeur (1992), Wang (1994)). The use of these methods as safety analysis techniques has been reviewed in Chapter 3.

In the hazard identification phase, the combined experience and insight of engineers is required to systematically identify all potential failure events at each required indenture level with a view to assessing their influences on system safety and performance. This is achieved using "brainstorming" techniques. The hazard identification phase can be further broken down into several steps as follows:

Problem definition - Define the bounds of study, generic vessel and generic stakeholder for the vessel.

Problem identification - The problem boundaries of a formal safety assessment study can be developed in the following manner:

- Range of the vessel.
- Geographic boundaries.
- Risks to be considered.
- Vessel systems.
- Relevant regulations.
- Measures of risk.

In addition, the following factors specifically related to the vessel are defined:

- The generic vessel.
- Vessel accident category.
- Vessel stakeholders.
- Vessel operational stages.

Hazard identification - The HAZard IDentification (HAZID) consists of determining which hazards affect the vessel's activities under consideration using "brainstorming" techniques. At the HAZID session the following information is gathered:

- Operational stage.
- Vessel systems.
- Hazards, causes and consequences.

Structuring HAZID output - The approach to structuring the HAZID output is to convert the information gathered at the HAZID meeting into hazard worksheets which record the causes, accident sub-categories, consequences and the source of information. These hazard worksheets provide a means for recording the output from the HAZID meeting and other hazards identified during the analysis period, that is, from incident databases or interviews with the vessel personnel.

Risk exposure groups - The next step is to group the causes into risk exposure groups. This can be achieved by using the guidewords taken from the risk exposure source given in MSC 68/14 (IMO (1993)). The groups are then further sub-divided, during the hazard-structuring phase into risk exposure sub-groups. An example of this can be found in (MSC (1997b)). In order to sort the large amount of information collected at the HAZID meeting, accident sub-categories are established for each accident category and all the identified consequences are grouped according to the contributing factors.

Hazard screening - The purpose of hazard screening during Step 1 is to provide a quick and simple way of ranking hazards. It is a process of establishing, in broad terms, the risks of all identified accident categories and accident sub-categories, prior to the more detailed quantification, which will be conducted in Step 2. Risk is a combination of the frequency of occurrence of an accident type with the severity of its consequences. Possible consequences may be loss of lives, environmental pollution or damage to ship/cargo or financial loss. Accordingly, risk can also be read as the estimated loss in a given period of time. Two approaches can be used for the assignment of screening risk level in order to check the robustness of the resulting hazard rankings and to assist in the resolution of the rankings in cases where several hazards have similar ranking levels. These are:

1. Risk matrix approach (Loughran et al. (2002)).
2. Cumulative loss approach (MSC (1997a)).

5.1.2 Step 2 - Risk Estimation

Information produced from the hazard identification phase will be processed to estimate risk. In the risk estimation phase, the likelihood and possible consequences of each System Failure Event (SFE) will be estimated either on a qualitative basis or a quantitative basis (if the events are readily quantified). The risk estimation phase can be further broken down into several steps as follows:

Structuring of Risk Contribution Tree (RCT) - The causes and outcomes that were identified in Step 1 are structured in Step 2 for its employment in various parts of the Risk Contribution Tree (RCT). An RCT is structured in two distinct ways. Below the accident category, the structure is a graphical representation of the accident sub-categories and of the combinations of contributory factors relevant to each accident sub-category. Its structure is similar to a Fault Tree in its use of logical symbols, and the term "Contribution Fault Tree" has therefore been employed. Above the accident category level, the structure is an event tree representation of the development of each category of accident into its final outcome. An example of an RCT is shown in Figure 5.3.

Structuring and quantification of influence diagrams - The purpose of influence diagrams is to identify the influences, which determine the likelihood of an accident, and to enable those influences to be quantified. It also provides information for use in Step 3 of the FSA process. An example of an influence diagram for a fire accident in given in Appendix 3. An influence diagram takes into account three different types of influence, which are due to:

1. Hardware failure.
2. Human failure.

3. External event.

Additionally, each influence diagram incorporates dimensions of design, operation and recovery[1].

Quantification of RCT - The quantification of the RCT is accomplished by using available historical data from the incident database and where such data is absent, expert judgement is used to complement the quantification. The level of potential consequences of a SFE may be quantified in economic terms with regard to loss of lives/cargo/property and the degradation of the environment caused by the occurrence of the SFE. Finally, the calculation of FN (i.e. frequency (*F*) – fatality (*N*)) curves and Potential Loss of Life (PLL) through the RCT is carried out.

5.1.3 Step 3 - Risk Control Options (RCOs)

The next step aims to propose effective and practical Risk Control Options (RCOs). Focusing on areas of the risk profile needing control, several RCOs are developed and recorded in a Risk Control Measure Log (RCML). Upon identifying all possible RCOs from the estimated risks, the RCOs in the RCML are used to generate a Risk Control Option Log (RCOL). The information in the RCOL will be used in Step 4 of the FSA process.

In general, RCO measures have a range of following attributes (MSA (1993)):

1. Those relating to the fundamental type of risk reduction (preventative or mitigating).
2. Those relating to the type of action required and therefore to the costs of the action (engineering or procedural).
3. Those relating to the confidence that can be placed in the measure (active or passive, single or redundant).

The main objective of an RCO is to reduce frequencies of failures and/or mitigate their possible consequences.

5.1.4 Step 4 - Cost-Benefit Analysis (CBA)

Upon gathering the various control options, the next step is to carry out a Cost-Benefit Analysis (CBA) on each option. CBA aims at identifying the benefits from reduced risks and costs associated with the implementation of each risk control option for comparison. The evaluation of costs and benefits may be conducted using various techniques (IMO (1993)). It should be initially carried out for the overall situation and then for those interested entities influenced by the problem consideration.

5.1.5 Step 5 - Decision-making

The final step is the decision-making phase, which aims at making decisions and giving recommendations for safety improvement. At this point, the various stakeholders' interest in the vessel under study is considered. The cost and benefit applicable to each stakeholder have to be determined in order to decide the best risk control option – each RCO will have a different

[1] Recovery refers to taking remedial action to recover from an error or failure before the accident occurs.

impact on the identified stakeholders, as such, the most effective RCO should strike a balance between the cost and benefit for each stakeholder. In reality, this is not always possible, hence, any imbalance has to be addressed and justified before the selected RCO is accepted as being the best option. The information generated in Step 4 of the FSA process can be used to assist in the choice of a cost-effective RCO. However, the cost factor may not be the only criterion that should be considered. As such, at this stage, certain multi criteria decision-making techniques should be employed to select the most favourable RCO (Wang et al. (1996), Pillay and Wang (2001), Wang et al. (2002)).

5.1.6 A Brief Discussion

There are many different types of ships such as fishing vessels, cruising ships, bulk carriers and containerships. It has been noted that different types of ships have different characteristics in terms of available failure data, the corresponding safety regulations, etc. As a result, the formal safety assessment framework described above should be applied on a flexible basis. For example, for fishing vessels, due to the poor safety culture in the fishing industry and lack of reliable failure data, the FSA framework described above may not be entirely applied. A modified FSA framework with a more qualitative nature may be more useful as will be described later. In contrast, the FSA framework described above may be relatively easily applied to containerships.

5.2 A Formal Safety Assessment Framework for a Generic Fishing Vessel

The proposed FSA framework for a generic fishing vessel by the authors, is based on the FSA methodology described in the previous section and can be developed into five steps for ease of understanding as follows (Loughran et al. (2002)):

1. Hazard identification.
2. Risk quantification.
3. Risk ranking.
4. Recommendations.
5. Decision-making.

These five steps are represented in a flowchart as seen in Figure 5.4. These steps are further complemented by the supporting techniques described in this book. The interaction of the proposed framework and the supporting techniques described in this book can be seen in Figure 5.5. This framework is aimed at enhancing fishing vessel safety, including protection of life, health, the marine environment and property, by using a systematic risk based analysis. It can be viewed as a simplified version of the framework discussed in the previous section.

5.2.1 Generic Fishing Vessel

A generic fishing vessel should be defined in order to describe the function, features, characteristics and attributes, which are common to all ships of the type, or relevant to the problem under study (MSC (1998a)). The generic vessel facilitates an understanding of the subject under study and can be used to help identify relevant accidents and accident sub-categories, leading to an enhancement of the HAZID structuring. The description of the generic fishing vessel can be divided into several aspects as seen in Figure 5.6 and explained as follows:

Power/Propulsion - Auxiliary power of fishing vessels is normally provided by two or more diesel-electric generator sets or possibly main engine driven alternators on smaller vessels. Power distribution is by series switchboards, distribution panels and cabling systems. Emergency power sources are normally battery based. Medium speed engines (via a reduction gearing system) normally provide the propulsion power.

Bunkering - Bunkering operation is normally undertaken with manual connection of fuel from shore to a receptor on the vessel. Fuel used for fishing vessels has a flash point of no less than 43 degrees Celsius.

Communications - These are pre-dominantly external communication components, which consist of VHF, MF, HF and Satcom systems with EPIRBs (Emergency Position Indicating Radio Beacon) and SARTs (Search and Rescue Transponder) for emergencies. Larger deep-sea fishing vessels have internal communication components such as the public address system and telephone system to particular crew or operational area.

Control - This covers the control of the entire ship. The bridge or wheelhouse is generally the central and often the only control centre on fishing vessels. The bridge has facilities for all round vision, communication, navigation, safety and ship control equipment. The main machinery spaces are periodically manned (during manoeuvring) and unmanned during fishing operations. Local control positions are available for all fishing gear with some limited remote controls on the bridge.

Emergency response/control - The fishing vessel is expected to be equipped to react to emergencies such as rescue from water (either man overboard or third parties). Most vessels carry on board first-aid kits to administer first aid in case of an accident.

Habitable environment - The crew of the fishing vessel are provided with a habitable environment. This may require consideration of ship motion, noise, vibration, ventilation, temperature and humidity. Most accommodation areas of the vessel are provided with intake and exhaust blowers. Where there is an engine control room fitted, it is provided with an air conditioning system as with the navigation bridge.

Manoeuvring - Fishing vessels do not particularly need an accurate and sensitive manoeuvring system. However, when carrying out pair trawling (where two or more vessels are moving closely together), it could be vital to avoid collisions and contacts. Rudders are used with conventional propeller propulsion systems. There are usually no bow or stern thrusters fitted on fishing vessels.

Mooring - Mooring during berthing operations is normally undertaken in a conventional manner using rope mooring lines, fairleads, bollards and winches.

Anchoring - Anchoring arrangements are provided for all fishing vessels and comprise light weight-high holding power anchors with wire or fibre ropes for the main anchor line.

Navigation - Fishing vessels are normally fitted with a magnetic compass, a speed and distance measurement device, a depth of water indicator, one or more radar and an electronic positioning system. Vessel fixing procedures using visually observed bearings are generally carried out on deep-sea fishing vessels and not on smaller coastal fishing vessels.

Payload - The payload of fishing vessels consists of both processed and pre-packed fish (vessels with fish factory on board) or loose fish stored in the cargo holds. The fishing gear on board the vessel is also considered to be part of the payload. Unloading is normally via shore cranes and forklifts - frozen fish packages are placed on pallets and then lifted by a shore crane from the ship to be placed on the docks. Once the fish pallets are on the dock, it is transferred either into a shore freezer holding area or directly onto a truck by the forklift.

Pollution prevention - Oily bilge water is stored on board and discharged to a shore receptacle when the vessel berths for unloading. Oily water separators are rarely provided for smaller coastal vessels. Engine exhaust gases are normally visually monitored.

Stability - The stability requirements of fishing vessels are normally assessed for a range of loading and operating conditions. They relate to intact and damage stability consideration including effects of wind, sea condition and loads on fishing gear during fishing operation.

Structure - The material used for the construction of a fishing vessel include wood, aluminium, fibre-reinforced plastics, high tensile steel and ferro - cement. The arrangements of aluminium and steel structures normally consist of shell plating supported by longitudinal members and, in turn by transverse frames. The structure must withstand the envisage forces imposed, such as sea forces, dead loads and cyclic forces.

The generic fishing vessel is epitomised to be a hypothetical vessel of any size and method of fishing. To summarise, it is an appraisal of the functions of operation that is necessary for any fishing vessel. Fishing being a combined production and transport operation, is cyclic with the following distinct phases of the life:

1. Design, construction and commissioning.
2. Entering port, berthing, un-berthing and leaving port.
3. Fish loading.
4. Fish unloading.
5. Passage.
6. Dry dock and maintenance period.
7. Decommissioning and scraping.

A generic fishing vessel may also be thought of as being a combination of hard and soft systems as listed below:

1. Communications
2. Control
3. Electrical
4. Human
5. Lifting
6. Machinery
7. Management system
8. Navigation
9. Piping and pumping
10. Safety

5.2.2 HAZID

The first step of the analysis is the hazard identification. This consists of determining which hazards affect the fishing vessel's activities under consideration using "brainstorming" techniques involving trained and experienced personnel. In the HAZID phase, various safety analysis methods described in Chapter 3 may be used individually or in a combination to identify the potential hazards.

In the HAZID meeting, accident categories are determined for safety analysis. As a guide, the accident categories determined by the Marine Accident Investigation Branch can be used (Loughran et al. (2002)). These categories can be seen in Chapter 2 (Section 2.5) and are summarised as follows:

1. Capsizing and listing
2. Collisions and contact
3. Fires and explosions
4. Heavy weather damage
5. Foundering and flooding
6. Loss of hull integrity
7. Machinery damage
8. Missing vessels
9. Stranding and grounding
10. Others

Having identified the accident categories, the causes are then grouped into the following risk exposure groups:

1. Human Errors

Human performance	-Communication
	-Navigation
	-Competency
	-Fishing
	-Anchoring
	-Mooring
	-Abandonment
Commercial pressures	-Manning
	-Finance
	-Company or firm procedures
Management systems	-Onboard management
	-Loading fish
	-Shore side systems

2. Hardware failures

-Material of construction
-Structure
-Propulsion
-Steering
-Piping and plumbing
-Control
-Electrical
-Refrigeration
-Safety systems
-Habitable environment
-Emissions control
-Bunkering and storage
-Diagnostics systems
-Maintenance systems

3. External events

Environment	-Pollution prevention
	-Climatic variations
Payload	-Fish handling, loading and storage
	-Crane/lifting mechanisms
	-Berthing

In order to sort the large amount of information collected at the HAZID meeting, a set of accident sub-categories is established as follows:

Collision and contact accident sub-category

- Berthed
- Starting up
- Loading and unloading in port
- Departing and manoeuvring close to the berth
- Manoeuvring in harbour and close to harbour
- Passage in open sea
- Loading fish at sea
- Entering harbour
- Arrival manoeuvring close to the berth
- Shutdown

- Abnormal operation
- Maintenance
- Anchored
- Dry-docked

Fire accident sub-category

- Engine room
- Fish room space
- Wheelhouse
- Accommodation
- Galley

Loss of hull integrity accident sub-category

- Hull plating
- Framing
- Bulkheads
- Welds and joints
- Penetrations
- Seals
- Appendages
- Opening or failure of doors
- Opening or failure of scuttles
- Others

5.2.3 Hazard Screening

The risk matrix approach is used in the hazard screening process. For each appropriate combination of the frequency (*F*) of the hazard and the severity (*S*) of the consequences in terms of human injuries/deaths, property damage/loss and the degradation of the environment, an assessment is carried out. The corresponding Risk Ranking Number (RRN) is then selected from the risk matrix table. This method allows for expert judgements where detailed data is unavailable. Ranking of the various accidents/hazards determines their order in relation to one another. In short, the RRN is indicative of the relative order of magnitude of risk.

Table 5.1 shows the risk matrix table that presents in a tabular format, a risk level related to the frequency and severity of a hazard. RRN ranges from 1 (least frequent and least severe consequence) to 10 (most frequent and most severe consequence).

Table 5.2 gives the interpretation of the frequencies *F1* to *F7* as determined by (Loughran et al. (2002), MSC (1998a)), in terms of a generic fishing vessel based on the following estimations:

1. Vessel life expectancy – 25 years.
2. Operational days per year – 250.
3. Operational hours per day – 13.
4. Major maintenance per year – 1.

Using the risk matrix approach, for each accident category, a ranked risk table is produced, listing all accident sub-categories against each generic location. An example of this is seen in Table 5.3. The number in the brackets, (x), is the corresponding RRN obtained from Table 5.1. Upon completing the risk table, the next task is to determine the "Equivalent Total" for each accident category.

5.2.4 "Equivalent Total"

The purpose for calculating the "Equivalent Total" is to provide a means of integrating the risks evaluated for each hazard of the accident sub-category. It will also provide a means of estimating each accident category to determine and justify the allocation of resources - to eliminate or reduce the risk.

Table 5.4 represents an example for a fire accident category. The data has been drawn from the MAIB reports and Tables 5.1 and 5.2 are used to assign the values of *S* and *F*, respectively. An RRN is assigned for each accident sub-category at different generic locations. This table can be generated for each accident category by analysing the incident/accident data in terms of its occurrence likelihood and severity of consequences.

Table 5.5 shows the number of times each RRN appears within an accident category. For example, RRN 4 appears 5 times (as highlighted in Table 5.4). As the RRN for the accident sub-category is considered for different generic locations, an "Equivalent Total" is calculated to give the accident category an index which will later be used to compare and rank it against other accident categories.

The calculation makes use of the fact that both the frequency and severity bands of the risk matrix are approximately logarithmic (e.g. a risk level of 6 is treated as 10^6) (MSC (1997b)). Using 7 as a base then:

"Equivalent Total" = 7 + Log (3.000 + 0.700 + 0/050 + 0.005) = 7+ Log (3.755) = 7.57

Alternatively using the risk ranking score of 4 as the base, then:

"Equivalent Total" = 4 + Log (3000 + 700 + 40 + 5) = 4 + Log (3755) = 7.57

or rounded off = 7.6

It can be noted that the risk ranking score does not change with the base chosen. Similarly, for each accident category an "Equivalent Total" can be calculated and the value obtained will give a direct indication of the areas needing attention. The higher the value of the "Equivalent Total", the higher the associated risk with reference to that category.

5.2.5 Recommendations

For particular risk factors, there is a range of RCOs. It is most cost effective to reduce risk factors at the early design stages. Additional costs are incurred in redesigning or modifying plant or processes once they are being used.

The various objectives and attributes of an RCO are already discussed in Section 5.1.3. At this stage, practical RCOs are recommended while considering the effectiveness of each option. An RCO could be in the form of a preventive measure - where the RCO reduces the probability of occurrence, and/or a mitigating measure - where the RCO reduces the severity of the consequences. Other factors that need to be considered are the cost of the RCO and the stakeholders who will be affected by its implementations.

Stakeholders can be defined to be any entity (e.g. person, organisation, company, nation state or grouping of these), who is directly affected by accidents or by the cost effectiveness of the industry. For any particular stakeholder, its stake in a generic vessel can be a definite committed monetary value such as an investment or payment. Stakeholders can be voluntary, involuntary or a mixture of both. In the decision making process, the stakeholders may be affected directly, indirectly or by representative groupings. The following stakeholders are identified for a generic fishing vessel:

- Classification society
- Coastal state
- Crew
- Designer/constructor
- Emergency services
- Flag state
- Insurance companies
- Other vessels
- Owner
- Port authority
- Port state
- Suppliers

The best RCO for the estimated risks of the generic vessel can be identified by determining the cost and benefit of each RCO with respect to each of the stakeholders mentioned above. Each RCO can then be represented by a Cost per Unit Risk Reduction (CURR). The CURR is given by:

$$CURR = \frac{Cost - Benefit}{Risk\ reduction} \tag{5.1}$$

where risk reduction is given in terms of the number of injuries.

When applying Equation (5.1), it can be considered that 50 minor injuries = 10 serious injuries = 1 life. Property damage/loss and the degradation of the environment can also be converted to the equivalent number of injuries.

Each RCO needs to be evaluated in accordance with the costs for its implementation and maintenance through the vessel's lifetime, as well as the benefits received for the same period. This evaluation is required to be carried out in two levels, primarily for the overall situation and then for each of the parties concerned and/or affected (stakeholders) by the problem under review (IMO (1997), MCA (1998), MSC (1998b), Wang (2001)). The cost and benefit for each RCO have to be calculated in terms of its Net Present Value (NPV). Hence the numerator in Equation (5.1) is represented as:

$$NPV = \sum_{t=1}^{n}[(C_t - B_t)(1+r)^{-t}] \tag{5.2}$$

where:

t = time horizon for the assessment, starting in year 1.

B = the sum of benefits in year t.

C = the sum of costs in year t.

r = the discount rate.

n = number of years in the vessel's lifetime.

The risk reduction is the difference between the risk level of the given event in the base case and the risk level of the given event following the adoption of the RCO. A negative CURR suggests that implementation would be financially beneficial (cost-effective). All that is left now, is to rank the RCOs using their CURR values and to recommend the most appropriate RCO for an accident category.

5.3 An Example

The example presented in this section is for a generic fishing vessel as defined in Section 5.2.1. For demonstration purposes, only three accident categories are considered, namely, collision/contact, fire and loss of hull integrity. Using the accident data provided in Chapter 2, by the MAIB reports and complementing this data with expert judgement - where the data was absent or incomplete, Tables 5.6, 5.7 and 5.8 are generated (Pillay (2001)). Expert judgements were drawn from ship owners and operators during a round table discussion. These tables represent the evaluation of the three accident categories identified. The RRN definition and interpretation of the values for frequency of occurrence are given in Tables 5.1 and 5.2, respectively.

In order to calculate the "Equivalent Total", the number of occurrence of each ranking score for the three accident categories is determined and summarised in Table 5.9.

Using 4 as the base score, the "Equivalent Total" for accident category collision/contact is given by:

$$Equivalent\,Total = 4 + \log(10843) = 8.0351$$

Using 4 as the base score, the "Equivalent Total" for accident category fire is given by:

$$Equivalent\,Total = 4 + \log(3755) = 7.574$$

Using 4 as the base score, the "Equivalent Total" for accident category loss of hull integrity given by:

$$
\begin{aligned}
EquivalentTotal &= 4 + \log(400 + 170 + 18) \\
&= 4 + \log(588) \\
&= 6.769
\end{aligned}
$$

The result of this analysis is presented in a tabular format to enable easy reading and is given in Table 5.10. A larger risk ranking number ("Equivalent Total") indicates a higher risk, but the values only represent the relative risk levels. Hence, this ranking gives an indication as to which areas of the generic vessel are of higher priority.

It is then required to determine the possible RCOs for the generic vessel considered. As data to quantify each RCO is difficult to obtain, hypothetical RCOs are considered for the demonstration of this method. The cost and benefit columns represent the cumulative values for all the stakeholders involved in the study. The views presented by each stakeholder, will considerably affect the outcome of the CURR. Considering the four RCOs given in Table 5.11 and the associated cost, benefit and risk reduction, a CURR for each RCO can be obtained. Note that the value for risk reduction represents the total number of equivalent deaths for the system under consideration.

Assuming that the time horizon for the safety assessment is for 25 years at a discount rate of 3%, and using Equations (5.1) and (5.2), the CURR calculation for each RCO is given as follows:

$$CURR_1 = \sum_{t=1}^{25}\left(\frac{(50000 - 25000)(1 + 0.03)^{-t}}{1}\right) = 435{,}328.7$$

$$CURR_2 = \sum_{t=1}^{25}\left(\frac{(10000 - 25000)(1 + 0.03)^{-t}}{1}\right) = -261{,}197.8$$

$$CURR_3 = \sum_{t=1}^{25}\left(\frac{(10000 - 15000)(1 + 0.03)^{-t}}{1}\right) = -87{,}065.0$$

$$CURR_4 = \sum_{t=1}^{25}\left(\frac{(30000 - 40000)(1 + 0.03)^{-t}}{1}\right) = -174{,}132$$

A large negative CURR suggests that this implementation would be financially beneficial. From the results obtained, it is determined that RCO 2 is the best option (from a cost-benefit point of view) and can be recommended for implementation.

5.4 Formal Safety Assessment of Containerships

5.4.1 Generic Containership

A generic model of containership can be developed according to the IMO's Interim Guidelines in (IMO (1997)), taking into consideration the particular systems and characteristics required for the transportation of containerised cargo. The relevant study carried out by the UK MCA on High Speed Passenger Craft (MCA (1998)) as well as on Bulk Carrier offer an equally useful guide for the development of our generic model.

The generic containership is not a "typical" vessel but a hypothetical one consisting of all technical, engineering, operational, managerial and environmental (physical, commercial and regulatory) networks that interact during the transportation of containerised cargo. This generic model can also be broken down to its component and more detailed levels. Thus the generic container ship can take the form shown in Figure 5.7. Breaking down the model to the four basic levels of the containership operation produces the "generic engineering and technical system model" (Figure 5.8) (MSC (1998b)), the "generic personnel sub-system" (Figure 5.9), the "generic operational and managerial infrastructure" (Figure 5.10), and the "generic environment of operation" (Figure 5.11).

Containerships follow the general pattern that all internationally trading cargo ships do, but they differentiate in various aspects, of which, the primary ones appear to be as follows:

i. Structure

The structure of a containership is typified by holds longitudinally divided in two sections (fore and aft), each being able to accommodate one 40 ft unit or two 20 ft ones in length. Holds are fitted with vertical "*L*" shaped guides (cell guides) used to guide and secure the units into their stowage position. Internally containership holds are box shaped surrounded by ballast, fuel tanks and void spaces.

ii. Strength and Stability

Containerships like most cargo vessels are equipped with means to calculate stability, shear-forces (SF) and bending moments (BM). The differentiating feature of containerships is the additional need for the calculation of torsion moments (TM). This need is generated by the uneven distribution of cargoes in cases where the vessel is partly loaded proceeding to various ports before completing its loading.

The existence of deck cargo reduces the stability of the vessel and calls for increased inherent or design stability of the vessel itself. It is not an uncommon phenomenon that a "Metacentric Height" (GM) is 6.5m for a Panamax size containership in "light ship" condition. The use of high-speed diesel engines increases the fuel consumption rate, which imposes the need for large fuel tanks, usually located at, or close to, the mid-section of the vessels. Thus, as fuel is consumed bending moments and shearing forces are increasing. It is noteworthy that many modern containerships are equipped with real-time stress monitoring equipment allowing for automated correction of excessive values using ballast.

iii. Cargo and Ballast Operations

This is one of the main differences between containerships and other cargo vessels. Loading and unloading cargo operations are carried out simultaneously and at very high rates. The cargo loaded and discharged is calculated based on the values declared by the shippers for each unit and by weighing the units upon their arrival at the terminal gate. Cargo operations are normally pre-planned by terminal personnel in simulated conditions and are subject to evaluation and acceptance by the ship's personnel. Real-time follow-up of the operation is carried out both onboard and ashore and the final figures of stability, stresses and cargo quantities are then calculated.

iv. Manoeuvrability, Power and Propulsion

Containerships are generally fitted with thrusters (bow and/or stern) and in several cases active rudders. This coupled by the advanced hydrostatic features (i.e. block coefficient) of these vessels, results in a high level of manoeuvrability at all speed levels. High speeds, nevertheless, tend to reduce the time available for reaction by operators, adversely affecting the human reliability in close quarters situations.

v. The Cargoes Carried

The majority of the cargoes carried are usually of high value, as opposed to bulk carriers and crude oil tankers, which tend to carry raw materials of relatively lower values. Containerised cargoes come in small parcels, while bulk cargoes (dry or liquid) come in larger ones. Goods travelling in a sealed container produce a problem of uncertainty as far as the characteristics of the cargo (i.e. quantity, quality security and inherent hazards) are concerned. The information for such features is received by the documents accompanying the sealed unit and is rarely crosschecked. Only in cases of suspected existence of undeclared dangerous goods does the law provide for ship personnel to demand inspection of the unit's contents.

Again due to the high loading rates and pressure in time, most of the paperwork is received "in good faith" and the burden of avoiding and in the worse case combating hazardous situations falls on the ship personnel. Cases of undeclared hazardous substances as well as poorly maintained containers and tanks, have been identified but rarely reported to the authorities, following a compromising agreement between carriers and cargo owners (Industrial Contact (2000), TSBC (1999)).

vi. Cargo Recipients (consignees)

Another characteristic that containerships have is the one connected with the cargo recipients (consignees). Unlike other ship types (i.e. bulk carriers, tankers) the number of cargo consignees is highly increased. Even within the same unit there may be more than one of recipients. This fact, combined with the high value of the cargoes carried and their hazardous nature, increases both the exposure of the carriers for possible damage and the difficulty in co-ordination and co-operation, between ship and cargo owners, during contingency situations.

vii. Ports and Terminals

Container-handling ports and terminals follow a distinct path, as far as their general layout and organisation are concerned. Container terminals have the ability to concurrently carry out loading and discharging operations, while terminals handling bulk cargoes tend to be specialised loading or discharging ones. In cases where bulk carrier terminals can handle both loading and discharging, the two operations are never carried out simultaneously.

5.4.2 A Formal Safety Assessment Framework for Containerships

By considering the characteristics of containerships, a formal safety assessment framework for containerships is described as follows:

5.4.2.1 Hazard Identification (HAZID)

The aim of this step is to identify the hazards related to a specific problematic area and generate a list of them, according to their occurrence likelihood and the severity of their consequences towards human life, property and the environment, in order to provide the base or the reference point for the next step. The following assumptions are applied:

1. The containership average lifetime: 25 years.
2. The average number of operational days per year: 330.
3. Operational hours per day: 24.
4. Major maintenance frequency: 1 every 2.5 years (30 months).

Again, "brainstorming" technique involving trained and experienced personnel are used for the whole process of hazard identification. The accident categories identified with regard to the containerships' operation include (Wang and Foinikis (2001)):

1. Cargo damage
2. Contact and/or collision
3. Explosion and fire (including flame and heat)
4. External hazards (i.e. heavy weather)
5. Flooding
6. Grounding and/or stranding
7. Hazards related to hazardous substances (including leakage, noxious fumes, etc)
8. Hazards related to human errors
9. Hazards related to loading/discharging operations (including ballast operations)
10. Loss of hull integrity
11. Machinery failure (including electronic devices, navigation equipment and safety systems)

The containership's compartments include:

1. Cargo spaces
2. Crew accommodation
3. Engine room
4. Galley
5. Navigation bridge
6. Provisions' storage spaces (including bonded stores)
7. Tunnels
8. Upper deck areas
9. Void spaces

The operational phases of a containership include:

1. Berthing and unberthing
2. Cargo and ballast operations
3. Coastal navigation
4. Decommissioning
5. Design – construction – commissioning
6. Entering and leaving port
7. Major maintenance (dry docking)
8. Open sea navigation
9. Planned maintenance (day-to-day onboard)

Once the hazards are identified with respect to each of above accident categories, compartments and operational phases, it is essential that they are "screened" so that they can be properly evaluated and the trivial ones to be excluded from further investigation. The "Risk Matrix Approach" (Loughran et al. (2002), MSA (1993), Wang et al. 1999)) can be used to combine frequency and severity rankings for the estimation of RRN values.

5.4.2.2 Risk Estimation

Following the study of the escalation of the basic or initiating events to accidents and their final outcomes, it is again necessary for an "influence diagram" to be constructed, in order to study how the regulatory, commercial, technical and political/social environments influence each accident category and eventually quantify these influences with regard to human and hardware failure as well as external events (MCA (1998), MSC (1998b), Wang et al. (1999), Billington (1999)). Again the various operational phases of the ship have to be taken into consideration and generic data or expert judgements to be used. A list of a containership's systems/compartments and operational phases can be shown in Table 5.12.

5.4.2.3 Ways of Managing Risks

The aim at this stage is to propose effective and practical RCOs to high-risk areas identified from the information produced by the risk estimation in the previous step (Wang, (2001), Wang and Foinikis (2001)). At this stage the implementation costs and potential benefits of risk control measures are not of concern.

Reducing the occurrence likelihood and/or the severity of the possible consequences of hazards can achieve risk reduction. There are three main methods used for risk reduction, namely the management, engineering and operational ones (Kuo (1998), Wang and Foinikis (2001)).

Managerial solutions involve activities related to the management of each organisation. The main objective of such activities is the development of a safety culture, while the key factor for their success is effective communication.

Engineering solutions involve the design and/or construction of the ship. Engineering solutions have the inherent advantages that they can be clearly identifiable (i.e. introduction of double hull in oil carriers) and address hazards in the early stages of a vessel's life.

Operational solutions involve the development and introduction of appropriate procedures for carrying out "risk-critical" tasks, as well as improving the effectiveness of personnel in these tasks. Thus safety procedures, safe working practices, contingency plans and safety exercises (drills) can be included. Such solutions address efficiently human error factors and ensure the existence of uniformity of the adopted safety standards.

The identified measures with the same effect, or applied to the same system, can then be grouped and it is up to the experts to estimate the effectiveness of each RCO. Selected RCOs can then be forwarded to the fourth step where their cost effectiveness is evaluated.

5.4.2.4 Cost-Benefit Assessment

Selected RCOs must also be cost-effective (attractive) so that the benefit gained will be greater than the cost incurred as a result of the adoption (IMO (1997), MSA (1993), Kuo (1998)). There are limitations in carrying out Cost benefit Analysis. The limitations come from imperfect data and uncertainty. It must also be pointed out that Cost Benefit Analysis, as suggested for use in FSA is not a precise science, but it is only a way of evaluation. Thus it cannot be used mechanistically, but only as a consulting instrument in decision-making.

Again, a "base case" is required in this stage. The RCO costs, benefits and the CURR can be estimated by comparing the base case with the one where the RCO is implemented. Equation (5.1) can be used to calculate CURR values.

Having estimated all costs-benefits and risk reduction values of each RCO, for both the overall situation and for each particular accident category, the next requirement is to list the findings with regard to their significance to the various stakeholders and their relative values.

5.4.2.5 Decision Making

The final step of FSA is "decision making", which aims at giving recommendations and making decisions for safety improvement taking into consideration the findings during the whole process. Thus the pieces of information generated in all four previous steps are used in selecting the risk control option which best combines cost effectiveness and an acceptable risk reduction, according to the set "risk criteria" by the regulators.

It is equally admitted, however, that the application of absolute numerical risk criteria may not always be appropriate, as the whole process of risk assessment involves uncertainties. Furthermore, opinions on acceptable numerical risk criteria may differentiate between individuals and societies with different cultures, experiences and mentalities.

The RCOs that could finally be adopted would be the ones with the best cost-effectiveness for the whole situation as well as for the particular stakeholders.

5.4.3 Evaluations and Recommendations in Formal Safety Assessment of Containerships

It becomes apparent that there is still plenty of room for improvement on containership safety. Areas on which such improvement can be achieved include, but not limited to (Wang and Foinikis (2001)):

- The vessels' strength and stability.
- Fire-fighting and life-saving equipment.
- Human reliability.
- Information availability, reliability and interchange.

Such areas are described in more detail as follows:

The Containership Hull Stresses

Mainly due to their configuration and the increased demand for full capacity utilisation, coupled by the subsequent increase in the vessels' sizes, containerships face the problem of increased structural stresses (i.e. bending moments, searing forces and torsion). The establishment of objectives aiming at the advancement of practical design strategies towards containership structures, optimal for both the operator and the operating environment, is considered crucial. Further research and testing towards that direction will greatly contribute to the rule-based treatment of the containership structural strength in the context of FSA.

In addition to the above, stress monitoring both in "harbour" and "open-sea" conditions, would provide a useful tool for the safe operation of large containerships producing information on both the current structural stress levels of the vessel and any possible deviations from the pre-calculated figures. Thus, "Real Time Stress Monitoring Systems", should not be considered "optional", but become compulsory for containership sizes of "Panamax" (3,000 TEU) and above.

The Containership Fire-fighting and Life-saving Equipment

The high concentration of dangerous goods with varying properties implies that apart from adequate contingency procedures containerships need to be fitted with the appropriate combating equipment although the available failure data do not show considerable fatalities, serious injuries or damage to the environment in such emergency situations.

The traditional combating methods (i.e. fixed and portable fire fighting arrangements) and materials (i.e. sea-water, chemical foam, CO_2 and personal protective equipment) which are used today are not designed to protect from conditions involving corrosive, toxic and biochemical substances or a chain reaction causing extensive fire and/or explosion. Since the introduction of specified combating materials for each particular type of cargo would not prove to be cost-effective, the introduction of advanced escape/evacuation systems and procedures should be considered.

The types of escape vehicles (i.e. lifeboats and life rafts) used on containerships currently follow the general pattern of dry cargo vessels, without taking into consideration the possibility of existence of corrosive, toxic or biochemical environments. Excluding their capacity, the choice of the type of lifeboats or life rafts is left on the shipping company's discretion. The compulsory inclusion, of protectively located and easily accessible lifeboats with "totally enclosed", "free fall", "self-righting" and "air tight" functions, equipped with "external sprinkler system" (as used in oil and gas carriers) for all containerships carrying dangerous goods would provide adequate protection to the evacuees.

Human Element

Considering the relevant statistics and failure data in hand, the human element appears to be the prominent factor for containership failures. The distribution of approximately 1:5 (21%) between shore based and ship operating personnel, suggests that the problem in hand is a multi-sided one.

Primarily, there is the need for adequate training of ship personnel specialising in the containership operation. Containerships should cease to be considered as simple "general dry cargo vessels", as dictated by their particular characteristics. Such characteristics include the increased ship speed, the long list of dangerous cargoes carried (e.g. explosives, biochemical, toxic, corrosive, nuclear etc.) and the often-marginal structural strength exploitation. The above suggest that personnel serving on containerships should be adequately qualified, with knowledge and skills exceeding the general ones offered by the various Nautical Academies. A similar requirement exists today for personnel serving on Oil and Gas Carriers. Thus, special courses and seminars should be introduced to provide containership personnel with the adequate theoretical and practical knowledge and the necessary documentation.

Other factors that diversely affect human reliability are the reduced port turnaround and the increased sea-passage time of this ship type. Containerships in very rare occasions have the opportunity for overnight stay in port, reducing the chances for crew recreation and thus increasing personal stress and fatigue. Measures such as in-built swimming pools, gymnasiums and recreation rooms and the introduction by the IMO of limits on the maximum amount of daily working hours per crewmember, have little effect in reducing crew stress and fatigue (Wang and Foinikis (2001)). Reductions in the contractual service time of crewmembers rest on each individual shipping company's discretion. It is believed that further consideration should be given to the matter and an international agreement be achieved.

Attention should also be paid to the required qualifications for shore-based personnel, as well as on the correct implementation of the relevant legislation regarding the proper inspection and documentation of the cargoes from the point of production to the point of loading. Better policing of the whole network will reduce incidents, which may prove to be disastrous for human lives, the environment and other property. Such incidents include, but not limited to, undeclared dangerous goods packed in inadequate containers, inaccurate or deliberately altered container weights and numbers, forged manifests and poorly if at all maintained reefer containers with inadequate settings.

Information Availability, Reliability and Interchange

Many of the weaknesses existing today in the shipping industry in general and the container sector in particular, would have been remedied if there had been an adequate flow of information amongst the parties concerned. Containerships and their owning/operating companies form a part of a multi-modal transportation system, which bases its successful function on an integrated logistics system and an electronic data interchange network. Thus, each company's existing infrastructure could easily be adapted to carry out the additional task of collection, processing, storing and interchange of safety information including failure rates at all sections.

It could also interact with regulatory bodies outside the shipping industry, responsible for land-based operations, and share the relevant data of non-compliance with established safety and quality standards for shore-based industries. This would eliminate a considerable percentage of errors attributable to factors not related to container shipping.

Until today, ship safety has been subject to sets of prescriptive rules and good practices established. Matters are usually resolved in an intuitive manner by ship personnel. The constantly evolving ship technology and the new hazardous cargoes carried impose new hazards in one form or another and call for equally advanced safety measures with the ability to follow up and adapt to the above evolutions. Possibly, the most illustrative example of fast evolution is the containership sector of the industry. Within only 44 years of life, containerships have moved from 58 to up to 7,000 TEU per vessel, from 13 to 27 knots and from simple dry general cargo to refrigerated, corrosive, toxic, explosive, biochemical and nuclear ones.

5.5 Discussions

FSA can be feasibly applied to ship design and operation, provided that several areas, which cause uncertainties, are further deliberated. These areas influence both the general principles of FSA and the specific requirements for a particular ship type, either directly or indirectly. The most prominent ones are analysed and alternative suggestions are described as follows:

5.5.1 The Brainstorming Process

Although the knowledge and expertise of the people involved in the "brainstorming" process is absolutely respectable, certain safety aspects may be overlooked as it might be considered "natural" from their point of view, while to a person outside the profession it might be something completely new and thus causing concern.

Since by definition the "brainstorming session" ought to be structured to encourage the unfettered thinking and participation of the people involved, the contribution by people with less expertise in the subject would be a positive one, as they might bring up safety issues, which otherwise would have been overlooked.

5.5.2 Need for Interaction with Other Industries' Safety and Quality Management Systems

FSA for ships should develop the ability to interact with regulatory bodies responsible for land-based operations. Sharing the relevant data of non-compliance with established safety and quality standards for shore-based industries would eliminate a considerable percentage of the uncertainty created in this direction.

5.5.3 Human Factor

Another important factor to be taken into consideration is human factor. Problems like differences in language, education, training, mentality, etc., have increased over the past years, especially with the introduction of multi-national crews. Such problems largely contribute to marine casualties. On the other hand, crew reductions have increased the workload of operators, which in connection with the reduced opportunities for port stay and recreation (especially with containerships) equally increases the probabilities for human errors.

It becomes apparent that FSA's success largely depends on two essential conditions. The first condition is the development of a safety culture at all levels of the industry's infrastructure, from company managers to vessel operators. The second one is the inclusion into the FSA framework itself of further guidance on how human factors would be integrated in a feasible manner.

5.5.4 The Availability and Reliability of Failure Data

Primarily, great attention should be paid to the data resources, as the various databases do not always use the same platform for data analysis. This is attributable to the fact that different organisations look into safety issues from a different prospective, which facilitates their own interests. In order to overcome the problems created by the availability and reliability of failure data, international co-operation and co-ordination are required with the intention that a new global database will be established, controlled and updated by an international regulatory body (i.e. IMO). Such a database should be easily accessible by both administrations and analysts/researchers.

It is noted that different types of ships have different levels of difficulties in terms of collecting and processing failure data. For example, the task of data collection and processing for containerships appears to be relatively easier than for other ship types. This is attributable to the fact that containerships and their owning/operating companies form a part of a multi-modal transportation network and therefore are highly computerised with the necessary infrastructure. With the adequate adaptations the existing infrastructure can be feasibly utilised for the purpose of FSA and failure data can be easily collected, processed and communicated both internally (i.e. company head offices, branches and ships) and externally (i.e. central international and national databanks, other industrial bodies). For fishing vessels, the necessary infrastructure for data collection is generally not ready yet and as a result the application of FSA may need more time and effort.

5.5.5 Cost Benefit Analysis

The use of cost benefit analysis as a platform on which a given option is finally selected for implementation is an appealing proposal. In practice, however, it can be quite complicated, especially in cases where human lives are involved. The fact that ships are manned with multi-national crews, usually officers from developed countries and crew from developing ones, and obliged to trade in all parts of the world, creates a difficulty in selecting the proper human life value for cost benefit analysis. Furthermore, the use of different values on different nationalities would have an adverse and undesirable effect on both international relations and working conditions onboard ships.

A feasible solution to this problem would, once more, involve an international agreement on a reliable method of estimating the current value of human life. The international regulatory bodies should not only be responsible for the initial deliberations, but also for the constant follow up of the international economic, political and social trends that influence that value.

5.6 Conclusion

The main intention of FSA (during the development stages of the approach) was to be applied to the regulatory regime for shipping. However, over the years, its potential has been recognised not only as a tool to develop safety rules and regulations but as a tool to identify safety related problems with design, operation and procedures of a maritime entity. The FSA approach has several benefits to offer the shipping industry, these benefits are summarised as follows:

1. FSA provides a consistent regime that addresses all aspects of safety (design and operation) in an integrated manner.
2. FSA is a pro-active approach. Hence, it enables hazards that have not yet given rise to accidents to be properly considered.
3. Owners and operators can ensure that safety investments are targeted where it will achieve the greatest benefit.
4. It provides a rational basis for addressing new risks posed by the changes in marine technology.

This Chapter has described a trial application of the proposed FSA technique for a generic fishing vessel. The application of FSA to containerships is also studied. Several problems have been identified with the use of the current approach as proposed by the MCA to the IMO, these include:

1. Reliable data is generally not available for ships. When it is available, there is a high level of uncertainty associated with the data.
2. The risk matrix approach is a simple subjective method to quantify the probability of occurrence and severity of the associated consequences, however, it lacks a formal way to quantifying expert judgement and opinion when using the risk matrix. This would entail that conflicting opinions of two different analysts on the severity of an accident could result in a deadlock.
3. It is difficult to quantify the costs and benefits of each RCO for each of the identified stakeholders. A more subjective approach may be needed to express the preference of one RCO over the others.
4. Human reliability can be considered in the FSA methodology, however, quantification may be impractical due to the lack of human reliability data associated with maritime tasks. As such there is a need to address this problem using a formal subjective approach.

The setbacks of the FSA methodology identified here are addressed by the development of various approaches that are presented in the following Chapters of this book.

5.7 References (Chapter 5)

1. Billington C.J., (1999), "Managing Risks in Ports", Proceeding of Managing Risk in Shipping Conference, The Nautical Institute, London, pp. 57-69.
2. Henley E.J. and Kumamoto H., (1996) "Probabilistic Risk Assessment and Management for Engineers and Scientists", 2nd Edition, IEEE Press, New York.
3. IMO, (1993) "Draft Guidelines for FSA Application to the IMO Rule-making Process", MSC 68/14, IMO Maritime Safety Committee, London.
4. IMO, (1997) "IMO/MSC Circular 829, "Interim Guidelines for the Application of Formal Safety Assessment to the IMO Rule-Making Process", IMO, London, 17th November.
5. Industrial Contact, (2000) "Accidents' and Near Misses", Reports Provided under Confidentiality, by Private Shipping Company. Information available from the authors of this book up on request.
6. Kuo C., (1998) "Managing Ship Safety", LLP, London.
7. Loughran C.G., Pillay A., Wang J., Wall A.D. and Ruxton T., (2002) "Formal Fishing Vessel Safety Assessment", Journal of Risk Research, Vol. 5, No. 1, pp. 3-21.
8. MCA, (1998) "Notes on the Experience Gained on Formal Safety Assessment", Informal Paper submitted by UK to IMO/MSC, 69th session, London, 12th February (IMO/MSC 69/INF14).
9. MSA, (1993) "Formal Safety Assessment", MSC 66/14, Submitted by the United Kingdom Marine Safety Agency to IMO Maritime Safety Committee.
10. MSC, (1997a) "Formal Safety Assessment: Trial Application to High-speed Passenger Catamaran Vessels", 68/Informal paper, 29th May, International Maritime Organization, London.

11. MSC, (1997b) "FSA Trial Application to High Speed Passenger Catamaran Vessel", MSC 68/14/2 & 68/INF., Maritime Safety Committee, International Maritime Organization, United Kingdom.

12. MSC, (1998a) "Formal Safety Assessment: Experience Gained from the Trial Application Undertaken by the United Kingdom", MSC 69/INF.14, Maritime Safety Committee, International Maritime Organization, 12th February,

13. MSC, (1998b) "Formal Safety Assessment of Bulk Carriers", MSC 70/Informal Paper, 27th November, Maritime Safety Committee, International Maritime Organization.

14. Pillay A. and Wang J., (2001) "Human Error Assessment and Decision-making Using Analytical Hierarchy Processing", Proceeding of the International Mechanical Engineering Congress and Exposition - ASME, November 11-16, New York.

15. Smith D.J., (1992) "Reliability, Maintainability and Risk", 4th Edition, Butterworths-Heineman Limited, Oxford, United Kingdom.

16. TSBC (Transportation Safety Board of Canada), (1999) "Statistics and Accident/Incident Reports for the Period 1995 – 1999", Safety Analysis and Communications Directorate, Quebec, Canada.

17. Villemeur A., (1992) "Reliability, Availability, Maintainability and Safety Assessment", John Wiley, Chichester, United Kingdom.

18. Wang J., (1994) "Formal Safety Assessment Approaches and Their Application to the Design Process", PhD Thesis, Engineering Design Centre, University of Newcastle upon Tyne, July, 258p.

19. Wang J., (2001) "Current Status of Future Aspects of Formal Safety Assessment of Ships", Safety Science, Vol. 38, pp. 19-30.

20. Wang J. and Ruxton T., (1998) "A Design for Safety Methodology of Large Engineering Systems", Journal of Engineering Design, Vol. 9, No. 2, pp. 159-170.

21. Wang J. and Foinikis P., (2001) "Formal Safety Assessment of Containerships", International Journal of Marine Policy, Vol. 25, pp. 143-157.

22. Wang J., Pillay A., Wall A. and Loughran C., (1999) "The Latest Development in Marine Safety Assessment", Proceeding of the 4th International Conference on Reliability, Maintainability & Safety, Shanghai, China, May 18-21, pp.711-719.

23. Wang J., Yang J.B. and Sen P., (1996) "Multi-person and Multi-attribute Design Evaluations Using Evidential Reasoning based on Subjective Safety and Cost Analyses", Reliability Engineering & System Safety, Vol. 52, No. 2, pp. 113-129.

24. Wang J., Pillay A., Wall A. and Ruxton T., (1999) "The Latest Development in Ship Safety Assessment", Proceeding of the 4th International Conference on Reliability, Maintainability and Safety (ICRMS '99), Shanghai, China, May 18-21, pp. 711-719.

25. Wang J., Sii H. S., Pillay A. and Lee J. A., (2002) "Formal Safety Assessment and Novel Supporting Techniques in Marine Applications", Proceeding of the International Conference on Formal Safety Assessment Conference, ISBN: 0 903055805, RINA, London, September 18-19, pp.123-142.

Table 5.1 Risk Matrix Table

		F1	F2	F3	F4	F5	F6	F7
S1	Minor Injuries	1	2	3	4	5	6	7
S2	Major injuries	2	3	4	5	6	7	8
S3	1 to 10 deaths	3	4	5	6	7	8	9
S4	> 10 deaths	4	5	6	7	8	9	10

Table 5.2 Key to Risk Matrix Table

Likely to happen on a vessel once per Frequency	General interpretation	Generic fishing vessel Interpretation
F1 10000 – 100000 years	Extremely remote to extremely improbable	Likely to happen every 20 years in the industry
F2 1000 – 10000 years	Remote to extremely remote	Likely to happen every 2 years in the industry
F3 100- 1000 years	Remote	Likely to happen 5 times per yr in the industry
F4 10 – 100 years	Reasonably probable to remote	Likely up to 3 times per vessel life
F5 1 – 10 years	Reasonably probable	Likely up to 15 times per vessel life
F6 yearly	Reasonably probable to frequent	Likely annually per vessel
F7 monthly	Frequent	Likely monthly per vessel

Table 5.3 Example of a Risk Table

Accident sub-category	Generic location				
	Berth	Harbour	Coastal	At sea	Dry dock
Engine room	F5 S1 (5)	F5 S2 (6)	F5 S3 (7)	....	.....
Bridge	F1 S2 (2)	F1 S3 (3)	....	....	.....
Cargo hold	.	.	....	....	.

Table 5.4 Fire Rankings Using Risk Matrix Approach - Expert Judgement

Accident sub-category	Berthing/ unberthing	Manoeuvring (harbour)	At sea (coastal)	At sea (open sea)	Dry dock maintenance
Fish Room Space	F4 S1=4	F4 S2 = 5	F4 S3 = 6	F4 S3 = 6	F4 S1 = 4
Galley	F4 S1 = 4	F4 S2 = 5	F4 S3 = 6	F4 S4 = 7	F4 S1 = 4
Crew accommodation	F4 S2 = 5	F4 S2 = 5	F4 S3 = 6	F4 S3 = 6	F4 S1 = 4
Bridge	F3 S1 = 3	F3 S1 = 3	F3 S1 = 3	F3 S1 =3	F3 S1 = 3
Engine room	F5 S1 = 5	F5 S2 = 6	F5 S3 = 7	F5 S3 = 7	F5 S2 = 6

Table 5.5 Number of Occurrences of Risk Ranking Scores

RRN	No. of occurrence for accident. sub category
4	5
5	5
6	7
7	3

Table 5.6 Collision/Contact Risk Ranking

Accident sub-category	Generic location				
	Berthing/ unberthing	Manoeuvring (harbour)	At sea (coastal)	At sea (open sea)	Dry dock maintenance
Berthed	F3 S2 (4)				
Loading/ unloading	F4 S2 (5)				
Departure	F5 S2 (6)				
Manoeuvring		F5 S2 (6)			
Passage open sea				F4 S3 (6)	
Loading fish at sea				F6 S3 (8)	
Entering harbour		F5 S2 (6)			
Manoeuvring close to berth	F5 S2 (6)				
Shutdown	F4 S2 (5)				
Abnormal operation	F4 S2 (5)	F4 S2 (5)	F4 S3 (6)	F4 S3 (6)	F4 S1 (4)
Maintenance					F3 S1 (3)
Anchored	F5 S2 (6)				
Dry docked					F4 S1 (4)

Table 5.7 Fire Risk Ranking

Accident sub-category	Generic location				
	Berthing/ unberthing	Manoeuvring (harbour)	At sea (coastal)	At sea (open sea)	Dry dock maintenance
Fish room space	F4 S1 (4)	F4 S2 (5)	F4 S3 (6)	F4 S3 (6)	F4 S1 (4)
Galley	F4 S1 (4)	F4 S2 (5)	F4 S3 (6)	F4 S4 (7)	F4 S1 (4)
Crew Accommodation	F4 S2 (5)	F4 S2 (5)	F4 S3 (6)	F4 S3 (6)	F4 S1 (4)
Bridge	F3 S1 (3)	F3 S1 (3)	F3 S1 (3)	F3 S1 (3)	F3 S1 (3)
Engine room	F5 S1 (5)	F5 S2 (6)	F5 S3 (7)	F5 S3 (7)	F5 S2 (6)

Table 5.8 Loss of Hull Integrity Risk Ranking

Accident sub-category	Generic location				
	Berthing/ unberthing	Manoeuvring (harbour)	At sea (coastal)	At sea (open sea)	Dry dock maintenance
Hull plating	F3 S1 (3)	F3 S2 (4)	F3 S2 (4)	F3 S2 (4)	F3 S1 (3)
Framing	F3 S1 (3)	F3 S2 (4)	F3 S2 (4)	F3 S2 (4)	F3 S1 (3)
Bulkheads	F3 S1 (3)	F3 S2 (4)	F3 S2 (4)	F3 S3 (5)	F3 S1 (3)
Welds and joints	F4 S1 (4)	F3 S2 (4)	F4 S2 (5)	F4 S2 (5)	F4 S1 (4)
Penetrations	F5 S1 (5)	F5 S1 (5)	F5 S2 (6)	F5 S2 (6)	F5 S1 (5)
Seals	F5 S1 (5)	F5 S1 (5)	F5 S1 (5)	F5 S1 (5)	F5 S1 (5)
Appurtenances	F4 S1 (4)	F4 S2 (5)	F4 S2 (5)	F4 S3 (6)	F4 S1 (4)
Doors	F4 S1 (4)	F4 S2 (5)	F4 S2 (5)	F4 S3 (6)	F4 S1 (4)
Windows	F4 S1 (4)	F4 S1 (4)	F4 S2 (5)	F4 S2 (5)	F4 S1 (4)

Table 5.9 Number of Occurrence of Each Ranking Score (Three Accident Categories)

RRN	***No. of occurrence for accident category***		
	Collision/contacts	Fire	Loss of hull integrity
4	3	5	18
5	4	5	17
6	8	7	4
7	-	3	-
8	1	-	-

Table 5.10 Summary of Analysis

Accident category	Collision/contact	Fire	Loss of hull integrity
Equivalent Total	8.035	7.574	6.769

Table 5.11 RCOs Determined for Generic Vessel

Risk Control Options	Cost (each year)	Benefit (each year)	Risk reduction (no. of injuries)
RCO 1	£50000	£25000	1
RCO 2	£10000	£25000	1
RCO 3	£10000	£15000	1
RCO 4	£30000	£40000	1

Table 5.12 A Ship's Systems/Compartments and Operational Phases

Ship's systems and compartments	**Ship's operational phases**
Navigation bridge	Design-construction-commissioning
Cargo spaces	Entering and leaving port
Engine room	Berthing-unberthing
Void spaces	Cargo operations
Crew accommodation	Coastal navigation
Passengers' accommodation (if applicable)	Open sea navigation
Galley	Dry-docking
Bonded stores & provision storage areas	Decommissioning
	Maintenance onboard/in port

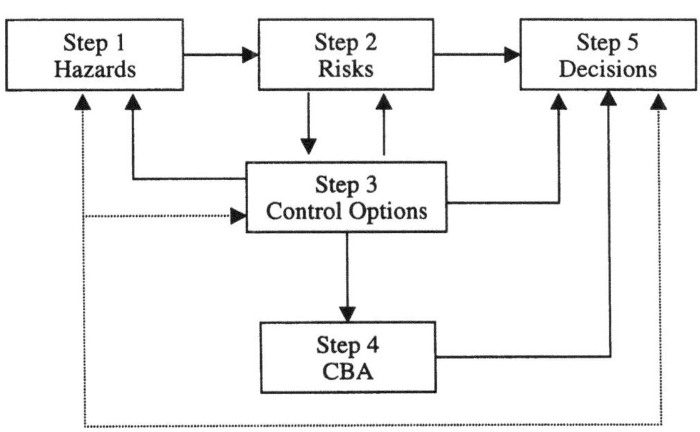

Figure 5.1 Flowchart of FSA process

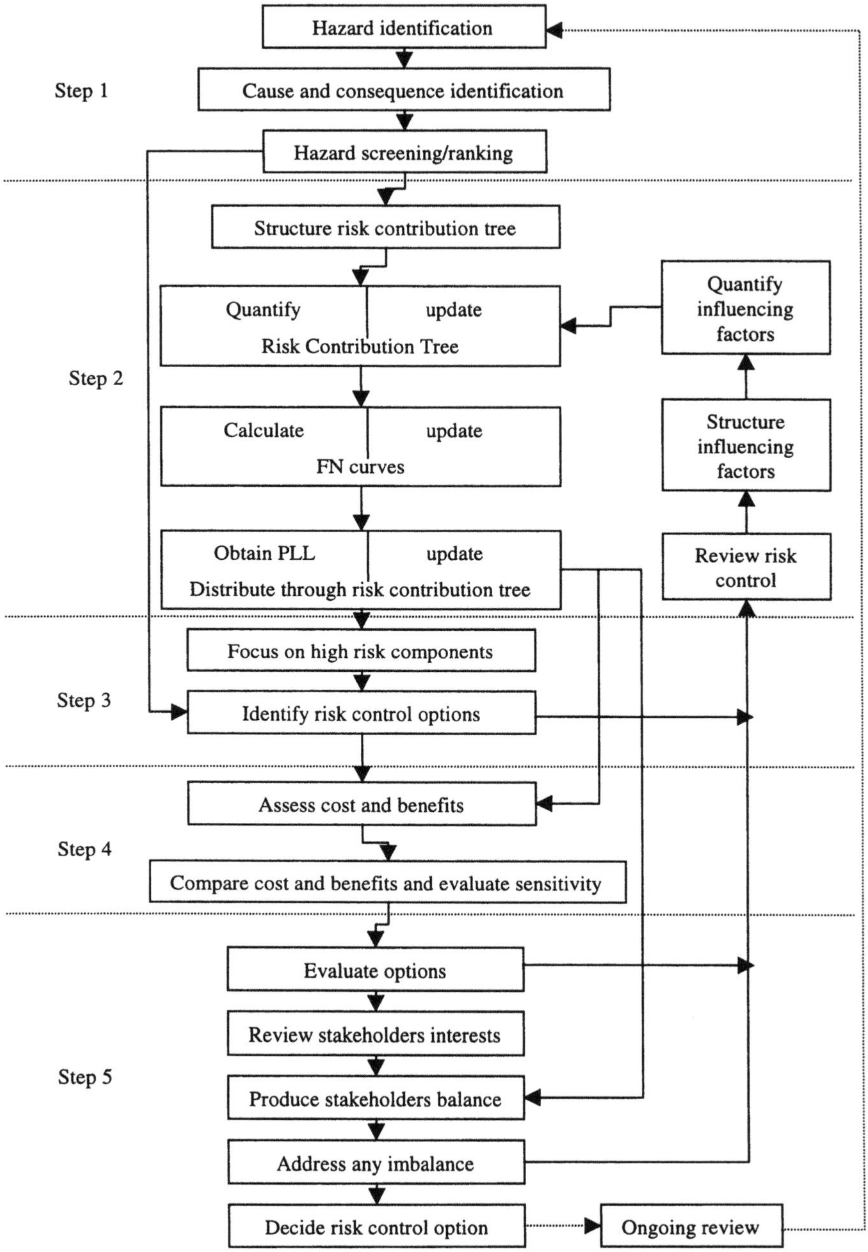

Figure 5.2 Detailed breakdown of FSA process

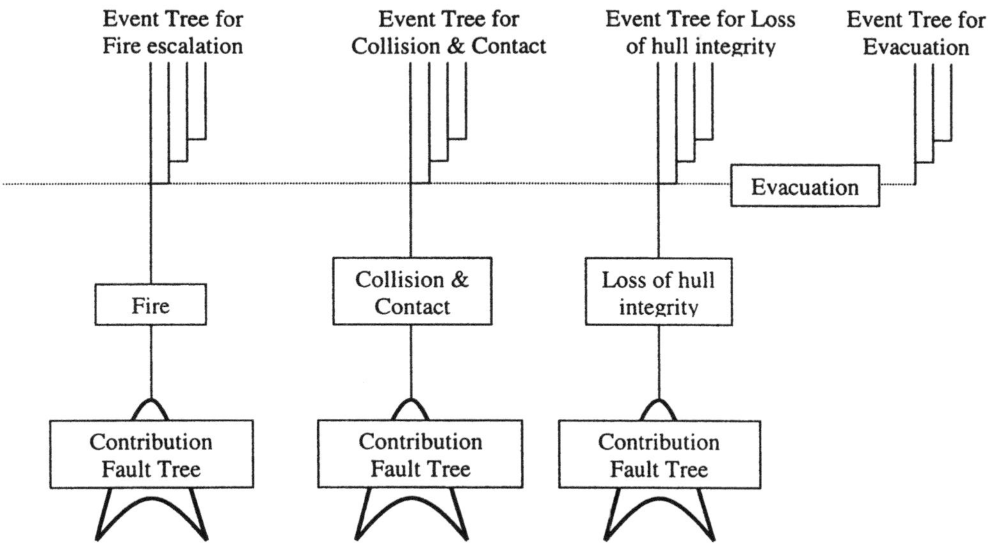

Figure 5.3 An example of Risk Contribution Tree (RCT)

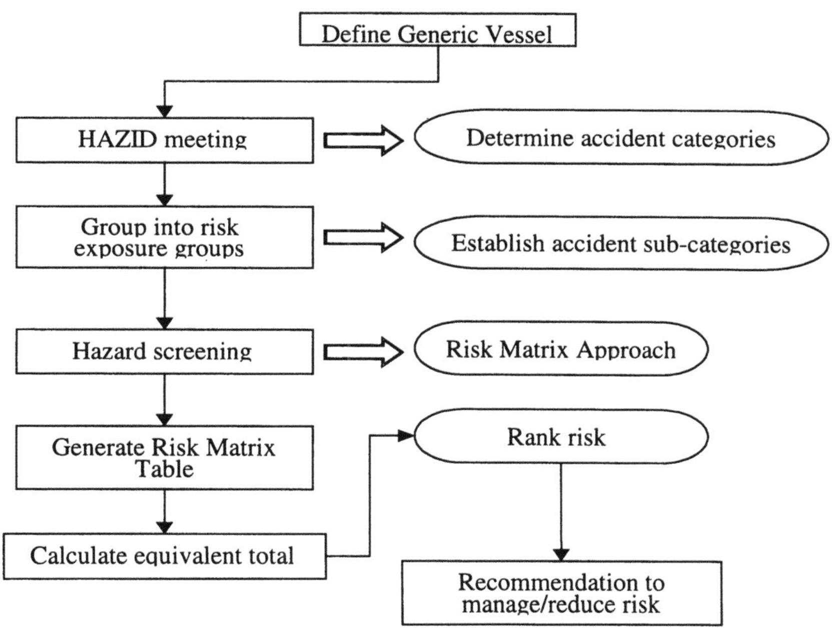

Figure 5.4 Flowchart of proposed approach

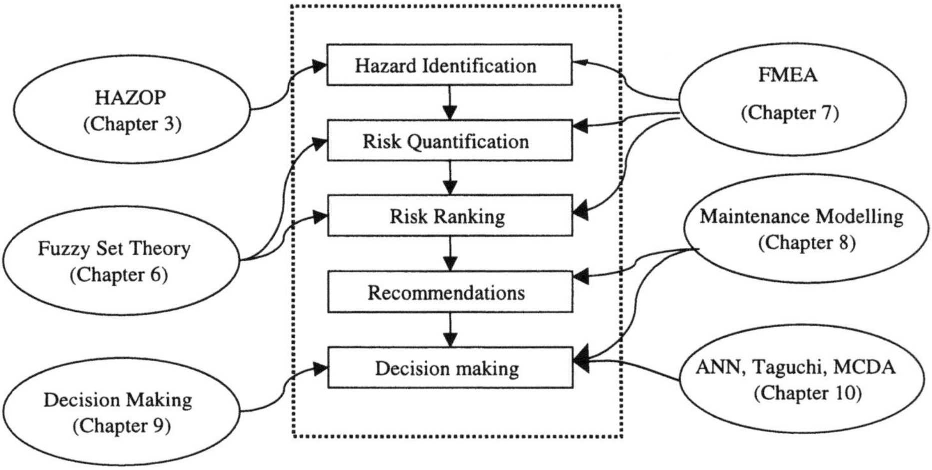

Figure 5.5 FSA framework of a generic fishing vessel

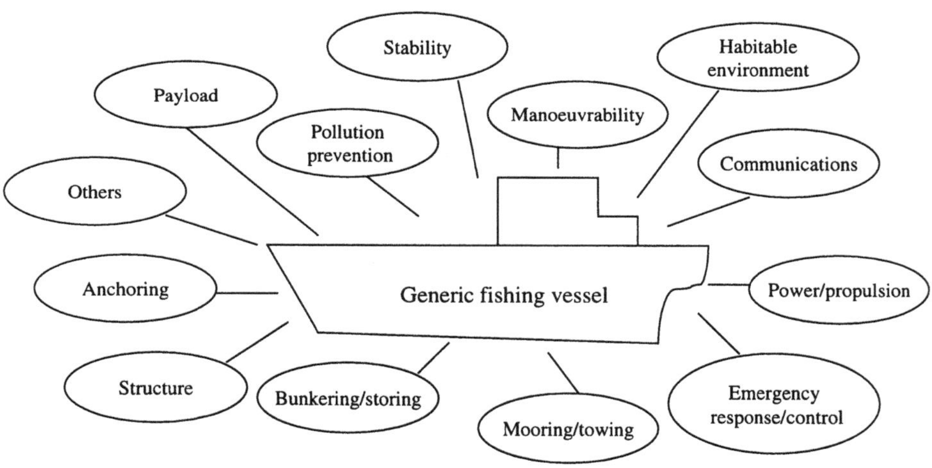

Figure 5.6 A generic fishing vessel

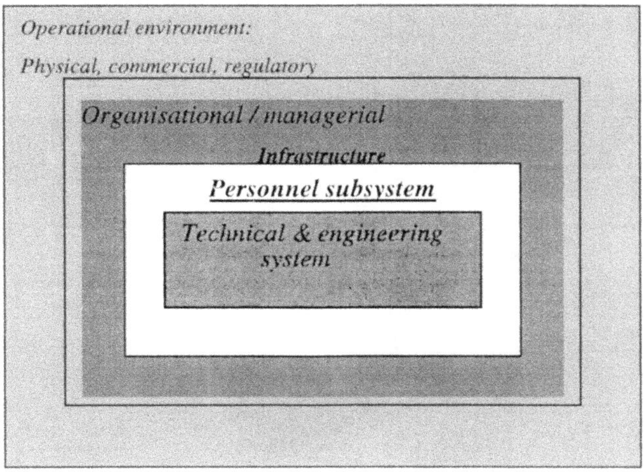

Figure 5.7 The component levels of the generic model of containership

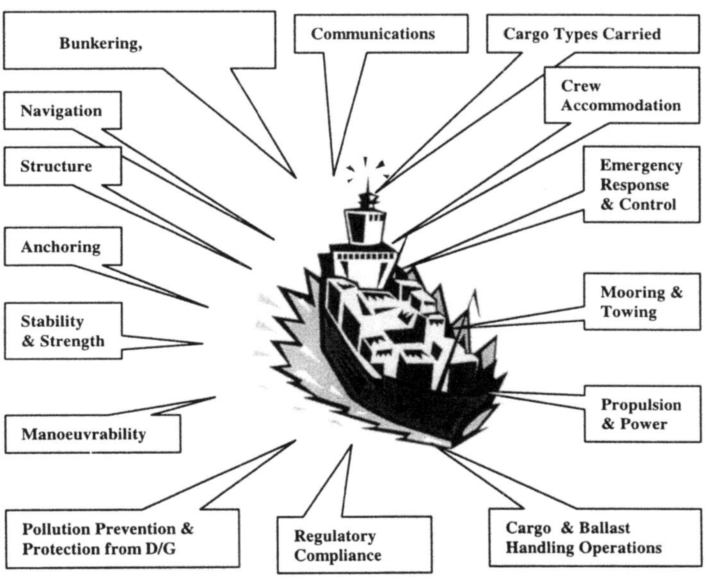

Figure 5.8 The generic engineering and technical system

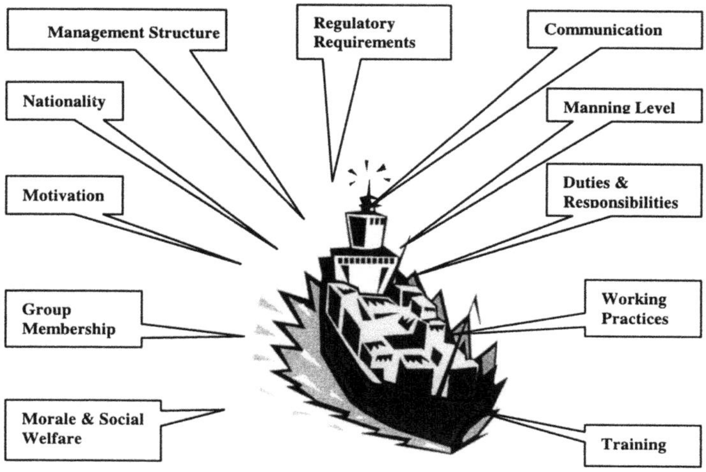

Figure 5.9 The generic personnel sub-system

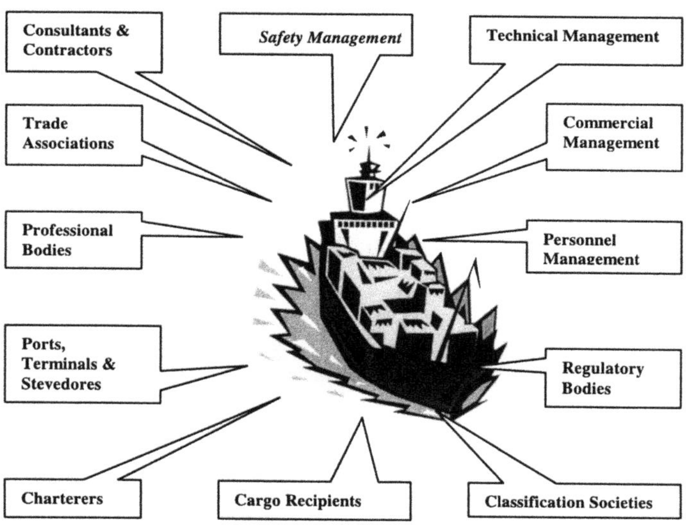

Figure 5.10 The generic organisational & managerial infrastructure

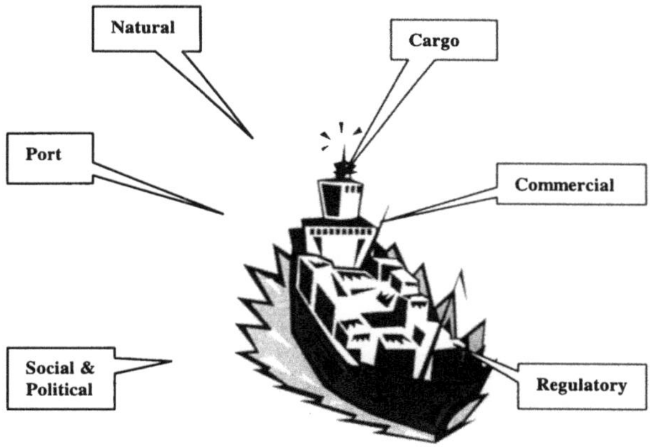

Figure 5.11 The generic environment of operation

Chapter 6

Risk Assessment Using Fuzzy Set Approach

Summary

The failure data available for fishing vessels are scarce and often accompanied with a high degree of uncertainty. For this reason the use of conventional probabilistic risk assessment methods may not be well suited. In this Chapter, an approach using Fuzzy Set Theory (FST) to model the occurrence likelihood and consequences of hazards is presented. The different ways uncertainties can manifest in an analysis are discussed and this is followed by a review of FST, identifying the various applications of the theory in the past. The approach described uses fault tree analysis to calculate the fuzzy probability of the system failure. The consequences of failure events are considered for four different categories. The risks associated with failure events are determined by combining their occurrence likelihood and possible consequences to produce a risk ranking. The application of this approach is demonstrated using a hydraulic winch operating system of a fishing vessel.

Keywords: Fault tree, fuzzy set, risk assessment, risk ranking.

6.1 Introduction

Where a major decision regarding cost or safety implication has to be made, it has become increasingly difficult to defend the traditional qualitative process called "engineering judgement". Thus, there has been a steady trend towards quantifying risks and/or costs, in particular the techniques of HAZard IDentification (HAZID), Quantitative Risk Assessment (QRA) and Cost-Benefit Analysis (CBA), have come very much to the fore.

QRA is a process of investigating potential accidents and expressing the results in terms of measures that reflect both the frequency and the potential loss severity of each type of accident that can occur (Henley and Kumamoto (1992)). The measures in most common use are Fatal Accident Rate (FAR)[1], Individual Risk Per Annum (IRPA)[2] and the FN curve.

Upon identifying the list of potential hazards and its contributing factors, which could be achieved by several methods including HAZard and OPerability studies (HAZOP) (Villemeur (1992)), Failure Mode and Effects Analysis (FMEA) (Military Standard (1980)), Fault Tree Analysis (FTA) (Henley and Kumamoto (1992)), etc, the next step is to quantify these events for the risk estimation phase. Quantification of risk considers two parameters, namely,

- Probability of failure event occurrence.

- Consequence severity.

These are the two parameters that are considered in many risk assessments utilised by the industry at present (Preyss (1995)). The frequencies of hazardous events are usually based on historical failure data. Often, little is known of the basis of the data or its processing and interpretation. The little that is known often raises doubts as to its quality, completeness and relevance. In the case of data relating to material or equipment failure, the attributes of the material or equipment are often not recorded and insufficient data is given in the context of its use. Almost invariably, failures are assumed to be random in time, that is, the observed number of failures is divided by an exposure period to give a failure rate and this is assumed to be age-independent. In reality, some modes of failure are more common in the earlier or later years of the life of a component or a system. Even where data is of high quality, sample sizes are often small and statistical uncertainties are correspondingly large. As such, a fuzzy set modelling approach may be more appropriate to model the probability of a hazardous event occurring.

The quantification of severity can be accomplished in several ways, subjective reasoning and expert judgement is one of the common methods. As many accidents in the marine industry, especially on fishing vessels, are rarely properly reported, it may be difficult to quantify the severity of an accident. Once again, the use of a fuzzy set modelling approach integrating expert knowledge may be well suited for this purpose.

Many fishing companies in the UK have very poor organisational structure and most are skipper owned vessels. This would entail that documented records on vessel, system and component would be difficult to come by and the availability of data for quantitative analysis is either unavailable or far from the ideal format. This has led to the need of developing a risk assessment method that could address the high level of uncertainty in the data.

6.2 Uncertainty

There is a close relationship between complexity and uncertainty and it is said that as the complexity increases, certainty decreases (Friedlob and Schleifer (1999)). Albert Einstein said that so far as mathematics refers to reality, it is not certain, and so far as mathematics is certain, it does not refer to reality (McNeill and Freiberger (1993)). In his "Law of Incompatibility", Zadeh states "As complexity rises, precise statements lose meaning, and meaningful statements lose precision" (McNeill and Freiberger (1993)). In 1965, while pondering this loss of precision, Zadeh conceived the notion of fuzzy logic, the first new method of dealing with uncertainty since the development of probability (Zadeh (1965)).

Uncertainty comes about when information is deficient, but information can be deficient in different ways. Uncertainty may be divided into several basic types (McNeill and Freiberger (1993), Klir and Yuan (1995), Klir (1989)):

- Fuzziness
- Ambiguity resulting from discord
- Ambiguity resulting from non-specificity

Fuzziness is uncertainty resulting from vagueness. Most natural language descriptors are vague and somewhat uncertain, rather than precise. Following are a few examples of fuzzy, uncertain events in engineering within a ship:

- Change lubricating oil within 100 days of operation.

- Filters should be cleaned when differential pressure is high.
- Maintain heavy fuel oil temperature above 90°C.

The vagueness in these operating instructions may lead the crew to use their own judgement to carry out the operation and hence there will be a non-uniform approach to maintenance and this could lead to failures within the operating system. From a safety assessment point of view, it would be difficult for the safety analyst to interpret these instructions and determine the interval of maintenance or the storage temperature of the heavy fuel oil.

Discord can be defined as a conflict or dissonance. For example, in a probability distribution, *P(x)*, each probability measure is for a specific alternative in a set of exhaustive, mutually exclusive alternatives. Each *P(x)* expresses the "degree of belief" (based on some evidence) that a particular alternative is the correct one. Thus, the beliefs expressed in a probability distribution may be in conflict with each other.

To illustrate this point, take the probability of failure of a component as an example. A 90% belief (probability) that the component will fail under certain conditions is in conflict with a 10% belief that the component will not fail. Probability theory can model only situations where there are no conflicting beliefs about mutually exclusive alternatives. If there are other aspects to the uncertainty (perhaps fuzziness), they are not captured in a probability theory model (Klir (1991)).

Non-specificity is a lack of informativeness resulting from not clearly stating or distinguishing alternatives. Non-specificity is characterised by cardinalities (sizes) of relevant sets of alternatives. The more possible alternatives a situation has, the less specific the situation is (a situation is completely specific if there is only one possible alternative) (Klir (1991)). Because each probability in a probability distribution is completely specific to a particular alternative, probability theory is not capable of conceptualising non-specificity (Klir (1991)). Figure 6.1 shows the types of uncertainty along with a brief description of each uncertainty (Klir and Yuan (1995)).

Uncertainty in a safety analysis can be caused by three main factors as listed below (Villemeur (1992)):

- *Uncertainties linked to the parameters* - for various reasons, the available information on dependability is uncertain: a small sample leading to a wide confidence interval, extrapolation of data from one installation to another, etc. Certain other parameters (delayed appearance of physical factors, time available after losing a system before undesirable effects ensue, etc.) connected with design or operation are also with uncertainties. Dependability is defined as the ability of an entity to perform one or several required functions under given conditions. This concept can encompass reliability, availability, maintainability, safety, durability, etc - or combinations of these abilities. Generally speaking, dependability is considered to be the science of failures and faults.

- *Uncertainties connected with modelling* - these are due to the use of an approximate dependability model. It is particularly true in the modelling of failures with a common cause, human error or software bugs. Generally, modelling can integrate all relevant variables without assessing their relationship in sufficient detail.

- *Uncertainties connected with the non-exhaustive nature of the analysis* - the analyst cannot be totally sure that his modelling has taken all important factors, relevant figures and significant interactions into account.

Analysing uncertainties therefore consists of identifying all the uncertainties and their repercussions on the assessment. Usually, only the first source of uncertainty is taken into account; an attempt is then made to assess the uncertainty of the final result.

6.3 Fuzzy Set Theory Background

Fuzzy Set Theory (FST) was formalised by Prof. Lofti Zadeh at the University of California in 1965. The significance of fuzzy variables is that they facilitate gradual transition between states and consequently, possess a natural capability to express and deal with observation and measurement uncertainties.

Traditional variables, which may be referred to as *crisp variables* do not have this capability. Although the definition of states by crisp sets is mathematically correct, in many cases, it is unrealistic in the face of unavoidable measurement errors. A measurement that falls into a close neighbourhood of each precisely defined border between states of a crisp variable is taken as evidential support for only one of the states, in spite of the inevitable uncertainty involved in decision. The uncertainty reaches its maximum at each border, where any measurement should be regarded as equal evidence for the two states on either side of the border. When dealing with crisp variables, the uncertainty is ignored; the measurement is regarded as evidence for one of the states, the one that includes the border point by virtue of an arbitrary mathematical definition. Bivalent set theory can be somewhat limiting if we wish to describe a 'humanistic' problem mathematically (Zadeh (1987)). For example, Figure 6.2 illustrates bivalent sets to characterise the temperature of a room.

The limiting feature of bivalent sets is that they are mutually exclusive - it is not possible to have a membership of more than one set. It is not accurate to define a transition from a quantity such as '*warm*' to '*hot*'. In the real world a smooth (unnoticeable) drift from '*warm*' to '*hot*' would occur. The natural phenomenon can be described more accurately by FST. Figure 6.3 shows how the same information can be quantified using fuzzy sets to describe this natural drift.

A set, A, with points or objects in some relevant universe, X, is defined as these elements of x that satisfy the membership property defined for A. In traditional 'crisp' sets theory each element of x either is or is not an element of A. Elements in a fuzzy set (denoted by ~, eg $\tilde{A}$) can have a continuum of degrees of membership ranging from complete membership to complete non-membership (Zadeh (1987)).

$\mu(x)$ gives the degree of membership for each element $x \in X$. $\mu(x)$ is defined on [0,1]. A membership of 0 means that the value does not belong to the set under consideration. A membership of 1 would mean full representation of the set under consideration. A membership somewhere between these two limits indicates the degree of membership. The manner in which values are assigned to a membership is not fixed and may be established according to the preference of the person conducting the investigation.

Formally $\tilde{A}$ is represented as the ordered pair $[x, \mu(x)]$:

$$\tilde{A} = \{(x, \mu(x)) | x \in X \text{ and } 0 \leq \mu(x) \leq 1\} \quad (6.1)$$

The use of a numerical scale for the degree of membership provides a convenient way to represent gradation in the degree of membership. Precise degrees of membership generally do not exist. Instead they tend to reflect sometimes subjective 'ordering' of the element in the universe.

Fuzzy sets can be represented by various shapes. They are commonly represented by S-curves, π-curves, triangular curves and linear curves. The shape of the fuzzy set depends on the best way to represent the data. In general the membership (often indicated on the vertical axis) starts at 0 (no membership) and continues to 1 (full membership). The domain of a set is indicated along the horizontal axis. The fuzzy set shape defines the relationship between the domain and the membership values of a set.

6.3.1 Types of Membership Functions

In principle a membership function associated with a fuzzy set Ã depends not only on the concept to be represented, but also on the context in which it is used. The graphs of the functions may have very different shapes and may have some specific properties (e.g. continuity). Whether a particular shape is suitable or not can be determined only in the application context (Klir and Yuan (1995)). In many practical instances, fuzzy sets can be represented explicitly by families of parameterised functions, the most common being:

Triangular functions

$$\tilde{A}(x) = \begin{cases} 0, \text{ if } x \le a \\ \dfrac{x-a}{m-a}, \text{ if } x \in [a,m] \\ 1, \text{ if } x = m \\ \dfrac{b-x}{b-m}, \text{ if } x \in [m,b] \\ 0, \text{ if } x \ge b \end{cases}$$

where *m* is a modal value and *a* and *b* denote lower and upper bounds, respectively, for non-zero values of $\tilde{A}(x)$. Sometimes it is more convenient to use the notation explicitly highlighting the membership function parameters, in this case it is given by:

$$A(x; a,m,b) = \max\{\min[(x-a)/(m-a),(b-x)/(b-m)],0\} \quad (6.2)$$

1) 2) Trapezoidal function.

$$\tilde{A}(x) = \begin{cases} 0, \text{ if } x < a \\ \dfrac{x-a}{m-a}, \text{ if } x \in [a,m] \\ 1, \text{ if } x \in [m,n] \\ \dfrac{b-x}{b-n}, \text{ if } x \in [n,b] \\ 0, \text{ if } x > b \end{cases}$$

Using equivalent notation, it is given by:

$$A(x; a,n,b) = \max\{\min[(x-a)/(m-a),1,(b-x)/(b-n)],0\} \quad (6.3)$$

Figure 6.4 shows an example of a parameterised trapezoidal function. This is a graphical representation of the explicit families of parameterised functions defining the bounds of the function. In this example, parameters *a, m, n* and *b* in Equation (6.3) are given by -2.5, 0, 2.5 and 5, respectively.

Fuzzy sets can be characterised in more detail by referring to the features used in characterising the membership functions that describe them (Kandel (1986), Dubois et al. (1993)).

The "Support" of a fuzzy set *A*, denoted by *Supp(A),* means that all elements of *X* belong to *A* to a non-zero degree (Kruse et al. (1994)). Formally, this is given by:

$$\text{Supp}(A) = \{ x \in X \mid A(x) > 0 \} \tag{6.4}$$

Alternatively, the "Core" of a fuzzy set *A* is the set of all elements of X that exhibit a unit level of membership in *A* (Kruse et al. (1994)), Formally, this is given by:

$$\text{Core}(A) = \{ x \in X \mid A(x) = 1 \} \tag{6.5}$$

Figure 6.5 shows a graphical representation of the "Support" and "Core" of a fuzzy set. The "Support" and "Core" of fuzzy sets may be viewed as closely related concepts in the sense that they identify elements belonging to the fuzzy set and they are both sets. All elements of a "Core" are sub-named by the "Support". Interval [*a,d*] is called the "Support" and interval [*b,c*] is called the "Core".

6.3.2 Representation Theorem

Any fuzzy set can be regarded as a family of fuzzy sets. This is the essence of an identity principle known also as the representation theorem. To explain this construction, it is required to define the notion of an α–cut of a fuzzy set. The α –cut of *A*, denoted by A_α, is a set consisting of those elements in the universe *X* whose membership values exceed the threshold level α. This is formally represented by:

$$A_\alpha = \{ x \mid A(x) \geq \alpha \} \tag{6.6}$$

In other words, A_α consists of elements of *x* identified with *A* to a degree of at least α. In particular, the highest level, $\alpha = 1$, determines a set of *x* totally belonging to *A*. Clearly the lower the level of α, the more elements are admitted to the corresponding α-cut, that is, if $\alpha_1 > \alpha_2$ then $A_{\alpha 1} \subset A_{\alpha 2}$. The representation theorem states that any fuzzy set *A* can be decomposed into a series of its α-cuts. This can be represented by:

$$A = \bigcup_{\alpha \in (0,1)} (\alpha A_\alpha) \text{ or equivalently}$$

$$A(x) = \sup_{\alpha \in [0,1]} [\alpha A_\alpha(x)] \tag{6.7}$$

Conversely, any fuzzy set can be "reconstructed" from a family of nested sets (assuming that they satisfy the constraint of consistency: if $\alpha_1 > \alpha_2$ then $A_{\alpha 1} \subset A_{\alpha 2}$). This theorem's importance lies in its underscoring of the very nature of the generalisation provided by fuzzy sets. Furthermore, the theorem implies that problems formulated in the framework of fuzzy sets (such as risk and reliability analysis) can be solved by transforming these fuzzy sets into their corresponding families of nested α-cuts and determining solutions to each using standard, non-fuzzy techniques. Subsequently, all the partial results derived in this way can be merged, reconstructing a solution to the problem in its original formulation based on fuzzy sets. By quantifying more levels of the membership values (that is the α-cuts), the reconstruction can be made more detailed. Figure 6.6 shows a diagrammatic representation of α-cuts. Clearly, the lower the level of α, the more elements are admitted to the corresponding α-cut, that is, if $\alpha_1 > \alpha_2$ then $A_{\alpha 1} \subset A_{\alpha 2}$.

6.3.3 Application of FST

Since FST was proposed almost four decades ago, it has found many useful applications. The linguistic approach based on fuzzy sets has given very good results for modelling qualitative information. It has been widely used in different fields, for example, information retrieval (Bordogna and Pasi (1993)), clinical diagnosis (Degani and Bortolan (1988)), marketing (Yager et al. (1994)), risk modelling in software development (Lee (1996a), Lee (1996b)), technology transfer strategy selection (Chang and Chen (1994)), education (Law (1996)), decision making (Bordogna et al. (1997)), environmental engineering (Deshpande (1999)), and many more. A review by Maiers and Sherif in 1985, covered over 450 papers addressing FST application in areas of automation control, decision making, biology and medicine, economics and the environment (Maiers and Sherif (1985)).

The use of FST in system safety and reliability analyses could prove to be a useful tool, as these analyses often require the use of subjective judgement and uncertain data. By allowing imprecision and approximate analysis, FST helps to restore integrity to reliability analyses by allowing uncertainty and not forcing precision where it is not possible. However, the theory can be difficult to use directly. The use of linguistic variables allows a flexible modelling of imprecise data and information. A linguistic variable differs from a numerical variable in that its values are not numbers but words or sentences in a natural or artificial language. Since words in general are less precise than numbers, the concept of a linguistic variable serves the purpose of providing a means of approximate characterisation of phenomena, which are too complex or ill defined to be amenable to description in conventional quantitative terms (Schmucker (1984)). More specifically, fuzzy sets, which represent the restriction associated with the values of a linguistic variable, may be viewed as summaries of various sub-classes of elements in a universe of discourse (a universe of discourse is the range of all possible values for an input to a fuzzy system). This is analogous to the role played by words and sentences in a natural language.

6.4 Risk Assessment Using FST

An approach for risk assessment using FST is divided into two main modelling categories, that is, probability (Part 1) and severity of consequences (Part 2) (Pillay (2001), Pillay and Wang (2002)). It involves several steps, which are represented in the flowchart shown in Figure 6.7. A combination of FST and expert judgement is used to accomplish the modelling of the two parameters. The outcome is used to rank the risk associated with failure events according to their priority. Part 1 of the approach uses Fault Tree Analysis (FTA) to identify the critical components of a system (Pillay et al. (2000)). Using this FTA, fuzzy arithmetic calculation is performed on the basic events to obtain the fuzzy probability estimates of the primary events. The results are left in the linguistic state to enable integration with the analysis of severity of consequences.

In Part 2 of the approach, the severity of a failure is assessed for its effect on four categories, as will be discussed later. The results of the analysis in Parts 1 and 2 are combined using the min-max inference rule to obtain a linguistic term for the risk. This linguistic term is then defuzzified using the weighted mean of maximum method to produce the risk ranking.

The first step of the described approach is to establish the type of data that is available for analysis. Depending on the size and organisational structure of the company, such data will vary in terms of its format and availability. For example, the data available from fishing vessels are most likely repair data that would just reflect the date the repair was carried out and the spares were consumed. Such data should be restructured to enable analysis using the fuzzy set approach.

The consequences of an event may not be documented in a format where it is readily useable for analysis. The severity of the consequences could be determined by the cost incurred from the result of the failure. This however may only be limited to equipment loss, production loss, environmental clean up cost, etc. The injury or loss of life (due to the failure of an equipment) is normally expressed in terms of number of casualties and the extent of the injury (bruises, required shore medical assistance, permanent disablement or death).

6.4.1 Part 1: Probability of Failure Event Occurrence

Constructing fault tree - Once the failure data has been gathered, it is grouped and sorted by its equipment/sub-system and finally the system to which the component belongs. The top event of the fault tree will be the failure of the equipment (e.g. main winch failure) while the initiating events and basic events will be the component failures (e.g. seal leakage, brake failure, control valve failure, etc). A full description of FTA has been provided in Section 3.8. It is best to construct a fault tree for equipment within a system separately as it enables data handling and analysis to be conducted. The individual fault trees can later be collated to analyse the system failure. Fault tree construction can be achieved with the use of computer software packages such as Fault Tree +V6.0 and AvSim+ (Isograph Limited (1995, 1998)).

Structure selection - In the structure selection phase, the linguistic variable is determined with respect to the aim of the modelling exercise. Informally, a linguistic variable is a variable whose values are words or sentences rather than numbers. Considering the available data at hand and the aim of this approach, the linguistic variable is determined to be the occurrence likelihood of an undesired critical event. The linguistic terms to describe this variable are then decided, for example, Very High, High, Moderate, Low and Remote.

Membership function and estimation - Six classes of experimental methods help to determine membership function: horizontal approach, vertical approach, pairwise comparison, inference based on problem specification, parametric estimation and fuzzy clustering (Pedrycz and Gomide (1998)). The method selected depends heavily on the specifics of an application, in particular, the way the uncertainty is manifested and captured during the sampling of data. The membership function chosen must be able to represent the available data in the most suitable and accurate manner. Due to the nature of the arithmetic involved, the shape of the membership function suited for the approach would either be triangular or trapezoidal, therefore the horizontal or vertical approach for function determination is applied (Pedrycz and Gomide (1998)). The vertical method takes advantage of the identity principle and 'reconstructs' a fuzzy set via identifying its α-cuts. After several levels of α are selected, the investigator is required to identify the corresponding subset of X whose elements belong to A to a degree not less than α. The fuzzy set is built by stacking up the successive α-cuts. Figure 6.8 shows an example of the stacking process of α-cuts.

Fuzzy calculation in fault trees - Given the critical event or undesired condition (top event), a fault tree can be developed using backward logic to create a network of intermediate events linked by logic operators (usually AND and OR operators) down to the basic events. The fault tree itself is the logic structure relating the top event to the primary events. These primary/basic events may be related to human error (operators, design or maintenance), hardware or software failures, environmental conditions or operational conditions.

The probability of an event defined as a fuzzy set is developed as below (Bowles and Paláez (1995)). Let S be a sample space and P a probability measure defined on S. Then,

$$P(S) = \int_S dP = 1$$

If $E \subseteq S$ is an event then

$$P(E) = \int_S C_E(x)dP$$

where $C_E(x) =$ 1 if $x \in E$

0 otherwise

Zadeh has observed that P(E) can be viewed as the expected value of the characteristic function that defines the set E (Zadeh (1987)). By analogy, he defines the probability of the fuzzy set $\tilde{A}$ as the expected value of the membership function for $\tilde{A}$:

$$P(\tilde{A}) = \int_S \mu_{\tilde{A}}(x)dP \tag{6.8}$$

On a discrete sample space, $S = \{x_1, x_2, x_3 \ldots\ldots x_n\}$, this is,

$$P(\tilde{A}) = \sum_{i=1}^{n} \mu_{\tilde{A}}(x_i)P(x_i) \tag{6.9}$$

Intuitively, Equations (6.8) and (6.9) define the probability of a fuzzy event as the summation over all elements, of the probability the event occurs weighted by the degree to which the element is a member of the event. Alternatively, it can be viewed as the probability of the possibility of the fuzzy event.

The following properties of the probability of ordinary events also hold for the probabilities of fuzzy events (Terano et al. (1992)).

$$P(\tilde{A}) \le P(\tilde{B}) \text{ if } \tilde{A} \subseteq \tilde{B}$$

$$P(\overline{\tilde{A}}) = 1 - P(\tilde{A})$$

$$P(\tilde{A} \cup \tilde{B}) = P(\tilde{A}) + P(\tilde{B}) - P(\tilde{A} \cap \tilde{B})$$

For most systems, the organisational structure can be described as either "parallel" or "series" or a combination of series and parallel as seen in Figures 6.9, 6.10 and 6.11.

System reliability can be analysed probabilistically as shown below:

Let P_i = probability of failure of component i. Then

R_i = reliability of component = $1 - P_i$

Let P_{sys} = system probability of failure. Then

R_{sys} = system reliability = $1 - P_{sys}$

For a parallel system, the system will work as long as at least one component is in operational order. If, as is traditionally assumed, components are independent and the system is either working or failed, the system probability failure (P_{sys}) is the product of the individual component failure probabilities:

$$P_{sys} = P_1 \bullet P_2 \bullet P_3 \bullet \ldots\ldots P_n \qquad (6.10)$$

Applying Equation (6.10) to a parallel system of two components *A* and *B* with fuzzy probabilities will give:

$$\tilde{P}_{sys} = \tilde{P}_A \tilde{P}_B \qquad (6.11)$$

where:

$\tilde{P}_{sys}$ is fuzzy system probability of failure.

$\tilde{P}_A$ and $\tilde{P}_B$ are fuzzy probabilities of failure events *A* and *B*.

In a series system, all constituent components must be operational in order for the system to work. Series systems are analysed in terms of their component reliabilities: $R_{sys} = R_1 \bullet R_2 \bullet R_3 \ldots\ldots \bullet R_n$. The analysis of a series system using reliabilities is identical to that of a parallel system using failure probabilities. In terms of failure probabilities for the series system:

$$P_{sys} = 1 - [(1 - P_1)(1 - P_2)(1 - P_3) \ldots\ldots (1 - P_n)] \qquad (6.12)$$

Applying Equation (6.12) to a series system of two components *A* and *B* with fuzzy probabilities will give:

$$\tilde{P}_{sys} = [1 - (1 - \tilde{P}_A)(1 - \tilde{P}_B)] \qquad (6.13)$$

When two basic events represent the input to an AND gate as shown in Figure 6.12, it can be assumed that these two events are in a parallel configuration. It denotes that the occurrence of both events will cause the AND gate to be operative and the probability will be given by Equation (6.11). For an OR gate with two basic events as its input as shown in Figure 6.13, it can be considered that the two events are in a series configuration. This denotes that if either events occur, the OR gate will be operative and the probability will be given by Equation (6.13) (Bowles and Paláez (1995)).

Fuzzy arithmetic operations - In standard fuzzy arithmetic, basic operations on real numbers are extended to those on fuzzy intervals. A fuzzy interval *A* is a normal fuzzy set on *R* (set of real numbers) whose α-cuts for all $\alpha \in (0,1]$ are closed intervals of real numbers and whose support is bounded by *A*.

Two common ways of defining the extended operation are based on the α-cut representation of fuzzy intervals and on the extension principle of FST (Kaufman and Gupta (1985), Klir and Yuan (1995)). When α-cut representation is employed, arithmetic operations on fuzzy intervals are defined in terms of arithmetic operations on closed intervals. To define the individual arithmetic operation specifically, let the symbols $[a_1^\alpha, a_2^\alpha]$ and $[b_1^\alpha, b_2^\alpha]$ denote for each $\alpha \in (0,1]$ the α-cuts of fuzzy intervals A and B, respectively. Using this notation, the individual arithmetic operations on the α-cuts of A and B are defined by the well known formulas from interval analysis (Kaufman and Gupta (1985), Klir and Yuan (1995)) given below:

$$A_\alpha + B_\alpha = [a_1^\alpha + b_1^\alpha, a_2^\alpha + b_2^\alpha] \quad (6.14)$$

$$A_\alpha - B_\alpha = [a_1^\alpha - b_2^\alpha, a_2^\alpha - b_1^\alpha] \quad (6.15)$$

$$A_\alpha \times B_\alpha = [a_1^\alpha b_1^\alpha, a_2^\alpha b_2^\alpha] \quad (6.16)$$

$$A_\alpha / B_\alpha = [a_1^\alpha / b_2^\alpha, a_2^\alpha / b_1^\alpha] \quad (6.17)$$

$$A_\alpha \pm k = [a_1^\alpha, a_2^\alpha] \pm k = [a_1^\alpha \pm k, a_2^\alpha \pm k] \quad (6.18)$$

$$A_\alpha \times k = [a_1^\alpha, a_2^\alpha] \times k = [ka_1^\alpha, ka_2^\alpha] \quad (6.19)$$

Equations (6.16) and (6.17) are true for all non-negative numbers. Figures 6.14 and 6.15 illustrate simple addition and subtraction operations of α-cuts of sets A and B, respectively.

6.4.2 Part 2: Severity of Consequences

List of consequences - When carrying out a comprehensive analysis, it is important that all the consequences of a failure are considered. It has been noted that due to the poor documentation of accidents on fishing vessels, the list of identifiable consequences is limited to the serious or life threatening ones, for example, death of a crew, complete loss of a vessel/equipment and so on. Therefore, expert judgement should be used to compile a list of consequences and complement the historical data. This can be achieved in the form of an FMEA (Smith (2002)). Upon being satisfied that all the consequences for each event/failure have been compiled, the analyst has to assign them into their respective groups. In the approach described here, four groups have been identified, that is, Personnel, Equipment, Environment and Catch. For each event or failure, a rating from 1 - 4 is given for each of the groups. The ratings describe the consequences of an event occurring in linguistic terms such as "Negligible", "Marginal", "Critical" and "Catastrophic". The significance of each of the ratings are listed and described as follows:

Personnel:

Effect of failure of the item on personnel (worst case always assumed)

Rating 1 = Negligible (No or little damage- bruises/cuts)

Rating 2 = Marginal (Minor injuries - treatable on board)

Rating 3 = Critical (Major injuries – requires professional attention)

Rating 4 = Catastrophic (Death/permanent disablement)

Environment:

Effect of failure of the item on the environment

Rating 1 = No effect (No or little effect)

Rating 2 = Marginal effect (Can be controlled by ship-staff)

Rating 3 = Critical effect (Requires shore assistance)

Rating 4 = Catastrophic effect (permanent damage to the environment)

Equipment:

Effect of failure on machinery or system in terms of down time if failure occurs and cost of repair

Rating 1 = Negligible (No or little attention needed - cleaning up/drying)

Rating 2 = Marginal (Minor repair – few hrs lost)

Rating 3 = Critical (Major repair – few days lost)

Rating 4 = Catastrophic (Destruction of equipment - total plant shutdown)

Catch:

Effect of failure on fishing operation in terms of catch effected:

Rating 1 = No effect (No or little effect)

Rating 2 = Marginal effect (Catch affected for a few hours)

Rating 3 = Critical effect (Catch affected for a few days)

Rating 4 = Catastrophic effect (No catch for a few months)

Calculate Total Score (Σx_{ij}) - Upon assigning a score for each group, a table is generated as shown in Table 6.1. From this table, a "Total Score" is calculated by summing the score of each individual group for an event. This total score will later be used to assign the membership function for that event using fuzzy rules.

Fuzzy rules - The fuzzy rules determining the membership function of each event are divided into 4 categories i.e. Hazard Class 1(HC_1), HC_2, HC_3 and HC_4. The maximum score of an event is used to assign that particular event to the appropriate hazard class. Therefore, if an event has a score of [2,2,1,1] for each group respectively, it would be assigned to HC_2 (the maximum score for that event is 2 for the Personnel and Environment categories).

Fuzzy rules are generated based on available historical data, experience and complemented by expert knowledge. Where possible, logbooks are analysed for casualty and accident reports to develop the following rules:

Hazard Class 1 (HC1)

If an event has a score of [1,1,1,1], which entails that for all categories considered, the effect of the failure is negligible, then the total effect of that failure on the system and environment should be negligible as well. Hence,

1) If $\Sigma X_{ij} = 4$, then 1.0 Negligible................(A.0)

Hazard Class 2 (HC_2)

The minimum score possible in the HC_2 category is 5, i.e. [2,1,1,1] or any variation of this score. The maximum possible score is 8, i.e. [2,2,2,2], therefore the range of membership function between these two extremities is assigned so as to ensure a smooth transition between limits to have overlapping of functions. Hence,

2) If $X_{ij\,max} = 2$, and $\Sigma X_{ij} = 5$ then 0.8 Negligible, 0.6 Marginal.........(B.0)

$\Sigma X_{ij} = 6$ then 1.0 Marginal, 0.2 Critical.............(B.1)

$\Sigma X_{ij} = 7$ then 0.5 Marginal, 0.8 Critical............(B.2)

$\Sigma X_{ij} = 8$ then 1.0 Critical, 0.2 Catastrophic.........(B.3)

The above rules can be represented graphically as seen in Figure 6.16.

Hazard Class 3 (HC_3)

The minimum score possible in the HC_3 category is 6, i.e. [3,1,1,1] or any variation of this score. The maximum possible score is 12, i.e. [3,3,3,3]. When assigning the linguistic membership function for HC_3, it is important to compare the values with those of the HC_2 to ensure that it does not contradict the rules generated for that hazard class. For the same total score in HC_2 and HC_3, the linguistic membership function for HC_3 (for that particular score) should logically reflect a more severe consequence. For example, for a total score of 7 for HC_2 and HC_3, which would have a combination of [2,2,2,1] and [3,2,1,1] respectively, using expert judgement, one would say that although both classes have the same total score, a total score of 7 for HC_3 would entail a more severe consequence. Hence the membership function for HC_3 and a total score of 7 is 0.8 Critical, 0.2 Catastrophic while the membership function for HC_2 with the same total score of 7 is 0.5 Marginal, 0.8 Critical (Pillay and Wang (2002)). Using this method, the rules for HC_3 are generated for the other values of its total scores and are reflected below:

3) If $X_{ij\,max} = 3$, and $\Sigma X_{ij} = 6$ then 0.5 Marginal, 1.0 Critical......................(C.0)

$\Sigma X_{ij} = 7$ then 0.8 Critical, 0.2 Catastrophic.................(C.1)

$\Sigma X_{ij} = 8$ then 0.5 Critical, 0.5 Catastrophic.................(C.2)

$\Sigma X_{ij} = 9$ then 0.2 Critical, 0.8 Catastrophic.................(C.3)

$\Sigma X_{ij} = 10$ then 1.0 Catastrophic..................................(C.4)

$\Sigma X_{ij} = 11$ then 1.0 Catastrophic..................................(C.5)

$\Sigma X_{ij} = 12$ then 1.0 Catastrophic..................................(C.6)

The above rules can be represented graphically as seen in Figure 6.17.

Hazard Class 4 (HC_4)

4) If $X_{ij\,max} = 4$, and $\Sigma X_{ij} \geq 7$ then Catastrophic....................................(D.0)

It is necessary to assign a hazard class for each event as the consequences of the event are considered for different groups. Grouping each event into a hazard class allows direct comparison with other events and enables the effects of a failure to be compared based on its linguistic terms assigned to it. For example, if an event *A* has a score of [3,3,1,1] with a total of 8 and event *B* has a score of [2,2,2,2] which also gives a total of 8, from experience and expert judgements, it can be said that event *A* is more serious in nature. Hence, it should be assigned a linguistic term which must be "more severe" compared to event *B*. To enable this distinction between events, which have the same total score, hazard classification is introduced, i.e. HC_1, HC_2, etc. Therefore, the membership functions for events *A* and *B* will be obtained from Rules (C.2) and (B.3), respectively. At this stage of the described approach, each event would be assigned occurrence likelihood and possible consequences. The next step would be to analyse these two parameters and provide a risk ranking number for each event.

6.4.3 Risk Assessment

The risk associated with an event increases as either its severity of the consequences or its occurrence probability increases. Judgement of the severity of possible consequences is, by its very nature, highly subjective. Using a priority matrix, the "riskiness" of an event can be obtained. The risk posed by the event is expressed in linguistic terms such as *'Very Important'*, *'Important'*, *'Moderate'* and *'Low'*. This matrix is based on the probability of occurrence and the severity of possible consequences. Table 6.2 displays the various combinations of these two parameters.

The interpretation of risk ranking is given as below:

Very Important	⇒	Needs immediate corrective action.
Important	⇒	Review and corrective action to be carried out.
Moderate	⇒	Review to be carried out and corrective action implemented if found to be cost effective.
Low	⇒	Review subject to availability of revenue and time.

From Table 6.2, a risk ranking in linguistic terms can be obtained for the failure events of a system/sub-system or component. For example, if the probability of a failure event is *'High'* and the severity is "*Marginal*", then the risk would be classified as *'Important'*.

Fuzzy set approach may provide a more flexible and meaningful way of assessing risk. The analysis uses linguistic variables to describe severity and probability of occurrence of the failure. These parameters are "fuzzified" to determine their degree of membership in each input class using the membership functions developed. The resulting fuzzy inputs are evaluated using the linguistic rule base to yield a classification of the "riskiness" of the failure and an associated degree of membership in each class. This fuzzy conclusion is then defuzzified to give a single crisp priority for the failure.

Figure 6.18 shows the membership function of the riskiness of an event on an arbitrary scale, which would later be used to defuzzify the fuzzy conclusion and rank the risk according to a priority number. The membership function used is a triangular function. Unlike the trapezoidal function, the membership value of 1 in the triangular function is limited to only one value of the variable on the x-axis.

6.4.4 Rule Evaluation and Defuzzification

Rules are evaluated using min-max inferencing to calculate a numerical conclusion to the linguistic rule based on their input value (Zadeh (1992)). The result of this process is called the fuzzy risk conclusion.

The "truth value" of a rule is determined from the conjunction (i.e. minimum degree of membership of the rule antecedents) (Zadeh (1973)). Thus the truth-value of the rule is taken to be the smallest degree of truth of the rule antecedents. This truth-value is then applied to all consequences of the rule. If any fuzzy output is a consequent of more than one rule, that output is set to the highest (maximum) truth-value of all the rules that include it as a consequent. The result of the rule evaluation is a set of fuzzy conclusions that reflect the effects of all the rules whose truth-values are greater than zero.

Consider the risk priority table (Table 6.2) where the probability of occurrence is "*High*", the severity is "*Marginal*" and their associated degrees of belief are 0.6 and 1.0, respectively. Thus the conclusion Riskiness = "*Important*" has a membership value of min (0.6,1.0) = 0.6. To establish how risky the hazard is, this fuzzy conclusion has to be defuzzified to obtain a single "crisp" result.

The defuzzification process creates a single assessment from the fuzzy conclusion set expressing the risk associated with the event, so that corrective actions can be prioritised. Several defuzzification techniques have been developed (Runkler and Glesner (1993)). One common technique is the weighted mean of maximum method, which is illustrated here. This technique averages the points of maximum possibility of each fuzzy conclusion, weighted by their degrees of truth. Hence, if the conclusion from the risk evaluation phase is, for example, 0.5 Low, 0.1 Low and 0.5 Mod, the maximum value for each linguistic term is taken. This reduces the conclusion to 0.5 Low and 0.5 Mod to be defuzzified.

The following is given to demonstrate how riskiness is obtained. Suppose event *A* has the following probability of occurrence and severity of consequences:

Probability of Occurrence – **Moderate** (0.6 High, 1.0 Moderate, 0.5 Low).

Severity – **Marginal** (1.0 Marginal).

Then from Table 6.2, event *A* will be denoted by the prefix **MM** and therefore is associated with a riskiness of "**Important**". However, considering all the membership functions of the two parameters, i.e. probability of occurrence and severity, the following terms of riskiness are generated:

0.6 High, 1.0 Marginal = HM = 0.6 Important

1.0 Moderate, 1.0 Marginal = MM = 1.0 Important

0.5 Low, 1.0 Marginal = LM = 0.5 Moderate

The riskiness is obtained as:

Riskiness = (0.5 "Moderate", 1.0 "Important")

From Figure 6.18, the support value for each linguistic term is obtained, where:

The support value for Moderate = 4

The support value for Important = 6

The support value represents an average value for a particular linguistic term. Taking the maximum value for each term of the riskiness, that is, 1.0 Important and 0.5 Moderate, the weighted mean is calculated as follows:

The weighted mean (Z) = [(1.0)(6) + (0.5)(4)]/(1.0+0.5) = 5.33

From this result the riskiness of event A can be prioritised as being "*Important*" with a support of 5.33.

6.5 Application to a Hydraulic Winch System

To demonstrate the described approach, the data from a fishing vessel is used as a test case (Pillay (2001)). The data collected for the test case is in the format of repair data. It includes:

- Voyage no (shows the date when the repair was carried out).

- Equipment repaired.
- Parts that were changed.
- Modifications that were made.
- Cause of failure (in some instances).

Specialists/contract workers carry out the repairs for this particular vessel in the floating dock. Should a failure occur during operation at sea, temporary repair is carried out by the crew and the equipment is kept operating in the "abnormal" condition. No records are kept of any temporary repairs done on board, however, a repair list is compiled by the Chief Engineer for the equipment to undergo permanent repair work at the next "docking".

In order to use this data for the modelling process, the following assumptions were made:

- Repairs and modifications are only carried out when the equipment/component has failed.
- Upon completion of repair, the equipment is assumed to be "same-as-new".

For this test case the trapezoidal function was selected and estimated. The boundaries of the trapezoidal function were determined for each set. These values being the values of x for the respective α-cuts are subjective and were predominantly based on the policies and attitude of the company and on what the company thought to be tolerable limits within which they wish to operate. To describe the probability of occurrence, linguistic terms such as "Very High", "High", "Moderate", "Low", and "Remote" are used. A range of limits quantifying the probability of occurrence is then assigned to each term. These limits are in the form of Mean Time Between Failure (MTBF). MTBF is given by:

$$MTBF = \frac{\sum t_i + \sum s_i}{n} \tag{6.20}$$

where:

t_i = time to failure.

s_i = survival time.

n = number of failures.

These limits are then converted into failure rates by the following formula:

$$\lambda = \frac{1}{MTBF} \tag{6.21}$$

A failure rate is calculated under the assumption that the mean down time and repair time are very small compared to the operating time. The MTBF is then converted to failure rate using Equation (6.21) and is reflected along an ordinal scale as shown in Table 6.3.

The membership function used (i.e. trapezoidal function) allows a membership value of 1 for a range of probabilities unlike the triangular function. This function is thought to model the probability of occurrence close to what it is in reality (Pillay (2001)). Figure 6.19 shows the membership function along with its ordinal scale. The limits and the centre point values of the ordinal scale are given by the dotted line and will be used to perform the fuzzy arithmetic.

The system used to demonstrate this approach is an operating system of a Gilson winch on board an ocean trawler. This trawler is a 1266 GRT (Gross Tonnage), deep-sea trawler with an L.O.A (length overall) of 60 meters (Pillay and Wang (2002)). The Gilson winch is hydraulically operated and is situated forward of the main winches. Unlike the main winches, it does not bear the load of the catch. It serves as an auxiliary winch to the main winches.

Table 6.4 shows the failure data of the primary/basic events for a Gilson winch failure. The data collected is over a period of 66 months (14 voyages), and from this data, the linguistic term for failure probability of each basic event is determined by identifying the number of occurrences per operating day(s) on the ordinal scale. The membership function is then determined by reading off the intersecting points on the y-axis.

The fault tree shown in Figure 6.20 is generated from the data collected for the failure of the Gilson winch. Each secondary or intermediate event (e.g. brake failure, clutch failure, hydraulic leakage, etc) is modelled by gathering the available failure data and then grouping them according to the component or system they affect. For example, the failure of the brake cylinder (GBCyl) and brake seal leakage (GBSeal) will cause the brake to fail. Hence, the brake failure (G.Brake) is the secondary event with the GBCyl and GBSeal being its basic events. To demonstrate the application of this approach with an example, the fault tree used only traces the path of failures that have been known to occur in the past, rendering the system inoperable.

Take two basic events from the fault tree in Figure 6.20, GBCyl and GBSeal as an example. The occurrence rates for GBCyl and GBSeal are 1 failure in 750 days and 1 failure in 300 days, respectively. Therefore event GBCyl would have a fuzzy probability of *Low* and GBSeal, *Moderate*. Performing the arithmetic operation using Equations (6.13), (6.15) and (6.16) on both these events will yield the result of 0.62 *High*, *Moderate* and 0.46 *Low* for the secondary event, brake failure (G.Brake). Figure 6.21 shows a graphical representation of this. This can be interpreted as the secondary event belonging to the linguistic term *High* with a membership of 62%, complete membership (100%) to *Moderate* and *Low* with a membership of 46%. Similarly, all the basic events in the fault tree are analysed in this manner producing an end result for the top event. The Gilson winch failure has a fuzzy failure probability of *HIGH* with a membership function of 0.9 Very High, 0.84 High and 0.1 Moderate. Although the membership to the *Very High* linguistic term is the highest, when the result is defuzzified to reflect the range of probability which it belongs to, it falls into the *High* category on the ordinal scale. It can therefore be stated that the failure rate of the Gilson winch lies between 2×10^{-1} and 2×10^{-2} (per operating day).

6.5.1 Severity of Consequences Modelling

The amount of data that was available on the consequences of a failure was scarce and difficult to come by. However, much of the data was collected in terms of cost and reports of accidents and incidents that led to injuries. Since there is no standard format for reporting an accident, the data was obtained from telexes, faxes, superintendent reports, Chief Engineer's logbook and various other sources. To complement the data, expert knowledge and judgement was used to assign ratings to each group (Personnel, Environment, Equipment and Catch). Table 6.5 shows the analyses of various failures in a Gilson winch system.

6.5.2 Risk Ranking of the Hydraulic Winch System

The probability of occurrence is determined for each basic event (Table 6.4) and the severity of the same basic events is as shown in Table 6.5. The risk estimation and ranking of these basic events can be carried out. For the pipe flange leak event, the probability of occurrence was determined to be 0.5 *Mod, Low* and 0.1 *Rem,* and the severity as 0.8 *Neg* and 0.6 *Marg*. Using the rule evaluation method described, which is summarised in Table 6.6, the linguistic term for risk is determined.

From Table 6.6, the risk evaluation for the pipe flange failure can be summarised as being (0.5 Low, 0.5 Imp, 0.8 Low, 0.6 Mod, 0.1 Low and 0.1 Low). Using the minimum-maximum inferencing, this can be reduced to 0.8 Low, 0.6 Mod and 0.5 Imp. The numbers 0.8, 0.6 and 0.5 represent the degree of belief and not the membership function of the particular linguistic term. Similarly, the risk evaluation for all other basic events is carried out. The results of the evaluation are shown in Table 6.7.

Weighted mean for event pipe flange leak is calculated as follows:

$$Z = \frac{(0.8\times 2)+(0.6\times 4)+(0.5\times 6)}{(0.8+0.6+0.5)} = 3.68$$

Therefore from Figure 6.18, the pipe flange leak event will be prioritised by "Moderate" with a support value of 3.68. Similarly, the weighted means can be calculated for all the other events within the system. Table 6.8 shows the results of these calculations. The risk ranking of the events associated with the Gilson winch can be easily obtained from Table 6.8.

6.6 Conclusion

Lack of reliable safety data and lack of confidence in safety assessment have been two major problems in safety analysis of various engineering activities. This is particularly true in FSA due to the fact that the level of uncertainty is high. In ship safety assessment it may often be difficult to quantify the probability of undesired events occurring and the associated consequences due to this very reason.

The described approach addresses these concerns and offers an alternative solution. Its application can be extended to sub-systems within an operating system to generate a list of components, which are ranked according to their priority for attention. This can help the owners and operators of ships to improve operating and maintenance strategies. This approach can be adopted within the FSA framework for generic ships and the results obtained from the analysis can be further utilised in Step 4 of the FSA (MSA (1993)). Due to the fact that precision is not forced, it would be appealing to use this approach in situations where reliable safety data is scarce and hard to come by.

6.7 References (Chapter 6)

1. Banard R.C., (1996) "A New Approach to Risk Assessment Integrating Scientific Evaluation and Economic Assessment of Cost and Benefits", Regulatory Toxicology and Pharmacology, Vol. 24, pp. 121-125.
2. Bordogna G. and Pasi G., (1993), "A Fuzzy Linguistic Approach Generalising Boolean Information Retrieval: a Model and Its Evaluation", Journal of the American Society for Information Science, Vol. 44, pp. 70-82.
3. Bordogna G., Fedrizzi M. and Pasi G., (1997), "A Linguistic Modelling of Consensus in Group Decision Making Based on OWA Operators", IEEE - Systems, Man and Cybernetics, Vol. 27, pp. 126-132.

4. Bowles J.B. and Paláez C.E., (1995), "Application of Fuzzy Logic to Reliability Engineering", Proceedings of the IEEE, Vol. 83, No. 3, pp. 435-449.
5. Chang P. and Chen Y., (1994), "A Fuzzy Multicriteria Decision Making Method for Technology Transfer Strategy Selection in Biotechnology", Fuzzy Sets and Systems, Vol. 63, pp. 131-139.
6. Degani R. and Bortolan G., (1988) "The Problem of Linguistic Approximation in Clinical Decision Making", International Journal of Approximate Reasoning, Vol. 2, pp. 143-162.
7. Deshpande A.W., (1999) "Application of Fuzzy Set Theory to Environmental Engineering Systems", BISC Seminar, 7th October, National Environmental Engineering Research Institute (NEERI), India.
8. Dubois D., Prade H. and Yager R.R., (1993) "Readings in Fuzzy Sets for Intelligent Systems", Morgan Kaufmann, San Mateo, California, USA.
9. Friedlob G.T. and Schleifer L.L.F., (1999) "Fuzzy Logic: Application for Audit Risk and Uncertainty", Managerial Auditing Journal, Vol. 14, No. 3, pp. 127-135.
10. Henley E.J. and Kumamoto H., (1992) "Probabilistic Risk Assessment", IEEE Press, New York.
11. Isograph Limited, (1995) "Fault Tree+ (1995), Version 6.0", Manchester, United Kingdom.
12. Isograph Limited, (1998) "AvSim+, Version 7.03", Manchester, United Kingdom.
13. Kandel A., (1986) "Fuzzy Mathematical Techniques with Applications", Addison Wesley, Reading, Massachusetts, USA.
14. Kaufman A. and Gupta M.M., (1985) "Introduction to Fuzzy Arithmetic", Van Nostrand Reinhold, New York.
15. Klir G.J., (1989) "Is There More to Uncertainty than Some Probability Theorist Might Have Us Believe?", International Journal of General Systems, Vol. 15, No. 4, pp. 360-372.
16. Klir G.J., (1991) "Principal of Uncertainty and Information Invariance", International Journal of General Systems, Vol. 17, pp. 258-269.
17. Klir G.J. and Yuan B., (1995) "Fuzzy Sets and Fuzzy Logic: Theory and Application", Prentice-Hall, New York.
18. Kruse R., Gebharat J. and Klawonn F., (1994) "Foundations of Fuzzy Systems", John Wiley & Sons, New York.
19. Law C.K., (1996) "Using Fuzzy Numbers in Educational Grading System", Fuzzy Sets and Systems, Vol. 83, pp. 311-323.
20. Lee H.M., (1996a) "Group Decision Making Using Fuzzy Set Theory for Evaluating the Rate of Aggregative Risk in Software Development", Fuzzy Sets and Systems, Vol. 80, pp. 261-271.
21. Lee H.M., (1996b) "Applying Fuzzy Set Theory for Evaluating the Rate of Aggregative Risk in Software Development", Fuzzy Sets and Systems, Vol. 80, pp. 323-336.
22. Loughran C.G., Pillay A. and Wang J., (1999) "Formal Safety Analysis of a Generic Fishing Vessel", International Conference of Fishing Vessel Safety, 12-15 April, Plymouth, United Kingdom.
23. Maiers J. and Sherif Y.S., (1985) "Applications of Fuzzy Set Theory", IEEE - Systems, Man and Cybernetics (SMC), Vol. 15, pp. 175-189.
24. MSA, (1993) "Formal Safety Assessment", MSC 66/14, Submitted by the United Kingdom Marine Safety Agency to IMO Maritime Safety Committee.
25. McNeill D. and Freiberger P., (1993) "Fuzzy Logic", Touchstone Press, New York.
26. Military Standard, (1980) "Procedures for Performing a Failure Mode, Effects and Criticality Analysis", MIL-STD-1629, November, AMSC Number N3074, Naval Ship Engineering Centre, Washington, USA.

27. Pedrycz W. and Gomide F., (1998) "An Introduction to Fuzzy Sets, Analysis and Design", MIT Press, USA.
28. Pillay A., (2001) "Formal Safety Assessment of Fishing Vessels", PhD Thesis, School of Engineering, Liverpool John Moores University, UK.
29. Pillay A. and Wang J., (2002) "Risk Assessment of Fishing Vessels Using Fuzzy Set Approaches", International Journal of Safety, Quality and Reliability, Vol. 9, No. 2, pp. 163-181.
30. Pillay A., Wang J., Wall, A. and Ruxton T., (2000) "Analysis of Fishing Vessel Failure Data Using Fuzzy Set Approach", Proceedings of the ESREL 2000 and SRA-Europe Annual Conference, Edinburgh, Scotland, UK, 14-17 May, Vol. 2, ISBN: 90 5809 1406, pp. 1005-1013.
31. Preyssl C., (1995) "Safety Risk Assessment and Management - the ESA Approach", Reliability Engineering & System Safety, Vol. 49, pp. 303-309.
32. Runkler T.A. and Glesner M., (1993) "A Set of axioms for Defuzzification Strategies toward a Theory of Rational Defuzzification Operators", Proceedings of the 2nd IEEE International Conference on Fuzzy Set Systems, New York, pp. 1161-1166.
33. Schmucker K.J., (1984) "Fuzzy Sets, Natural Language Computations and Risk Analysis, Computer Science Press, Rockville, Madison, USA.
34. Smith D.J., (2002) "Reliability, Maintainability and Risk", 6th Edition, Butterworths-Heineman Limited, Oxford, United Kingdom.
35. Terano T., Asai K. and Sugeno M., (1992) "Fuzzy Systems Theory and Application", Academic Press, United Kingdom.
36. Villemeur A., (1992) "Reliability, Availability, Maintainability and Safety Assessment", John Wiley, Chichester, United Kingdom.
37. Yager R.R., Goldstein L.S. and Mendels E., (1994) "FUZMAR: An Approach to Aggregating Market Research Data Based on Fuzzy Reasoning", Fuzzy Sets and Systems, Vol. 68, pp. 1-11.
38. Zadeh L.A., (1965) "Fuzzy Sets", Information and Control, Vol. 8, pp. 338-353.
39. Zadeh L.A., (1973) "The Outline of a New Approach to the Analysis of Complex Systems and Decision Process", IEEE Transactions On System, Man and Cybernetics (SMC), Vol.3, pp. 28-44.
40. Zadeh L.A., (1987) "Fuzzy Sets and Applications: Selected Papers", John Wiley, New York.
41. Zadeh L.A., (1992) "The Calculus of Fuzzy If/Then Rules", AI Experts, Vol. 7, pp. 23-27.

Table 6.1 Event Score

	Personnel	Environment	Equipment	Catch	Total Score (ΣX_{ij})
Failure Y_1	X_{11}	X_{21}	X_{31}	X_{41}	ΣX_{i1}
Failure Y_2	X_{12}	X_{22}	X_{32}	X_{42}	ΣX_{i2}
Failure Y_3	X_{13}	X_{23}	X_{33}	X_{43}	ΣX_{i3}

Table 6.2 Probability and Consequence Matrix

		Severity of Occurrence			
		NEG	***MARG***	***CRIT***	***CAT***
Probability of Occurrence	***REMOTE***	**RN**	**RM**	**RC**	**RCAT**
	LOW	**LN**	**LM**	**LC**	**LCAT**
	MODERATE	**MN**	**MM**	**MC**	**MCAT**
	HIGH	**HN**	**HM**	**HC**	**HCAT**
	VERY HIGH	**VHN**	**VHM**	**VHC**	**VHCAT**

Very **Important**

Moderate **Low**

Table 6.3 Probability Range for Linguistic Terms

Probability (Linguistic term)	*MTBF range (days)*	*Failure rate (ordinal scale)*
Very High	1 to 5	1 to 2×10^{-1}
High	5 to 50	2×10^{-1} to 2×10^{-2}
Moderate	50 to 500	2×10^{-2} to 2×10^{-3}
Low	500 to 2000	2×10^{-3} to 5×10^{-4}
Remote	2000 to 10000	5×10^{-4} to 1×10^{-5}

Table 6.4 Probabilities of Basic Events for a Gilson Winch Failure

Basic Events	*MTBF (days)*	*Linguistic term*	*Membership function*
Pipe flange leak	900	Low	0.5 Mod, 1.0Low, 0.1 Rem
Pipe	450	Moderate	0.6 High, 1.0 Mod, 0.5 Low
Control valve fail	900	Low	0.5 Mod, 1.0Low, 0.1 Rem
Filter choke	40	High	0.72 V.High, 1.0 High, 0.18 Mod
Brake cylinder fail	750	Low	0.5 Mod, 1.0 Low, 0.1 Rem
Brake seal fail	300	Moderate	0.6 High, 1.0 Mod, 0.5 Low
Clutch cylinder fail	900	Low	0.5 Mod, 1.0 Low, 0.1 Rem
Clutch seal leak	900	Low	0.5 Mod, 1.0 Low, 0.1 Rem
Air cylinder fail	900	Low	0.5 Mod, 1.0 Low, 0.1 Rem

Table 6.5 Gilson Winch Severity of Consequence

	Personn.	*Environ.*	*Equip.*	*Catch*	*Total*	*HC*	**Membership function**
Pipe Flange leak	1	2	1	1	5	2	0.8 Neg, 0.6 Marg.
Pipe leak	1	2	1	1	5	2	0.8 Neg, 0.6 Marg.
Control v/v fail	1	1	2	3	7	3	0.8 Crit., 0.2 Cat.
Filter choke	1	1	1	3	6	3	0.5 Marg., 1.0 Crit.
Brake cyl fail	1	1	3	3	8	3	0.5 Crit., 0.5Cat.
Brake seal leak	1	1	2	2	6	2	1.0 Marg., 0.2 Crit.
Clutch cyl fail	1	1	3	3	8	3	0.5 Crit., 0.5Cat.
Clutch seal leak	1	1	2	2	6	2	1.0 Marg., 0.2 Crit.
Air cyl fail	1	1	1	1	4	1	1.0 Neg.

Table 6.6 Risk Evaluation for Pipe Flange Failure

Probability of occurrence	*Severity*	*Risk*
0.5 Moderate	0.8 Negligible	0.5 Low
0.5 Moderate	0.6 Marginal	0.5 Important
1.0 Low	0.8 Negligible	0.8 Low
1.0 Low	0.6 Marginal	0.6 Moderate
0.1 Remote	0.8 Negligible	0.1 Low
0.1 Remote	0.6 Marginal	0.1 Low

Table 6.7 Risk Evaluation of Gilson Winch Basic Events

Events	*Occurrence*	*Severity*	*Risk*
Pipe Flange leak	0.5 Mod, 1.0 Low, 0.1 Rem	0.8 Neg, 0.6 Marg	0.8 Low, 0.6 Mod, 0.5 Imp
Pipe leak	0.6 High, 1.0 Mod, 0.5 Low	0.8 Neg, 0.6 Marg	0.8 Low, 0.6 Mod, 0.6 Imp
Control v/v fail	0.5 Mod, 1.0 Low, 0.1 Rem	0.8 Crit, 0.2 Cat	0.2 Mod, 0.8 Imp, 0.2 V.Imp
Filter choke	0.72 V.High, 1.0 High, 0.18 Mod	1.0 Crit, 0.5 Marg	0.5 Imp, 1.0 V.Imp
Brake cyl fail	0.5 Mod, 1.0 Low, 0.1 Rem	0.5 Crit, 0.5 Cat	0.1 Mod, 0.5 Imp, 0.5 V.Imp
Brake seal leak	0.6 High, 1.0 Mod., 0.5 Low	1.0 Marg, 0.2 Crit	0.5 Mod, 1.0 Imp, 0.2 V.Imp
Clutch cyl fail	0.5 Mod, 1.0 Low, 0.1 Rem	0.5 Crit, 0.5 Cat	0.1 Mod, 0.5 Imp, 0.5 V.Imp
Clutch seal leak	0.5 Mod, 1.0 Low, 0.1 Rem	1.0 Marg, 0.2 Crit	0.1 Low, 1.0 Mod, 0.5 Imp
Air cyl fail	0.5 Mod, 1.0 Low, 0.1 Rem	1.0 Neg	1.0 Low

Table 6.8 Defuzzified Ranking of a Gilson Winch Failure Events

Event	*Risk (Linguistic term)*	*Support value*
Filter choke	Very Important	7.33
Clutch cyl fail	Important	6.72
Brake cyl fail	Important	6.72
Control v/v fail	Important	6.00
Brake seal leak	Important	5.65
Clutch seal leak	Moderate	4.50
Pipe leak	Moderate	3.68
Pipe flange leak	Moderate	3.68
Air cyl fail	Low	2.00

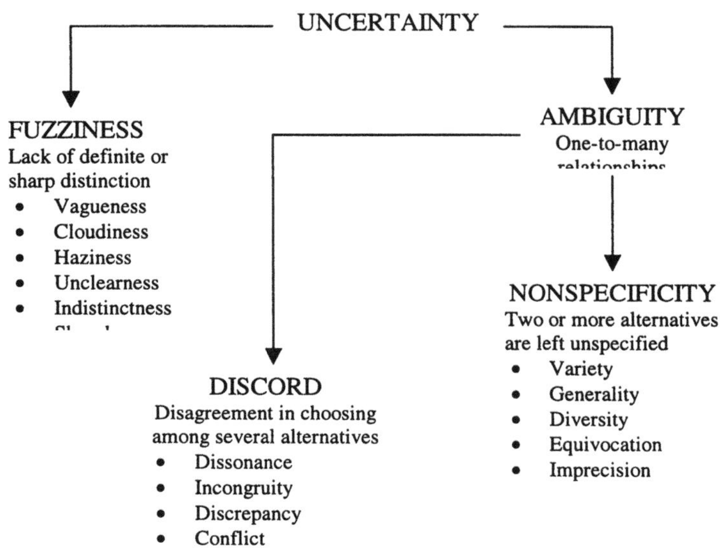

Figure 6.1 Types of uncertainty

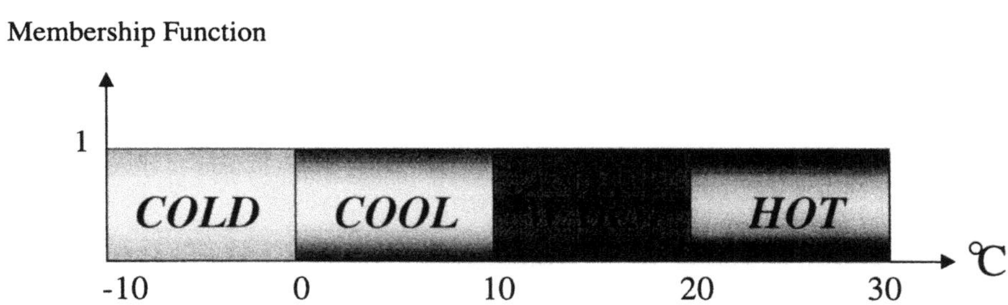

Figure 6.2 Bivalent set to characterise room temperature

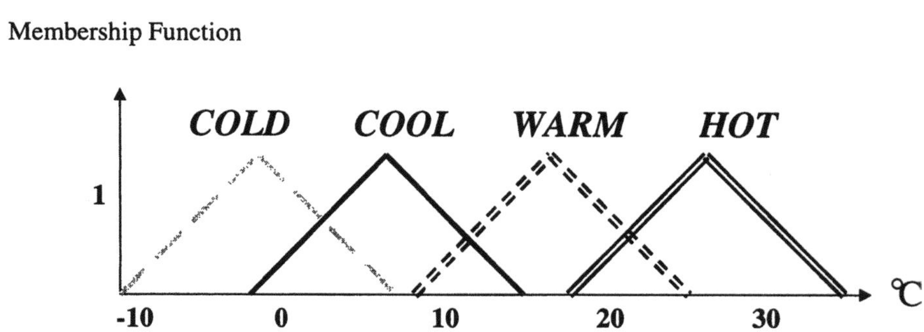

Figure 6.3 Fuzzy set to characterise room temperature

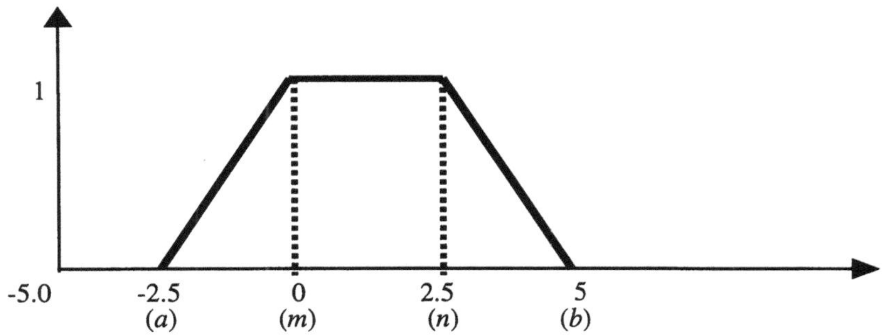

Figure 6.4 Parameterised trapezoidal function

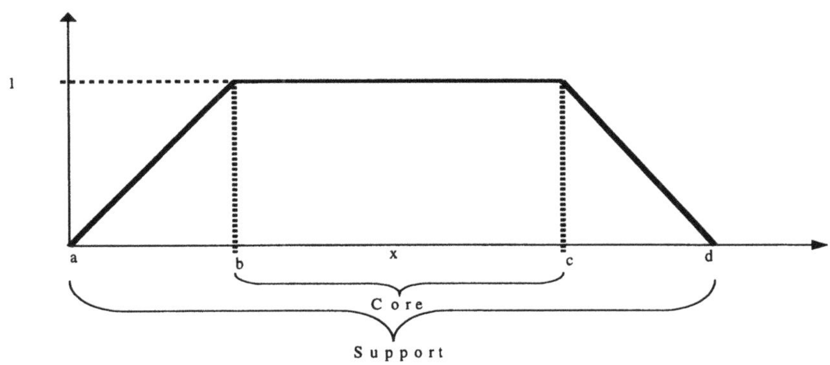

Figure 6.5 Representation of fuzzy set "Support" and "Core"

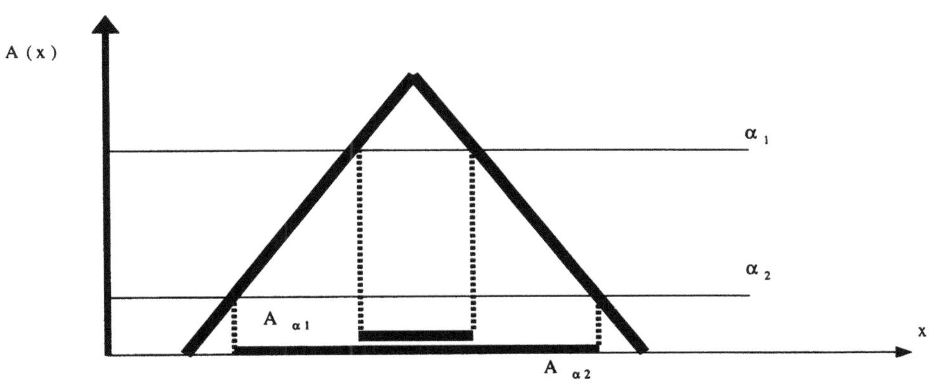

Figure 6.6 Representation of α-cuts on a fuzzy set

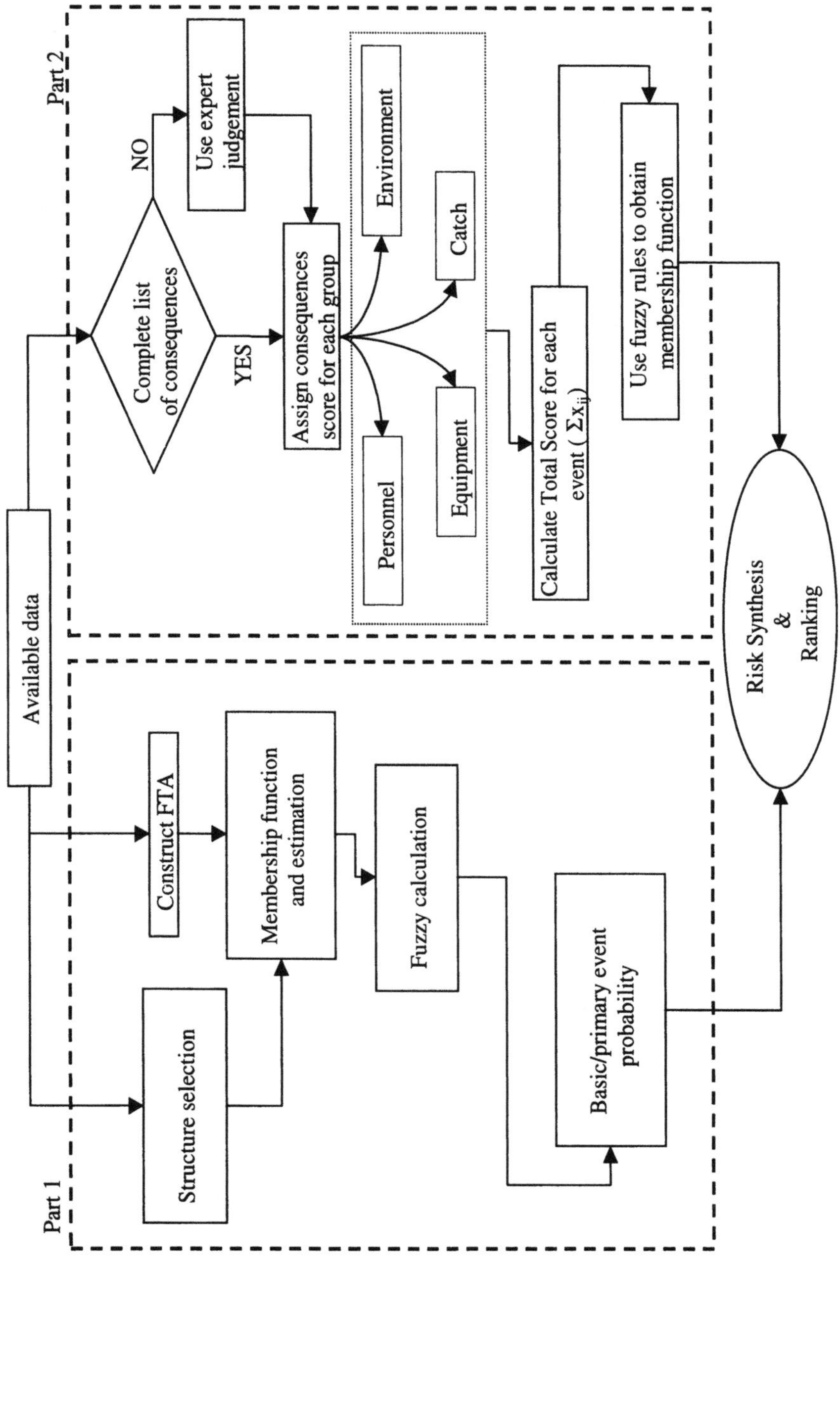

Figure 6.7 Flowchart of the approach

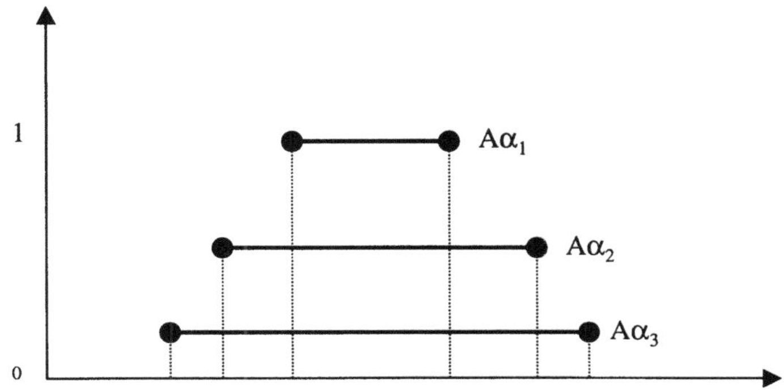

Figure 6.8 Vertical approach for function determination

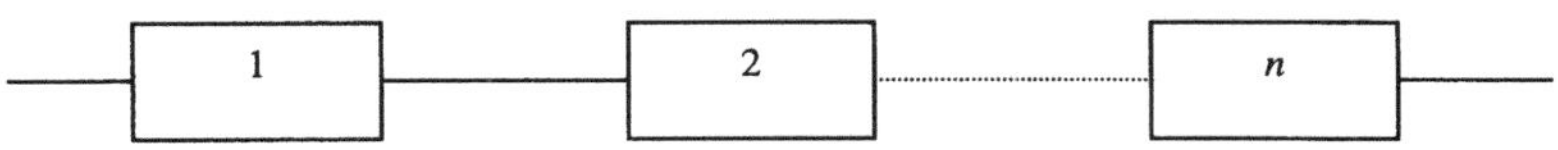

Figure 6.9 Series system

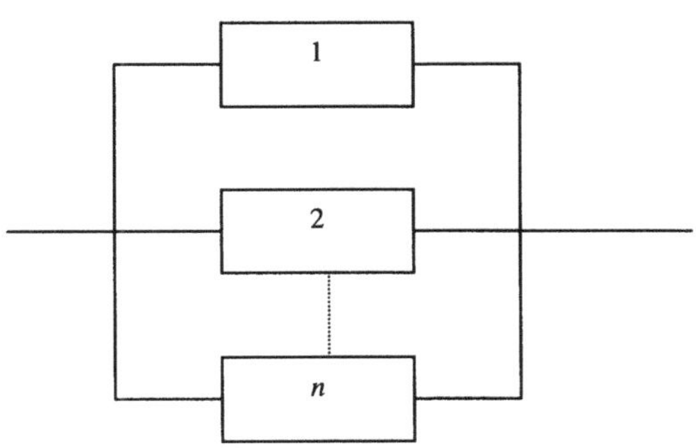

Figure 6.10 Parallel system

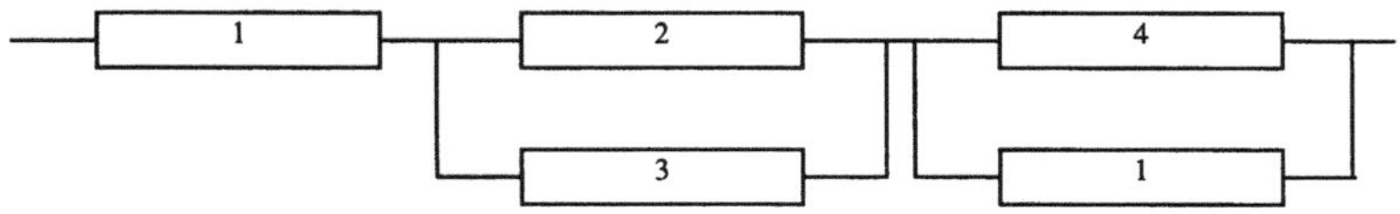

Figure 6.11 Series-parallel system

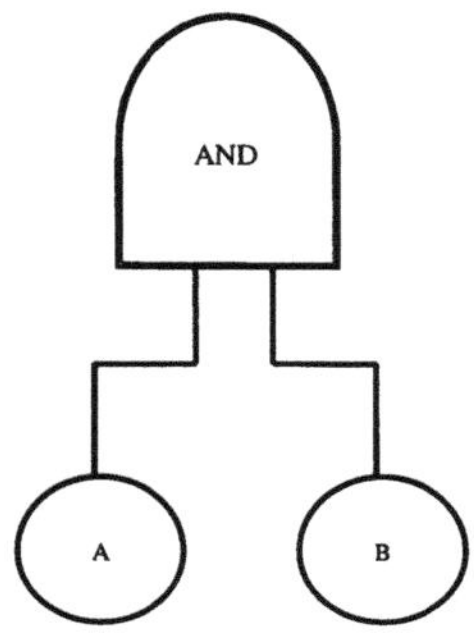

$$\tilde{P}_{sys} = \tilde{P}_A \tilde{P}_B$$

Figure 6.12 AND gate

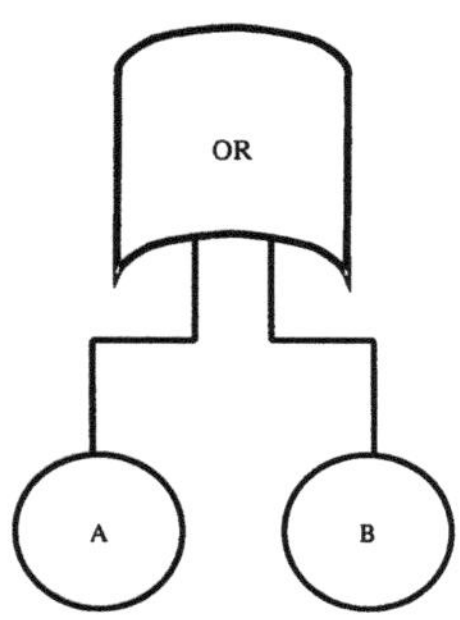

$$\tilde{P}_{sys} = [1-(1-\tilde{P}_A)(1-\tilde{P}_B)]$$

Figure 6.13 OR gate

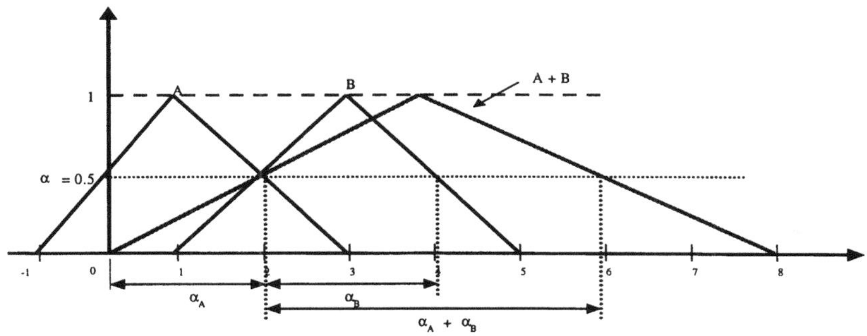

Figure 6.14 Addition operation on α-cut

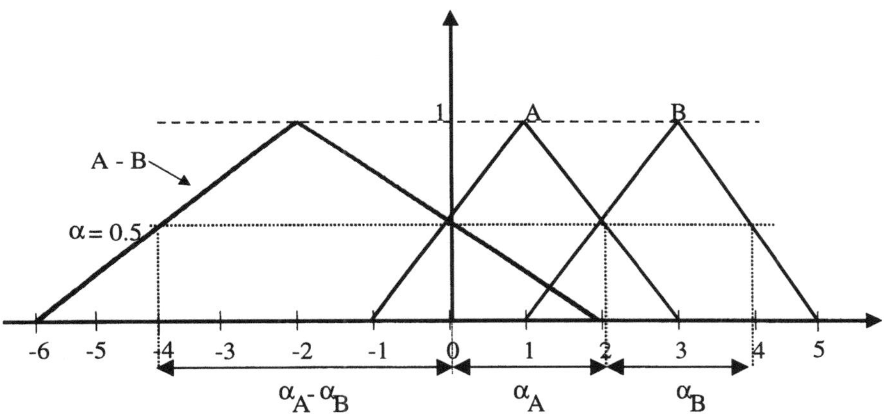

Figure 6.15 Subtraction operation on α-cut

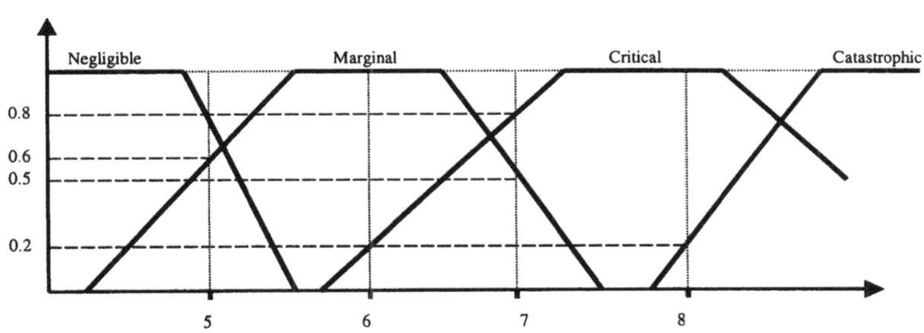

Figure 6.16 Hazard class 2

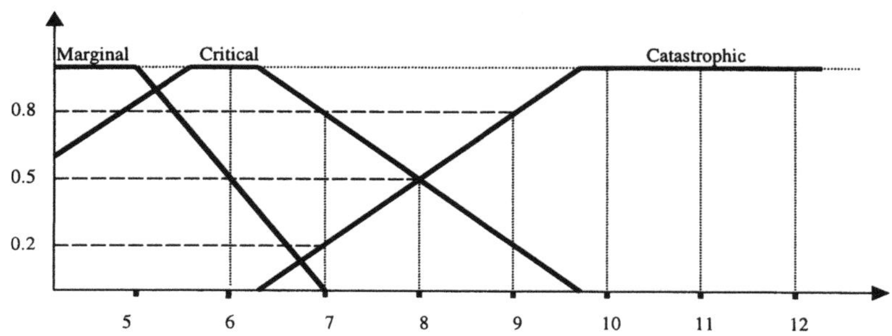

Figure 6.17 Hazard class 3

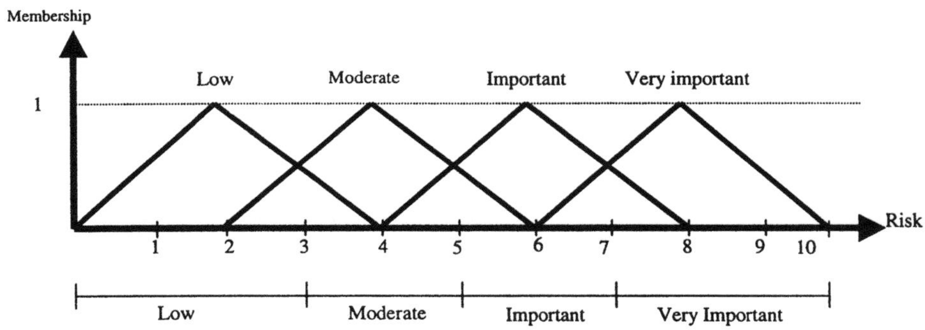

Figure 6.18 Membership function of riskiness

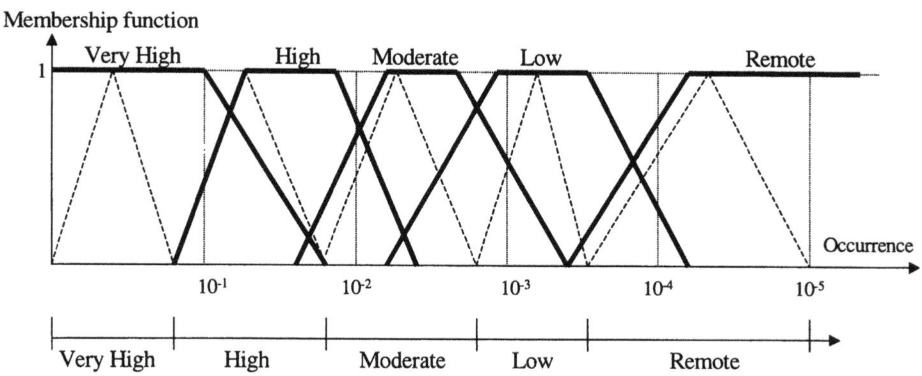

Figure 6.19 Membership function and ordinal scale

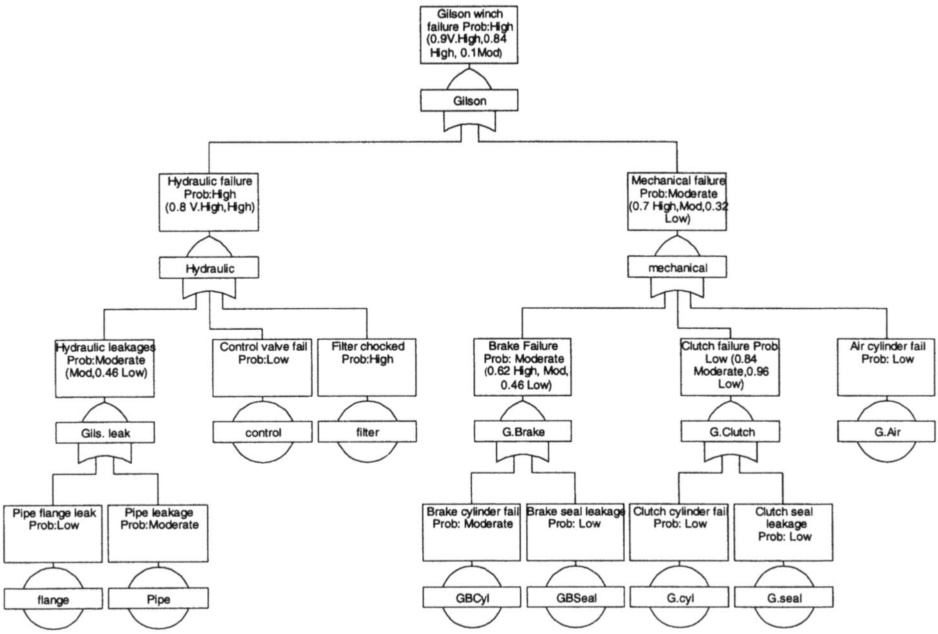

Figure 6.20 Fault tree of Gilson winch failure

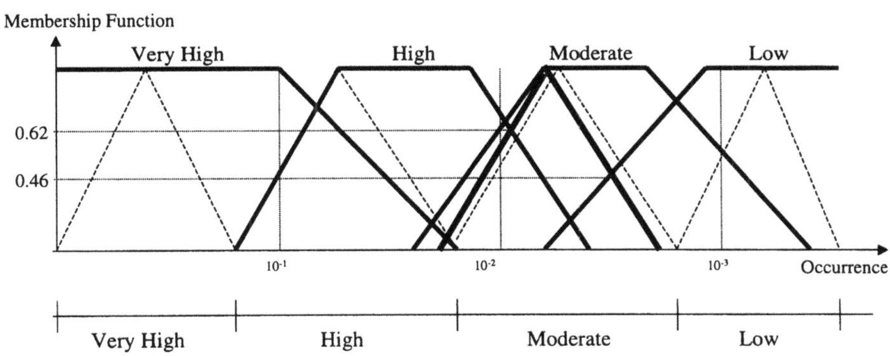

Figure 6.21 Graphical representation of fuzzy arithmetic operation on two basic events

Chapter 7

Modified Failure Mode and Effects Analysis

Summary

The marine industry is recognising the need for powerful techniques that can be used to perform risk analysis of marine systems. One technique that has been applied in both national and international marine regulations and operations is Failure Mode and Effects Analysis (FMEA). This risk analysis tool assumes that a failure mode occurs in a system/component through some failure mechanism. The effect of this failure is then evaluated. A risk ranking is produced in order to prioritise the attention for each of the failure modes identified. The traditional method utilises the Risk Priority Number (RPN) ranking system. This method determines the RPN by finding the multiplication of factor scores. The three factors considered are probability of failure, severity and detectability. Traditional FMEA has been criticised to have several weaknesses. These weaknesses are addressed in this Chapter. A new approach, which utilises the fuzzy rules base and grey relation theory, is presented.

Keywords: Failure mode and effects analysis, fuzzy set, grey theory, risk ranking.

7.1 Introduction

FMEA is intended to provide information for making risk management decisions. Detailed procedures on how to carry out an FMEA and its various application in the different industries have been documented in (Stamatis (1995)). Over the years several variations of the traditional FMEA have been developed. Russomano and Price discussed the use of knowledge base system for the automation of the FMEA process (Russomano et al. (1992), Price et al. (1992, 1995)). The use of a causal reasoning model for FMEA is documented in (Bell et al. (1992)). An improved FMEA methodology, which uses a single matrix to model the entire system and a set of indices derived from probabilistic combination to reflect the importance of an event relating to the indenture under consideration and to the entire system, was presented by Kara-Zaitri (Kara-Zaitri et al. (1991, 1992)). A similar approach was made to model the entire system using a fuzzy cognitive map (Pelaez and Bowles (1996)).

Many FMEAs have a quantitative objective, that is, to predict the likelihood of certain types of system failures. This requires good information on the statistical distribution of component failures. It also requires knowledge of dependency relationships among components under normal operations and under external perturbations.

FMEA can also be used as part of a qualitative analysis (or a semi-quantitative analysis). It attempts to identify critical components whose failure will lead to accident, injury, and/or property loss. The goal is to make systems safer or more reliable by:

1. Evaluating the effects of component failures on system performance.
2. Identifying those components that are critical to safety.
3. Developing system enhancements or administrative changes to improve safety and/or system reliability.

The major safety-related objectives of FMEA include:

1. Analysis of the system to determine effects of component failures on system performance and safety.
2. Identification of components that are critical to safety (identifying where component failure would compromise system operation, resulting in injuries, property damage, or other losses).
3. Redesigning the system to improve "passive" reliability and safety.
4. Improving maintenance routines to reduce the likelihood of component failures.

FMEA is used to assist analysts to perform hazard analyses and it is regarded as a supplement rather than a replacement for hazard analyses. Safety analysts can use FMEA to verify that all safety critical hardware has been addressed in the hazard analyses. The FMEA for hardware systems is an important technique for evaluating the design and documenting the review process. All credible failure modes and their resultant effects at the component and system levels are identified and documented. Items that meet defined criteria are identified as critical items and are placed on the Critical Item List (CIL). Each entry of the CIL is then evaluated to see if design changes can be implemented so that the item can be deleted from the CIL. Items that cannot be deleted from the CIL must be accepted by the programme/project, based on the rationale for acceptance of the risk. The analysis follows a well-defined sequence of steps that encompass: (1) failure mode, (2) failure effects, (3) causes, (4) detectability, (5) corrective or preventive actions, and (6) rationale for acceptance.

The general process of FMECA and criticality analysis has been described in Chapter 3. In an FMEA, Risk Priority Number (RPN) can also be used to model each failure mode in order to rank all the failure modes. Such a process can be divided into several steps as seen in Figure 7.1. These steps are briefly explained as follows:

1. Develop a good understanding of what the system is supposed to do when it is operating properly.
2. Divide the system into sub-systems and/or assemblies in order to "localise" the search for components.
3. Use Piping and Instrument Diagrams (P&ID), Process Flow Diagrams (PFD), schematics and flow charts to identify components and relations among components.
4. Develop a complete component list for each assembly.
5. Identify operational and environmental stresses that can affect the system. Consider how these stresses might affect the performance of individual components.
6. Determine failure modes of each component and the effects of failure modes on assemblies, sub-systems, and the entire system.

7. For each failure mode, establish detectability (dependent upon several elements including alarm/monitoring devices in place)
8. Categorise the hazard level (severity) of each failure mode (several qualitative systems have been developed for this purpose).
9. Estimate the probability. In the absence of reliable statistical information, this can also be done using qualitative estimates.
10. Calculate the RPN: the RPN is given as the multiplication of the index representing the probability, severity and detectability.
11. Determine if action needs to be taken depending on the RPN.
12. Develop recommendations to enhance the system performance. These fall into two categories:
    - Preventive actions: avoiding a failure situation.
    - Compensatory actions: minimising losses in the event that a failure occurs.
13. Summarise the analysis: this can be accomplished in a tabular form.

7.2 Some Weaknesses of FMEA

The traditional FMEA has been a well-accepted safety analysis method, however, it suffers from several weaknesses. One of the critically debated weaknesses, is the method that the traditional FMEA employs to achieve a risk ranking. The purpose of ranking risk in order of importance is to assign the limited resources to the most serious risk items. Traditional FMEA uses an RPN to evaluate the risk level of a component or process. The RPN is obtained by finding the multiplication of three factors, which are the probability of failure (S_f), the severity of the failure (S) and the probability of not detecting the failure (S_d). Representing this mathematically will give:

$$RPN = S_f \times S \times S_d \tag{7.1}$$

Tables 7.1, 7.2 and 7.3 list the scales used to measure the three factors given in Equation (7.1) (Pillay (2001), Pillay and Wang (2003)).

These tables show that the traditional FMEA uses five scales and scores of one to ten, to measure the probability of occurrence, severity and the probability of detection. Though this simplifies the computation, converting the probability into another scoring system, and then finding the multiplication of factor scores may cause problems. From Tables 7.1 and 7.3 it can be seen that the relation between S_f and the probability scale is non-linear, while it is linear for that between S_d and the probability scale.

The most critically debated disadvantage of the traditional FMEA is that various sets of S_f, S and S_d may produce an identical value of RPN, however, the risk implication may be totally different (Gilchrist (1993), Ben-Daya and Raouf (1993)). For example, consider two different events having values of 2, 3, 2 and 4, 1, 3 for S_f, S and S_d respectively. Both these events will have a total RPN of 12 ($RPN_1 = 2 \times 3 \times 2 = 12$ and $RPN_2 = 4 \times 1 \times 3 = 12$), however, the risk implications of these two events may not necessarily be the same. This could entail a waste of resources and time or in some cases a high risk event going unnoticed.

The other prominent disadvantage of the RPN ranking method is that it neglects the relative importance among S_f, S and S_d. The three factors are assumed to have the same importance. This may not be the case when considering a practical application of the FMEA process.

An approach using fuzzy rule base and grey relation theory is described to address these problems. A fuzzy rule base is used to rank the potential causes identified within the FMEA, which would have identical RPN values but different risk implications. The approach then extends the analysis to include weighting factors for S_f, S and S_d using defuzzified linguistic terms and grey relation analysis. The background of fuzzy set theory has been explained in Chapter 6 and the principle of grey relation theory will be briefly described in Section 7.3.

7.3 Background of Grey Theory

Grey system theory was proposed and developed by Deng in 1982 (Deng (1982, 1989)). In grey systems, the information, such as operation, mechanism, structure and behaviour, are neither deterministic nor totally unknown, but are partially known. It explores system behaviour using relation analysis and model construction. It also deals with making decisions characterised by incomplete information (Shih et al. (1996), Wu et al. (1984)).

Grey system theory has been widely used in many fields, such as optimisation (Zheng and Lewis (1993)), engineering (Wu and Ouhyoung (1994)), geomechanics (Zheng (1988)), economy (Jianer (1987)), history (Junjiang (1986)), geography (Li (1991)) and management (Deng (1986)).

The use of grey theory within the FMEA framework can be accomplished (Chang et al. (1999)). The method involves several steps, which are briefly discussed here. First, a comparative series, which reflects the various linguistic terms and decision factors of the study, is generated. The linguistic terms describing the decision factors are, for example, *Low, Moderate, High,* etc. The comparative series can be represented in a form of a matrix as shown in Equation (7.2). This matrix shows the failure modes, $\{x_1\ x_2\ \ldots\ x_n\}$ and the linguistic terms describing each decision factor of the failure mode, $\{x_1(1)\ x_1(2)\ \ldots\ x_1(K)\}$, $\{x_2(1)\ x_2(2)\ \ldots\ x_2(K)\}$, etc.

$$x = \begin{bmatrix} x_1 \\ x_2 \\ \cdot \\ \cdot \\ x_n \end{bmatrix} = \begin{bmatrix} x_1(1) & x_1(2) & . & . & . & x_1(K) \\ x_2(1) & x_2(2) & . & . & . & x_2(K) \\ \cdot & \cdot & . & . & . & \cdot \\ \cdot & \cdot & . & . & . & \cdot \\ x_n(1) & x_n(2) & . & . & . & x_n(K) \end{bmatrix} \quad (7.2)$$

where:

K = the number of the decision factors.

n = the number of failure modes considered.

The standard series is an objective series that reflects the ideal or desired level of all the decision factors and can be expressed as $x_0 = [x_0(1)\ x_0(2)\ \ldots\ x_0(K)]$. This could be assumed to be the lowest level of the linguistic terms describing the decision factors. The difference between the two series (comparative and standard series) is calculated. The grey relation coefficient is obtained using Equation (7.3):

$$\gamma(x_0(k), x_i(k)) = \frac{\min_i \min_k |x_0(k) - x_i(k)| + \zeta \max_i \max_k |x_0(k) - x_i(k)|}{|x_0(k) - x_i(k)| + \zeta \max_i \max_k |x_0(k) - x_i(k)|} \quad (7.3)$$

where ζ is an identifier, $\zeta \in (0,1)$, only affecting the relative value of risk without changing the priority (Hong (1986)).

To find the degree of relation, the weighting coefficients (β_k) of the decision factors must first be decided. For the application of the grey theory to FMEA, β_k should be set to suit the intention of the FMEA and comply with Equation (7.4).

$$\sum_{k=1}^{K} \beta_k = 1 \quad (7.4)$$

The degree of relation, $\Gamma(x_0, x_i)$, can then be calculated using Equation (7.5).

$$\Gamma(x_0, x_i) = \sum_{k=1}^{K} \beta_k \gamma\{x_0(k), x_i(k)\} \quad (7.5)$$

The degree of relation in FMEA denotes the relationship between the potential causes and the optimal value of the decision factors. The higher the value obtained from Equation (7.5), the smaller the effect of the identified events.

7.4 Fuzzy Rule Based Method

The aim of this method is to develop a method that does not require a utility function to define the probability of occurrence (S_f), severity (S) and detectability (S_d) considered for the analysis and to avoid the use of the traditional RPN. This is achieved by using information gathered from experts and integrating them in a formal way to reflect a subjective method of ranking risk.

The flowchart in Figure 7.2 illustrates a fuzzy set method for the modified FMEA process (Pillay (2001), Pillay and Wang (2003)). The first step is to set up the membership function of the three categories, that is, probability of occurrence (S_f), severity (S) and detectability (S_d). Once the membership functions of these three categories have been developed, the FMEA is carried out in the traditional manner with the use of brainstorming techniques (Brahm and Kleiner (1996), Van Gundy (1998)). Each of the failure modes is then assigned a linguistic term representing the three linguistic variables, (probability of occurrence, severity and detectability). Using the fuzzy rule base generated, these three variables are integrated to produce linguistic terms representing the *priority for attention*. This term represents the risk ranking of all the failure modes identified for the components. Once a ranking has been established, the process then follows the traditional method of determining the corrective actions and generating the FMEA report.

7.4.1 Fuzzy Membership Function

The fuzzy membership function is developed for each of the three variables using multiple experts. These experts should be appropriately chosen so as to ensure realistic and non-biased membership functions (Kuusela et al. (1998)). The application of the modified FMEA requires experts who are familiar with the operation and management circumstances of the industry.

Using the selected experts, the fuzzy sets and membership functions can be generated as explained here.

Assume that E experts are asked for some $x \in X$ to evaluate the proposition "x belongs to A" as either true or false, where A is a fuzzy set on X that represents a linguistic term associated with a given linguistic variable. Given a particular element $x \in X$, let $a_i(x)$ denote the answer of expert i ($i \in E$). Assume that $a_i(x) = 1$ when the proposition is valued by expert i as true, and $a_i(x) = 0$ when it is valued as false (Klir and Yuan (1995)). Then,

$$A(x) = \frac{\sum_{i=1}^{E} a_i(x)}{E} \tag{7.6}$$

may be viewed as a probabilistic interpretation of the constructed membership function. When the experts have different degrees of competencies, C_i, with regard to the model being constructed, Equation (7.6) is modified to give:

$$A(x) = \sum_{i=1}^{E} C_i a_i(x) \tag{7.7}$$

where

$$\sum_{i=1}^{E} C_i = 1 \tag{7.8}$$

The degree of competency for each of the experts should be determined based on their experience and knowledge of the system and should be agreed upon by all the experts involved in the study.

In the fuzzy rule base analysis, the linguistic variables are determined to be the probability of occurrence (S_f), the severity (S) and the detectability (S_d). Each of the three linguistic variables has five linguistic terms describing them. These linguistic terms are *Remote, Low, Moderate, High* and *Very High* (for simplicity, the term *Negligible* for the Severity category is substituted by *Remote*). The interpretations of these linguistic terms are given in Table 7.4. This information can also be represented graphically as seen in Figure 7.3, where it was developed by a collective agreement between the analysts involved in the study (Pillay (2001), Pillay and Wang (2003)). Each expert was asked for the values (on the x-axis) that they thought belonged to the appropriate linguistic term. The membership functions for the linguistic terms used were determined using Equation (7.7).

7.4.2 Fuzzy Rule Base Development

Fuzzy logic systems are knowledge-based or rule-based systems constructed from human knowledge in the form of fuzzy *IF-THEN* rules (Wang (1997)). An important contribution of fuzzy system theory is that it provides a systematic procedure for transforming a knowledge base into non-linear mapping. A fuzzy *IF-THEN* rule is an *IF-THEN* statement in which some words are characterised by continuous membership functions.

IF-THEN rules have two parts: an antecedent that is compared to the inputs and a consequent, which is the result/output. The input of the fuzzy rules base is the probability of occurrence, severity and detectability. The output is assigned a linguistic variable, *priority for attention*, and is described linguistically by *Low, Fairly Low, Moderate, Fairly High* and *High.*

In order to generate the fuzzy rule base for the FMEA, the selected experts are asked to group the various combinations of linguistic terms describing the three factors considered into a category reflecting the *priority for attention*. Since there are three factors and five linguistic terms describing each factor, the total number of rules are 125. However, some of these rules can be combined to reduce the number of rules of the fuzzy rule base. A typical rule from the rule base would read as:

"If failure probability is *Remote*, severity is *Remote* and detectability is *Low*, then *priority for attention* is *Low.*"

Using Equation (7.7), the membership function for the rules in the fuzzy rule base can be determined. The rule base is then used in the FMEA to ascertain the priority for attention for each of the potential failure modes identified.

7.4.3 Ranking the Priority for Attention

The defuzzification process creates a single assessment from the fuzzy conclusion set expressing how corrective actions can be prioritised. Several defuzzification techniques have been developed (Runkler and Glesner (1993)). One common technique is the weighted mean of maximum method (WMoM), which is illustrated here. This technique averages the points of maximum possibility of each fuzzy conclusion, weighted by their degrees of truth.

Assume the output of the FMEA is assigned a linguistic variable, *priority for attention*, and is described linguistically by *Low, Fairly Low, Moderate, Fairly High* and *High.* The support value for each of these linguistic terms is determined by taking the weighted average of the support values given by each expert. Suppose the support values for the five linguistic terms are calculated on an arbitrary scale of 1 to 10 and are defined as follows: *Fairly Low* - 0.055, *Low* - 0.461, *Moderate* - 0.911, *Fairly High* - 2.041 and *High* - 7.111 (Pillay (2001), Pillay and Wang (2003)).

Suppose the potential cause identified in the FMEA has the following probability of occurrence, severity and detectability: Probability of Occurrence – ***Remote***, Severity – ***Remote,*** and Detectability - ***Moderate.*** Referring to the rule base that will be introduced in Table 7.11, the priority of attention is, for example, ***Low, 0.06 Fairly Low*** with a support value of 0.055 and 0.461, respectively. Using the WMoM method, the weighted mean, (Z), can be calculated as:

$$Z = [(1.0)(0.055) + (0.06)(0.461)]/(1.0+0.06) = 0.078$$

From this result the *priority for attention* of this particular event can be numerically expressed as being 0.078. This method of defuzzification has been discussed in Chapter 6. Similarly all the potential failure modes identified in the FMEA can be analysed in this manner to produce a ranking such that the highest value of the defuzzified conclusion reflects the highest priority for attention.

7.5 Grey Theory Method

The flowchart in Figure 7.4 illustrates the grey theory method to rank the events, which are identified in the FMEA process (Pillay (2001), Pillay and Wang (2003)). The first step is to set up the membership function of the three categories (probability of occurrence (S_f), severity (S)

and detectability (S_d)). This can be carried out as explained in Section 7.3. In order to preserve consistency in the analysis, the membership functions estimated earlier are preserved and applied here. Hence, each of the linguistic variables, that is, the probability of occurrence, severity and detectability will have five linguistic terms describing them. Upon identifying all the failure modes and causes of failure using brainstorming techniques (as used in the traditional FMEA process), the probability of occurrence, severity and detectability are assigned linguistic terms accordingly.

Upon assigning the appropriate linguistic term to describe each linguistic variable (for each event), the next step requires a crisp number to be produced representing each of the linguistic terms assigned. In short, the application of these fuzzy sets with grey theory requires the defuzzification of the membership functions obtained in Figure 7.3 (Chang et al. (1999)). The defuzzified values of each of the linguistic terms are used to generate the comparative series, which is represented in the form of a matrix.

At this stage, the standard series for the variables is generated by determining the optimal level of all three variables for the events in the FMEA. The difference between the standard and comparative series is obtained and the results are used to determine the grey relation coefficient.

Using the value of the grey relation coefficient and introducing a weighting factor for all three linguistic variables, the degree of grey relation of each event can be calculated. This degree represents the ranking order of each event identified in the FMEA.

Chen and Klien have proposed an easy defuzzification method for obtaining the crisp number of a fuzzy set as shown here in Equation (7.9) (Chen and Klien (1997)).

$$K(x) = \frac{\sum_{i=0}^{I}(b_i - c)}{\sum_{i=0}^{I}(b_i - c) - \sum_{i=0}^{I}(a_i - d)} \quad (7.9)$$

where $K(x)$ is the defuzzified crisp number. As an example, consider the defuzzification of the linguistic term *Moderate* as seen in Figure 7.5. This linguistic term can be defuzzified to produce a crisp value as seen below:

$$K(x) = \frac{[b_0 - c] + [b_1 - c]}{\{[b_0 - c] + [b_1 - c]\} - \{[a_0 - d] + [a_1 - d]\}}$$

$$= \frac{[8-0]+[6-0]}{\{[8-0]+[6-0]\} - \{[4-10]+[6-10]\}} = 0.583$$

The values of c and d will remain the same for the defuzzification of all linguistic terms. The values a_0 and b_0 are rating values at the extreme limits of each linguistic term where the membership function is 0 and a_1 and b_1 are the rating values when the membership function is 1 (for a triangular membership function).

7.5.1 Comparative Series

An informative series with n components or decision factors can be expressed as $x_i = (x_i(1), x_i(2), \ldots x_i(k), \ldots) \in X$, where $x_i(k)$ denotes the kth factors of x_i. If all information series are comparable, the *n* information series can be described for the three linguistic variables as the following matrix (Deng (1989)):

$$x = \begin{bmatrix} x_1 \\ x_2 \\ . \\ . \\ x_n \end{bmatrix} = \begin{bmatrix} x_1(1) & x_1(2) & x_1(3) \\ x_2(1) & x_2(2) & x_2(3) \\ . & . & . \\ . & . & . \\ x_n(1) & x_n(2) & x_n(3) \end{bmatrix}$$

For the application of this matrix in an FMEA study, the value of $x_i(k)$ represents the defuzzified crisp number describing each linguistic variable considered for the identified failure modes. For example, consider three failure events, *A, B* and *C*, where the linguistic terms have been assigned for the three variables considered as seen in Table 7.5 and assume that the values in brackets represent the defuzzified value for the associated linguistic term. The information in Table 7.5 can be represented in a matrix form to reflect the comparative series.

$$\begin{bmatrix} A \\ B \\ C \end{bmatrix} = \begin{bmatrix} \text{Remote} & \text{Remote} & \text{High} \\ \text{Moderate} & \text{Very High} & \text{Low} \\ \text{Remote} & \text{Low} & \text{Remote} \end{bmatrix} = \begin{bmatrix} 0.196 & 0.196 & 0.370 \\ 0.583 & 0.952 & 0.804 \\ 0.196 & 0.370 & 0.952 \end{bmatrix}$$

7.5.2 Standard Series

The standard series for the decision factors are generated by determining the optimal level of all factors for the events in the FMEA. From a safety point of view, the lowest level of all the factors is desired. Hence, the standard series $x_0 = [x_0(1)\ x_0(2)\ x_0(3)]$ = [*Remote Remote Very High*] = [0.196 0.196 0.196].

7.5.3 Difference

The difference between the comparative and standard series, D_0, is calculated and reflected in a form of a matrix as seen below:

$$D_0 = \begin{bmatrix} \Delta_1(1) & \Delta_1(2) & \Delta_1(3) \\ . & . & . \\ \Delta_n(1) & . & \Delta_n(3) \end{bmatrix}$$

where:

$\Delta_i(k) = \| x_0(k) - x_i(k) \|$ (i = 1, 2, ..., n; k = 1, 2, 3).

$[x_0(1)\ x_0(2)\ x_0(3)]$ is the standard series.

$[x_i(1)\ x_i(2)\ x_i(3)]$ is a comparative series.

For the example used in Table 7.5, the difference of the comparative and standard series can be calculated as seen below:

$$D_0 = \begin{bmatrix} \|0.196-0.196\| & \|0.196-0.196\| & \|0.196-0.370\| \\ \|0.196-0.583\| & \|0.196-0.952\| & \|0.196-0.804\| \\ \|0.196-0.196\| & \|0.196-0.370\| & \|0.196-0.952\| \end{bmatrix} = \begin{bmatrix} 0 & 0 & 0.174 \\ 0.387 & 0.756 & 0.608 \\ 0 & 0.174 & 0.756 \end{bmatrix}$$

7.5.4 Grey Relation Coefficient

The grey relation coefficient, $\gamma\{x_0(k), x_i(k)\}$ (k = 1, 2, 3), is calculated using Equation (7.3) for each of the failure events identified in the FMEA. In the example used in Table 7.5, the grey relation coefficient can be calculated as shown here, assuming that $\zeta = 0.5$:

Using,

$$\gamma(x_0(k), x_i(k)) = \frac{\min_i \min_k |x_0(k) - x_i(k)| + \zeta \max_i \max_k |x_0(k) - x_i(k)|}{|x_0(k) - x_i(k)| + \zeta \max_i \max_k |x_0(k) - x_i(k)|}$$

for event A, the grey relation coefficient for the probability of occurrence, γ_f, is given as:

$$\gamma_f = \frac{0 + [(0.5)(0.756)]}{0 + [(0.5)(0.756)]} = 1.000$$

Similarly, the grey relations for the other two linguistic variables (Severity (γ_s) and Detectability (γ_d)), can be calculated as follows:

$$\gamma_s = \frac{0 + [(0.5)(0.756)]}{0 + [(0.5)(0.756)]} = 1.000$$

$$\gamma_d = \frac{0 + [(0.5)(0.756)]}{0.174 + [(0.5)(0.756)]} = 0.684$$

The grey relation coefficients for events B and C are calculated in the same way. The results of these calculations are summarised as seen in Table 7.6.

7.5.5 Grey Relation

The next step is to decide upon the weighting coefficients to obtain the degree of grey relation. Depending on the objective of the analysis and the reliability of the data available, the weighting coefficients (β_k), for the linguistic variables S_f, S and S_d are to be determined. The weighting coefficients will have a large influence on the final ranking of the failure events. As such, they must be carefully selected and agreed upon by all the experts involved in the study.

The degree of grey relation is calculated using Equation (7.5) for each failure event incorporating the weighted variables. For example, assume that the values of β_f, β_s and β_d are 0.4, 0.4 and 0.2, respectively, the degree of grey relation in the example shown in Table 7.5 can be calculated as seen here:

Using Equation (7.5), the grey relation for event A can be calculated as:

$$\Gamma_A = \{[(0.4)(1)] + [(0.4)(1)] + [(0.2)(0.684)]\} = 0.9368$$

The degrees of grey relation for events B and C are calculated in the same way. The results of these calculations are summarised as seen in Table 7.7.

The identified failure events in the FMEA are ranked according to the ascending order of the degree of relation. This entails that the failure mode with the smallest degree of grey relation gets the highest priority for attention. For the example in Table 7.7, failure event *B* would be at the top of the list for priority for attention, this will be followed by events *C* and *A*. The summary of the results for this example is shown in Table 7.8.

7.6 Application to Fishing Vessels

The application of the fuzzy rule base and grey theory to FMEA is demonstrated for an ocean going fishing vessel. The FMEA in this example is limited to a few systems and not all failure modes are considered. The FMEA for fishing vessels investigates four different systems, that is, the structure, propulsion, electrical and auxiliary systems. Each of the systems is considered for different failure modes that could lead to an accident with undesired consequences. The effect of each failure mode at both the system and vessel levels is studied along with the provisions that are in place/available to mitigate or reduce the risk. For each of the failure modes, the system is investigated for any alarms or condition monitoring arrangements, which are in place.

A traditional FMEA using the RPN ranking system is carried out in the first instance. This analysis is summarised in Table 7.9. In Table 7.9, S_f represents the probability of occurrence, S the severity, and S_d the detectability. The values for S_f, S and S_d are obtained by using the values detailed in Tables 7.1, 7.2 and 7.3, respectively. The same pool of experts that carried out the analysis for the fuzzy rule based method and grey theory method is used for the traditional FMEA analysis. This ensures the consistency in the opinion of each expert.

7.6.1 Fuzzy Rule Base Application

The fuzzy rule base is developed in such a way so as to enable comparison with the traditional FMEA method. Hence, in fuzzy terms, the linguistic variables are determined to be the probability of occurrence, severity and detectability. Each of these variables can be described in linguistic terms as: *Remote, Low, Moderate, High* and *Very High*. The interpretations of these linguistic terms have already been given in Table 7.4.

The membership functions of the five linguistic terms are as shown in Figure 7.6. The linguistic terms for detectability will be in reverse order but with the same membership function. The triangular membership function is chosen so as to ensure a smooth transition from one linguistic term to the other. This is in parallel with the ability of the experts to represent certain sets of data in this fashion. Apart from that, the triangular membership function facilitates easy defuzzification of each linguistic term. The membership function for each linguistic term is evaluated for its limits on an arbitrary scale from 0 to 1.

The experts for this study were carefully selected to ensure a well-balanced fuzzy rule base. The expertise and knowledge of the five experts selected along with the degree of competency, C_i, are tabulated in Table 7.10 (Pillay (2001), Pillay and Wang (2003)). The degrees of competency assigned to the experts do not reflect their personal competency in their respective field, but instead they represent their knowledge and experience in dealing with safety assessments of fishing vessels and the fishing industry. The degree of competency for each of the experts was decided and agreed upon by all the experts.

The selected experts were asked to assign linguistic terms describing the *priority for attention* for different combinations of the linguistic terms describing the three linguistic variables (probability of occurrence, severity and detectability). Upon receiving the feedback from each of the experts and applying Equation (7.7) with the values from Table 7.10, the membership function for the linguistic variable *priority for attention* is determined and graphically represented in Figure 7.7. Although the membership function for the *priority for attention* is triangular in shape, it can be noted that the membership functions for the linguistic terms are not symmetrical. This is due to the difference in opinions of individual experts. However, the graph still provides a smooth transition between states.

The support value for each of these linguistic terms is determined by taking the weighted average of the support values given by each expert. Using the information presented in Figure 7.7, the support value is assumed to be represented on the *x*-axis when the membership function for the particular linguistic term reaches 1. Hence, the support values for the linguistic terms describing the *priority for attention* can be summarised as:

Fairly Low - 0.055

Low - 0.461

Moderate - 0.911

Fairly High - 2.041

High - 7.111

The fuzzy rule base is generated based on the membership function derived from the experts (Figures 7.6 and 7.7). A total of 125 rules are generated. However, these rules are combined (where possible) and the total number of rules in the fuzzy rule base is reduced to 35. For example, consider the following three rules:

Rule 1: if probability of occurrence is *Moderate,* severity is *Low* and detectability is *Low* then *priority for attention* is 0.66 *Moderate*, 0.94 *Fairly High.*

Rule 2: if probability of occurrence is *Low,* severity is *Moderate* and detectability is *Low* then *priority for attention* is 0.66 *Moderate*, 0.94 *Fairly High*

Rule 3: if probability of occurrence is *Moderate,* severity is *High* and detectability is *High,* then *priority for attention* is 0.66 *Moderate*, 0.94 *Fairly High*

Rules 1, 2, and 3 can be combined to read:

"if probability of occurrence is *Moderate,* severity is *Low* and detectability is *Low* or any combination of the three linguistic terms assigned to these variables, then *priority for attention* is 0.66 *Moderate*, 0.94 *Fairly High*".

The degrees of belief 0.66 and 0.94, depend heavily upon the opinion of the experts involved in the study, as such, it can be assumed that these figures only represent the average values for all the opinions of the experts.

This method of rule reduction assumes that the probability of occurrence, severity and detectability have the same importance. Using this method to reduce the number of rules in the fuzzy rule base, a final set of rules is generated as shown in Table 7.11.

Using the same data from the traditional FMEA, and expressing the three variables considered linguistically with the aid of the membership function in Figure 7.6 and the fuzzy rule base in

Table 7.11, give the results of the modified FMEA. These results are then defuzzified using the WMoM method to obtain a ranking as shown in Table 7.12.

From Table 7.12, considering the first event (component - rudder bearing and failure mode - seizure), the three variables are linguistically described as:

Probability of occurrence (S_f) = *Remote*

Severity (S) = *High*

Detectability (S_d) = *High*

Using the fuzzy rule base generated in Table 7.11, *Rule 7* will apply to the first event. This rule is interpreted to read as, "if the probability of occurrence is *Remote,* severity is *High* and detectability is *High*, then *priority for attention* is *0.58 Low, 0.68 Fairly low*". The conclusion *0.58 Low, 0.68 Fairly low* can be defuzzified using the WMoM method to produce a crisp number as shown here:

$$Z = \frac{(0.58 \times 0.055) + (0.68 \times 0.461)}{(0.58 + 0.68)} = 0.274$$

where the support value for *Low* is 0.055 and *Fairly low* is 0.461 (as determined earlier).

The *priority for attention* for the first event can be represented numerically by ***0.274.*** Similarly, all other events are analysed and the corresponding priorities for attention are obtained such that the higher the value of the defuzzified results, the higher the priority in the ranking series. From the analysis and the results presented in Table 7.12, the failure event with the highest priority is failure component - hydraulic, failure mode - system loss, with a defuzzified result of 5.353. The lowest in the series is identified to be failure component - shaft & propeller, failure mode - propeller blade failure, with a defuzzified result of 0.055.

7.6.2 Grey Theory Application

There are many similarities in the data required to carry out the FMEA using grey theory, as it is to analyse it using a fuzzy rule base. Hence, the linguistic terms and membership functions generated for the fuzzy rule base application can be used in the grey theory method. The three variables are identical, these are the probability of occurrence (S_f), severity (S), and detectability (S_d). These three variables are described linguistically as *Remote, Low, Moderate, High* and *Very High.* The meaning of each of these terms is tabulated in Table 7.4 and graphically represented in Figure 7.6. These linguistic terms are defuzzified using Equation (7.9) to produce a crisp number. The result of the defuzzification is tabulated as seen in Table 7.13.

The data from the FMEA in Tables 7.9 and 7.12 is used here to demonstrate the application of the grey theory method. The same data is used for all three methods (traditional FMEA, fuzzy rule base and grey theory), to enable comparisons of the results. The comparative series is generated based on the linguistic terms assigned to each event for the three variables considered and is represented in a matrix linguistically and then converted by defuzzification to express it numerically as seen in the matrix below:

$$x = \begin{bmatrix} \text{Rem} & \text{High} & \text{High} \\ \text{Rem} & \text{High} & \text{High} \\ \text{Rem} & \text{High} & \text{High} \\ \text{High} & \text{High} & \text{Mod} \\ \text{Mod} & \text{High} & \text{Mod} \\ \text{Rem} & \text{High} & \text{V.High} \\ \text{Rem} & \text{High} & \text{V.High} \\ \text{Rem} & \text{Low} & \text{High} \\ \text{Low} & \text{Rem} & \text{High} \\ \text{Rem} & \text{Low} & \text{V.High} \\ \text{Low} & \text{Rem} & \text{High} \\ \text{High} & \text{Low} & \text{Mod} \\ \text{High} & \text{Low} & \text{Mod} \\ \text{Low} & \text{Mod} & \text{High} \\ \text{Low} & \text{Low} & \text{High} \\ \text{Rem} & \text{High} & \text{High} \\ \text{Low} & \text{High} & \text{Mod} \\ \text{Rem} & \text{Mod} & \text{Mod} \\ \text{Mod} & \text{Low} & \text{High} \\ \text{High} & \text{High} & \text{Low} \\ \text{High} & \text{Low} & \text{Mod} \end{bmatrix} = \begin{bmatrix} 0.196 & 0.804 & 0.370 \\ 0.196 & 0.804 & 0.370 \\ 0.196 & 0.804 & 0.370 \\ 0.804 & 0.804 & 0.583 \\ 0.583 & 0.804 & 0.583 \\ 0.196 & 0.804 & 0.196 \\ 0.196 & 0.804 & 0.196 \\ 0.196 & 0.370 & 0.370 \\ 0.370 & 0.196 & 0.370 \\ 0.196 & 0.370 & 0.196 \\ 0.370 & 0.196 & 0.370 \\ 0.804 & 0.370 & 0.583 \\ 0.804 & 0.370 & 0.583 \\ 0.370 & 0.583 & 0.370 \\ 0.370 & 0.370 & 0.370 \\ 0.196 & 0.804 & 0.370 \\ 0.370 & 0.804 & 0.583 \\ 0.196 & 0.583 & 0.583 \\ 0.583 & 0.370 & 0.370 \\ 0.804 & 0.804 & 0.804 \\ 0.804 & 0.370 & 0.583 \end{bmatrix}$$

The standard series is taken to be the lowest level of the linguistic terms describing all the three variables. These are *Remote* for the probability and severity and *Very High* for the detectability. When the linguistic term *Remote* is defuzzified, the crisp number obtained is 0.196, representing the average value. As such value 0 (lowest possible value) is used to represent linguistic term *Remote* in the standard series. Value 0 is also used to represent linguistic term *Very High* for the detectability in the standard series. A matrix representing the standard series is generated as shown below:

$$x_0 = [\text{Rem Rem V.High}] = [0\ 0\ 0]$$

The difference between the comparative series and standard series is then calculated and expressed as a matrix. Since all entries for the matrix representing the standard series was determined to be 0, the difference between the comparative and standard series would be equal to the comparative series (considering that $\Delta_i(k) = \| x_0(k) - x_i(k) \|$).

Using the values obtained from the difference of the standard and comparative series, the grey relation coefficient, $\gamma\{x_0(k), x_i(k)\}$, is calculated using Equation (7.3) for each variable of the events identified in the FMEA. Take $[x_1(1)\ x_1(2)\ x_1(3)]$ = [Rem High High] as an example. Equation (7.3) can be simplified and is represented by Equation (7.10) (Pillay (2001), Pillay and Wang (2003)):

$$\gamma(x_o(k), x_1(k)) = \frac{\Delta_{\min} - \zeta\Delta_{\max}}{\Delta_1(k) - \zeta\Delta_{\max}} \tag{7.10}$$

where Δ_{min} = 0.196, Δ_{max} = 0.804 and ζ = 0.5. ζ is an identifier, $\zeta \in (0,1)$, only affecting the relative value of risk without changing the priority. Generally, ζ can be set to 0.5 (Deng (1989)).

One of the objectives of applying an FMEA study to fishing vessels is to identify areas where safety features are lacking in the system. These include interlocks, alarms, auto cut-off/shutdown, condition monitoring and redundancy features. Due to the organisational and operating nature of fishing vessels, incorporating/improving safety features may be the easiest and most effective way to improve the operational safety of the vessel. As such the weighting coefficients (β_k) for the decision factors, S_f, S and S_d should be such that $\beta_{Sd} > \beta_S > \beta_{Sf}$. This would entail giving more preference to the detectability factor in the analysis. Hence, The weighting coefficients (β_k), are set to be 0.2, 0.3 and 0.5 for the probability of occurrence, severity and detectability, respectively.

Consider the first event where S_f, S and S_d are assigned *Remote, High* and *High* for the probability of occurrence, severity and detectability, respectively. The grey relation coefficients γ_f, γ_s and γ_d are calculated as shown below:

$$\gamma_f = \frac{0.196 + [(0.5)(0.804)]}{0.196 + [(0.5)(0.804)]} = 1$$

$$\gamma_s = \frac{0.196 + [(0.5)(0.804)]}{0.804 + [(0.5)(0.804)]} = 0.496$$

$$\gamma_d = \frac{0.196 + [(0.5)(0.804)]}{0.370 + [(0.5)(0.804)]} = 0.775$$

Substituting these values and the weighting coefficients into Equation (7.5) will give the degree of relation for the first event as follows:

$$\Gamma(x_0, x_i) = \{[(0.2)(1)] + [(0.3)(0.496)] + [(0.5)(0.775)]\} = 0.736$$

Similarly, the degree of relation is calculated for all the events identified in the FMEA to produce a ranking that determines the *priority for attention*. The complete analysis of the test case using grey theory is tabulated as seen in Table 7.14.

7.7 Analysis of Results

The results obtained for the FMEA using the fuzzy rule based method and grey theory method are collated with the results obtained from the traditional FMEA using the RPN method and are given in Table 7.15. From Table 7.15, consider events 1 and 11, where the RPN value is 24. From Table 7.9, it can be seen that the values of S_f, S and S_d are 1, 8 and 3 for event 1 and 4, 2 and 3 for event 11, hence an RPN value of 24 is obtained. Although the RPN for both events are the same, the risk levels are different. This difference is obvious when the fuzzy rule based method and grey theory method are applied. The results show that event 1 has a higher priority compared to event 11. However, the traditional RPN method puts these two events as having the same priority.

The ranking produced using the fuzzy rule based method and grey theory method do not differentiate events that have the same linguistic terms describing the factors considered. For example, for events 1, 2 and 3, where S_f, S and S_d are assigned *Remote, High* and *High,* respectively, the defuzzified ranking is 0.274 and the degree of grey relation is 0.736 for all three events. This entails that these three events should be given the same *priority for attention.* The RPN method however, produces a result of 24, 32 and 64 for events 1, 2 and 3, respectively. This denotes that event 3 has the highest priority followed by events 1 and 2.

The effects of the weighting coefficient introduced in the grey theory method can be clearly seen in the results obtained for events 17 and 21, where S_f, S and S_d are assigned *Low, High* and *Moderate* and *High, Low* and *Moderate,* respectively. Using the fuzzy rule base to analyse these two events produces a defuzzified ranking of 1.575, however, when using the grey theory method (incorporating the weighted coefficient), the grey relation ranking is 0.607 and 0.635 for events 17 and 21, respectively. This entails that event 17 should be given a higher priority compared to event 21.

7.8 Conclusion

When conducting an FMEA for safety assessment purposes, precision should not be forced where data is unreliable and scarce. Hence, to ask an analyst or an expert to assign scores ranging from 1 to 10 (as done in the RPN method) for the different factors considered could produce a false and unrealistic impression. The use of linguistic terms allows for the experts to assign a more meaningful value for the factors considered.

The advantages of the described fuzzy rule based method and grey theory method for application to FMEA of ships can be summarised as follows (Pillay (2001), Pillay and Wang (2003)):

1. It can be used for systems where safety data is unavailable or unreliable, as it does not force precision.
2. It provides an organised method to combine expert knowledge and experience for use in an FMEA study.
3. The use of linguistic terms in the analysis enables the experts to express their judgements more realistically and hence improving the applicability of the FMEA.
4. The flexibility of assigning weight to each factor in the FMEA provides a means of specifically identifying weak areas in the system/component studied.

The described method using fuzzy rule base (without the weighting factors of the linguistic variables) could be suitable for use in Step 1 of the FSA process (at the hazard-screening phase) as discussed in Chapter 5. During the hazard-screening phase, only a relative ranking order is needed. This will distinguish the hazards with a high-risk level from those with a low-risk level.

The described method using grey theory (with the weighting factors of the linguistic variables) would be suitable for use in Step 2 of the FSA (risk estimation phase) as discussed in Chapter 5. At this stage of the FSA, a more detailed analysis of each hazard is required to produce a ranking order that would determine the allocation of the limited resources. As the described method provides the analyst with the flexibility to decide which factor is more important to the analysis, the outcome of the analysis will provide valuable information for the decision making process.

The traditional FMEA, the fuzzy rule based method and the grey theory approach may complement each other to produce a risk ranking with confidence.

7.9 References (Chapter 7)

1. Bell D., Cox L., Jackson S. and Schaefer P., (1992) "Using Causal Reasoning for Automated Failure & Effects Analysis (FMEA)", Proceedings Annual Reliability and Maintainability Symposium, pp. 343-353.

2. Ben-Daya M. and Raouf A., (1993) "A Revised Failure Mode and Effects Analysis Model", International Journal of Quality and Reliability Management, Vol. 3, No. 1, pp. 43-47.
3. Brahm C. and Kleiner B.H., (1996) "Advantages and Disadvantages Of Group Decision-Making Approaches", Team Performance Management, Vol. 2, No. 1, pp. 30-35.
4. Chang C.L, Wei C.C. and Lee Y.H., (1999), "Failure Mode and Effects Analysis Using Fuzzy Method and Grey Theory", Kybernetes, Vol. 28, No. 9, pp. 1072-1080.
5. Chen C.B. and Klien C.M., (1997) "A Simple Approach to Ranking a Group of Aggregated Fuzzy Utilities", IEEE Transactions on System, Man, Cybernetics - Part B: Cybernetics, Vol. 27, No. 1, pp. 26-35.
6. Deng J., (1982) "Control Problems of Grey Systems", Systems and Control Letters, Vol. 1, No. 5, pp. 288-294.
7. Deng J., (1986) "The Application of Grey System to Quantitative Investment Strategy", Journal of Investment and Reform, Vol. 3, pp. 21-25.
8. Deng J., (1989) "Introduction to Grey System Theory", The Journal of Grey System, Vol. 1, No. 1, pp. 1-24.
9. Gilchrist W., (1993) "Modelling Failure Modes and Effects Analysis", International Journal of Quality and Reliability Management, Vol. 10, No. 5, pp. 16-23.
10. Hong G., (1986) "Identification Coefficient of Relational Grade of Grey Systems", Fuzzy Mathematics, Vol. 2, pp. 55-58.
11. Jianer H., (1987) "Application of Grey Statistics to Investment Decision Making of Forestry", Journal of Economic Problems of Forestry, Vol. 2, pp. 44-51
12. Junjiang H., (1986) "On the Applications of Grey System Theory and Methods to History Studies", Journal of Social Sciences, Vol. 10, pp. 64-68.
13. Kara-Zaitri C., Keller A.Z. and Fleming P.V., (1992) "A Smart Failure Mode and Effect Analysis Package", Proceedings Annual Reliability and Maintainability Symposium, pp. 414-421.
14. Kara-Zaitri C., Keller A.Z., Barody I. and Fleming P.V., (1991) "An Improved FMEA Methodology", Proceedings Annual Reliability and Maintainability Symposium, pp. 248-252.
15. Klir G.J. and Yuan B., (1995) "Fuzzy Sets and Fuzzy Logic: Theory and Application", Prentice Hall PTR, New Jersey, USA.
16. Kuusela H., Spence M.T. and Kanto A. J., (1998) "Expertise Effects on Pre-Choice Decision Processes and Final Outcomes: a Protocol Analysis", European Journal of Marketing, Vol. 32, No. 5/6, pp. 559-576.
17. Li H., (1991) "Grey System Theory for Population Geographers", The Geographical Approach to Fertility, edited by Jurgen Bahr and Paul Gans, Universitat Kiel, Geographisches Institut: Kiel, Germany, pp.29-39.
18. Pelaez C.E. and Bowles J.B., (1996) "Applying Fuzzy Cognitive-Maps Knowledge-Representation to Failure Mode Effects Analysis", Journal of Information Sciences, Vol. 88, No.1, pp. 177-199.
19. Pillay A., (2001) "Formal Safety Assessment of Fishing Vessels", PhD Thesis, School of Engineering, Liverpool John Moores University, UK.
20. Pillay A. and Wang J., (2003) "A Risk Ranking Approach Incorporating Fuzzy Set Theory and Grey Theory", Engineering Reliability and System Safety, Vol. 79, No. 1, pp. 61-67.
21. Price C.J., Hunt J.E., Lee M.H. and Ormsby R.T., (1992) "A Mel-based Approach to the Automation of Failure Mode Effects Analysis for Design", Proc. IMechE, Part D, Journal of Automobile Engineering, Vol. 206, pp. 285-291.

22. Price C.J., Pugh D.R., Wilson M.S. and Snooke N., (1995) "The Flame System: Automating Electrical Failure Mode & Effects Analysis (FMEA)", Proceedings Annual Reliability and Maintainability Symposium, pp. 90-95.
23. Runkler T.A and Glesner M., (1993) "A Set of Axioms for Defuzzification Strategies toward a Theory of Rational Defuzzification Operators", Proceedings of the 2nd IEEE International Conference on Fuzzy Set System, New York, pp. 1161-1166.
24. Russomano D.J., Bonnell R.D. and Bowles J.B., (1992) "A Blackboard Model of an Expert System for Failure Mode and Effects Analysis", Proceedings Annual Reliability and Maintainability Symposium, pp 483-489.
25. Shih K.C., Wu K.W. and Huang U.P., (1996) "Information of Grey Relation", Chuan-Hua Co. Ltd., Taiwan.
26. Stamatis D. H., (1995) "Failure Mode and Effects Analysis - FMEA from Theory to Execution", ASQC Quality Press, New York, USA.
27. Van Gundy A., (1998) "Techniques of Structured Problem Solving", Van Nostrand, New York.
28. Wang L.X., (1997) "A Course in Fuzzy Systems and Control", Prentice-Hall, ISBN 0 13 540882 2.
29. Wu H.H., Deng J.L. and Wen K.L., (1984) "Introduction to Grey Analysis", Kao-Li Co. Ltd., Taiwan.
30. Wu J.R. and Ouhyoung M., (1994) "Reducing the Latency in Head-mounted Display by a Novel Prediction Method Using Grey System Theory", Computer Graphics Forum, Vol. 13, No. 3, pp. 503-512.
31. Zheng Y., (1988) "Grey Prediction Method of Ground Settlements", Proceedings of the 3rd Chinese National Symposium on Numerical Analyses and Analytical Methods in Geomechanics, Guangdong, China, November, Vol. 2, pp. 469-484.
32. Zheng Y. and Lewis R.W., (1993) "On the Optimisation Concept of Grey Systems", Journal of Applied Mathematical Modelling, Vol. 17, No. 7, pp. 388-392.

Table 7.1 Traditional FMEA Scale for Probability of Occurrence (S_f)

Probability of occurrence	*Rating*	*Possible failure rate (operating days)*
Remote	1	< 1:20000
Low	2	1:20000
	3	1:10000
Moderate	4	1:2000
	5	1:1000
	6	1:200
High	7	1:100
	8	1:20
Very High	9	1:10
	10	1:2

Table 7.2 Traditional FMEA Scale for Severity (S)

Severity	*Rating*
Negligible	1
Low	2
	3
Moderate	4
	5
	6
High	7
	8
Very High	9
	10

Table 7.3 Traditional FMEA Scale for Detectability (S_d)

Detectability	*Rating*	*Probability (%) of detection*
Very High	1	86-100
High	2	76-85
	3	66-75
Moderate	4	56-65
	5	46-55
	6	36-45
Low	7	26-35
	8	16-25
Remote	9	6-15
	10	0-5

Table 7.4 Interpretations of the Linguistic Terms

Linguistic term	*Probability of occurrence*	*Severity*	*Detectability*
Remote	It would be very unlikely for these failures to be observed even once	A failure that has no effect on the system performance, the operator will probably not notice	Defect remains undetected until the system performance degrades to the extent that the task will not be completed
Low	Likely to occur once, but unlikely to occur more frequently	A failure that would cause slight annoyance to the operator, but that would cause no deterioration to the system	Defect remains undetected until system performance is severely reduced
Moderate	Likely to occur more than once	A failure that would cause a high degree of operator dissatisfaction or that causes noticeable but slight deterioration in system performance	Defect remains undetected until system performance is affected
High	Near certain to occur at least once	A failure that causes significant deterioration in system performance and/or leads to minor injuries	Defect remains undetected until inspection or test is carried out
Very High	Near certain to occur several times	A failure that would seriously affect the ability to complete the task or cause damage, serious injury or death	Failure remains undetected, such a defect would almost certainly be detected during inspection or test

Table 7.5 Example of Comparative Series

Failure events	*Probability of occurrence*	*Severity*	*Detectability*
A	*Remote* (0.196)	*Remote* (0.196)	*High* (0.370)
B	*Moderate* (0.583)	*Very High* (0.952)	*Low* (0.804)
C	*Remote* (0.196)	*Low* (0.370)	*Remote* (0.952)

Table 7.6 Example of Grey Relation Coefficient

Failure event	γ_f	γ_s	γ_d
A	1	1	0.684
B	0.494	0.333	0.383
C	1	0.684	0.333

Table 7.7 Example of Degree of Grey Relation

Failure Events	Degree of grey relation
A	0.9368
B	0.4074
C	0.7402

Table 7.8 Example of Ranking for Failure Events Using the Degree of Grey Relation

Failure events	*Probability of occurrence*	*Severity*	*Detectability*	*Degree of grey relation*	*Ranking (priority for attention)*
A	*Remote*	*Remote*	*High*	0.9368	**3**
B	*Moderate*	*Very High*	*Low*	0.4074	**1**
C	*Remote*	*Low*	*Remote*	0.7402	**2**

Table 7.9 Traditional FMEA for a Fishing Vessel

Descrip.	*Comp.*	*Failure Mode*	*Failure effect (System)*	*Failure effect (Vessel)*	*Alarm*	*Provision*	S_f	*S*	S_d	*RPN*
Structure	Rudder bearing	Seizure	Rudder jam	No steering ctrl.	No	Stop vessel	1	8	3	24
Structure	Rudder Bearing	Breakage	Rudder loose	Reduced steering ctrl.	No	Stop vessel	1	8	3	24
Structure	Rudder structure	Structural failure	Function loss	Reduced steering	No	Use beams	2	8	4	64
Propulsion	Main Engine	Loss of output	Loss of thrust	Loss of speed	Yes	None	8	8	5	320
Propulsion	Main Engine	Auto shutdown	M/E stops	Loss of speed	Yes	Anchor	6	8	6	288
Propulsion	Shaft & propeller	Shaft breakage	Loss of thrust	Loss of speed	No	Anchor	2	8	1	16
Propulsion	Shaft & propeller	Shaft seizure	Loss of thrust	Loss of speed	Yes	Anchor	2	9	2	36
Propulsion	Shaft & propeller	Gearbox seizure	Loss of thrust	Loss of speed	Yes	Anchor	1	4	3	12
Propulsion	Shaft & propeller	Hydraulic failure	Cannot reduce thrust	Cannot reduce speed	No	Anchor	3	2	3	18
Propulsion	Shaft & propeller	Prop. blade failure	Loss of thrust	Loss of speed	No	Slow steaming	1	2	4	8
Air services	Air receiver	No start air press.	Cannot start M/E	No propulsion	Yes	Recharge receiver	4	2	3	24
Electrical Sys.	Power genera-tion	Generator fail	No elec.power	Some system failures	Yes	Use st-by generators	9	3	7	189
Electrical Sys.	Main switch board	Complete loss	Loss of main supply	No battery charging	Yes	Use emergency 24v	8	3	6	144
Electrical Sys.	Emer. S/B	Complete loss	Loss of emer.supp.	No emergency supp.	No	Use normal supply	3	7	4	84
Electrical Sys.	Main batteries	Loss of output	Loss of main 24v	Loss of main low volt.	Yes	Use emergency 24v	3	3	4	36
Electrical Sys.	Emer. batteries	Loss of output	Loss of emer.supp.	No emer.supp.	No	Use normal supply	1	8	3	24
Auxiliary Sys.	Fuel System	Contamina-tion	M/E and Gen stop	Vessels stops	Yes	Anchor	4	8	5	160
Auxiliary Sys.	Fuel system	No fuel to M/E	M/E stops	Vessel stops	No	Anchor	2	7	7	98
Auxiliary Sys.	Water system	No cooling water	Engine overheat	M/E auto cut-out	Yes	Use st-by pump	7	2	4	56
Auxiliary Sys.	Hydraulic	System loss	No hydraulics	No steering	Yes	Stop vessel	9	8	9	648
Auxiliary Sys.	Lube oil system	Loss of pressure	Low pressure cut-off	M/E stops	Yes	Use st-by pump	9	3	6	162

Table 7.10 Selected Experts and Assigned Degree of Competency

Expert	*Expertise and knowledge*	C_I
Expert 1	Safety analyst (marine & offshore).	0.3
Expert 2	Marine surveyor (fishing vessels).	0.3
Expert 3	Superintendent engineer (fishing vessels).	0.2
Expert 4	Marine operations engineer (merchant vessels).	0.1
Expert 5	Statistician.	0.1

Table 7.11 Reduced Rules for the Fuzzy Rule Base

Rule No	*Probability of occurrence*	*Severity*	*Detectability*	*Priority for attention*
1	Rem	Rem	V.High to High	1.0 Low
2	Rem	Rem	Mod	1.0 Low, 0.06 F.Low
3	Rem	Rem	Low	0.86 Low, 0.14 F.Low
4	Rem	Rem	Rem	0.78 Low, 0.2 F.Low
5	Low	Rem	High	1.0 Low, 0.16 F.Low
6	Low	Rem	Mod	0.86 Low, 0.48 F.Low
7	Low	Rem	Low	0.58 Low, 0.68 F.Low
8	Mod	Rem	Mod	0.5 Low, 0.92 F.Low
9	Mod	Rem	Low	0.8 F.Low, 0.4 Mod
10	Mod	Rem	Rem	0.92 F.Low, 0.8 Mod
11	High	Rem	Mod	0.74 F.Low, 0.4 Mod
12	High	Rem	Low	0.48 F.Low, 0.92 Mod
13	High	Rem	Rem	0.88 Mod, 0.1 F.High
14	V.High	Rem	High	0.48 Low, 0.88 F.Low
15	V.High	Rem	Rem	0.82 Mod, 0.36 F.High
16	Low	Low	High	0.86 Low, 0.78 F.Low
17	Low	Low	Mod	0.4 F.Low, 0.58 Mod
18	Low	Low	Low	0.8 F. Low, 0.92 Mod
19	Low	Low	Rem	0.92 F.Low, 0.7 Mod
20	Mod	Low	Mod	0.94 F.Low, 0.46 Mod
21	Mod	Low	Low	0.66 Mod,0.94 F.High
22	Mod	Low	Rem	0.92 Mod, 0.92 F.High
23	High	Low	Low	0.58 Mod,0.88 F.High
24	High	Low	Rem	0.72 F.High, 0.22 High
25	V.High	Low	Rem	0.98 F.High, 0.38 High
26	Mod	Mod	Mod	0.92 Mod,0.84 F.High
27	Mod	Mod	Low	0.4 Mod,0.66 F.High
28	Mod	Mod	Rem	0.94 F.High,0.56 High
29	High	Mod	Low	0.88 F.High,0.62 High
30	High	Mod	Rem	0.74 F.High,0.9 High
31	V.High	Mod	Rem	0.58 F.High,0.6 High
32	High	High	Low	0.52 F.High,0.98 High
33	High	High	Rem	0.3 F.High,0.42 High
34	V.High	High	Rem	1.0 High
35	V.High	V.High	Rem	1.0 High

Table 7.12 Modified FMEA Using Fuzzy Rule Base

Descript.	*Component*	*Failure Mode*	S_f	*S*	S_d	*Priority for attention*	*Defuzzified ranking*
Structure	Rudder bearing	Seizure	Rem	High	High	0.58 Low,0.68 F.Low	0.274
Structure	Rudder Bearing	Breakage	Rem	High	High	0.58 Low,0.68 F.Low	0.274
Structure	Rudder structure	Structural failure	Rem	High	High	0.58 Low,0.68 F.Low	0.274
Propulsion	Main Engine	Loss of output	High	High	Mod	0.88 F. High, 0.62 High	4.136
Propulsion	Main Engine	Auto shutdown	Mod	High	Mod	0.4 Mod, 0.66 F.High	1.614
Propulsion	Shaft & propeller	Shaft breakage	Rem	High	V.High	0.86 Low, 0.14 F.Low	0.112
Propulsion	Shaft & propeller	Shaft seizure	Rem	High	V.High	0.86 Low, 0.14 F.Low	0.112
Propulsion	Shaft & propeller	Gearbox seizure	Rem	Low	High	1.0 Low,0.16 F.Low	0.111
Propulsion	Shaft & propeller	Hydraulic failure	Low	Rem	High	1.0 Low,0.16 F.Low	0.111
Propulsion	Shaft & propeller	Prop. Blade failure	Rem	Low	V.high	1.0 Low	0.055
Air services	Air receiver	No start air press.	Low	Rem	High	Low,0.16 F.Low	0.111
Electrical sys.	Power generation	Generator fail	High	Low	Mod	0.66 Mod,0.94 F.High	1.575
Electrical sys.	Main switch board	Complete loss	High	Low	Mod	0.66 Mod,0.94 F.High	1.575
Electrical sys.	Emergency S/B	Complete loss	Low	Mod	High	0.4 F.Low, 0.58 Mod	0.727
Electrical sys.	Main batteries	Loss of output	Low	Low	High	0.86 Low, 0.78 F.Low	0.248
Electrical sys.	Emergency batteries	Loss of output	Rem	High	High	0.58 Low,0.68 F.Low	0.274
Auxiliary sys.	Fuel sys.	Contamination	Low	High	Mod	0.66 Mod,0.94 F.High	1.575
Auxiliary sys.	Fuel sys.	No fuel to M/E	Rem	Mod	Mod	0.5 Low,0.92 F.Low	0.318
Auxiliary sys.	Water sys.	No cooling water	Mod	Low	High	0.4 F.Low, 0.58 Mod	0.727
Auxiliary sys.	Hydraulic	Sys. loss	High	High	Low	0.52 F.High,0.98 High	5.353
Auxiliary sys.	Lube oil sys.	Loss of pressure	High	Low	Mod	0.66 Mod, 0.94 F.High	1.575

Table 7.13 Defuzzified Crisp Number for Linguistic Terms Describing the Variables

Linguistic Term	Defuzzified crisp number
Remote	0.196
Low	0.370
Moderate	0.583
High	0.804
Very High	0.952

Table 7.14 Modified FMEA Using Grey Theory

Description	Component	Failure Mode	S_f	γ_f	S	γ_s	S_d	γ_d	Grey Relation
Structure	Rudder bearing	Seizure	Rem	1.000	High	0.496	High	0.775	0.736
Structure	Rudder bearing	Breakage	Rem	1.000	High	0.496	High	0.775	0.736
Structure	Rudder structure	Structural failure	Rem	1.000	High	0.496	High	0.775	0.736
Propulsion	Main Engine	Loss of output	High	0.496	High	0.496	Mod	0.607	0.552
Propulsion	Main Engine	Auto shutdown	Mod	0.607	High	0.496	Mod	0.607	0.574
Propulsion	Shaft & propeller	Shaft breakage	Rem	1.000	High	0.496	V.High	1.000	0.849
Propulsion	Shaft & propeller	Shaft seizure	Rem	1.000	High	0.496	V.High	1.000	0.849
Propulsion	Shaft & propeller	Gearbox seizure	Rem	1.000	Low	0.775	High	0.775	0.820
Propulsion	Shaft & propeller	Hydraulic failure	Low	0.775	Rem	1.000	High	0.775	0.843
Propulsion	Shaft & propeller	Prop. blade failure	Rem	1.000	Low	0.775	V.high	1.000	0.933
Air services	Air receiver	No start air press.	Low	0.775	Rem	1.000	High	0.775	0.843
Electrical Systems	Power generation	Generator fail	High	0.496	Low	0.775	Mod	0.607	0.635
Electrical Systems	Main switch board	Complete loss	High	0.496	Low	0.775	Mod	0.607	0.635
Electrical Systems	Emergency S/B	Complete loss	Low	0.775	Mod	0.607	High	0.775	0.725
Electrical Systems	Main batteries	Loss of output	Low	0.775	Low	0.775	High	0.775	0.775
Electrical Systems	Emergency batteries	Loss of output	Rem	1.000	High	0.496	High	0.775	0.736
Auxiliary Systems	Fuel system	Contamination	Low	0.775	High	0.496	Mod	0.607	0.607
Auxiliary Systems	Fuel System	No fuel to M/E	Rem	1.000	Mod	0.607	Mod	0.607	0.686
Auxiliary Systems	Water system	No cooling water	Mod	0.607	Low	0.775	High	0.775	0.741
Auxiliary Systems	Hydraulic	Sys. Loss	High	0.496	High	0.496	Low	0.496	0.496
Auxiliary Systems	Lube oil system	Loss of pressure	High	0.496	Low	0.775	Mod	0.607	0.635

Table 7.15 Ranking Comparison

ID	Component	Failure Mode	RPN	Fuzzy rule base	Grey Theory	Ranking (RPN)	Ranking (Rule base)	Ranking (Grey theory)
1	Rudder bearing	Seizure	24	0.274	0.736	15	11	10
2	Rudder Bearing	Breakage	32	0.274	0.736	14	11	10
3	Rudder structure	Structural failure	64	0.274	0.736	10	11	10
4	Main Engine	Loss of output	320	4.136	0.552	2	2	2
5	Main Engine	Auto shutdown	288	1.614	0.574	3	3	3
6	Shaft & propeller	Shaft breakage	16	0.112	0.849	19	16	19
7	Shaft & propeller	Shaft seizure	36	0.112	0.849	12	16	19
8	Shaft & propeller	Gearbox seizure	12	0.111	0.820	20	18	16
9	Shaft & propeller	Hydraulic failure	18	0.111	0.843	18	18	17
10	Shaft & propeller	Prop. blade fail	8	0.055	0.933	21	21	21
11	Air receiver	No start air press.	24	0.111	0.843	15	18	17
12	Power generation	Generator fail	189	1.575	0.635	4	4	5
13	Main switch board	Complete loss	144	1.575	0.635	7	4	5
14	Emer. S/B	Complete loss	84	0.727	0.725	9	8	9
15	Main batteries	Loss of output	36	0.248	0.775	12	15	15
16	Emer. batteries	Loss of output	24	0.274	0.736	15	11	10
17	Fuel System	Contamination	160	1.575	0.607	6	4	4
18	Fuel system	No fuel to M/E	98	0.318	0.686	8	10	8
19	Water system	No cooling water	56	0.727	0.741	11	8	14
20	Hydraulic	System loss	648	5.353	0.496	1	1	1
21	Lube oil system	Loss of pressure	162	1.575	0.635	5	4	5

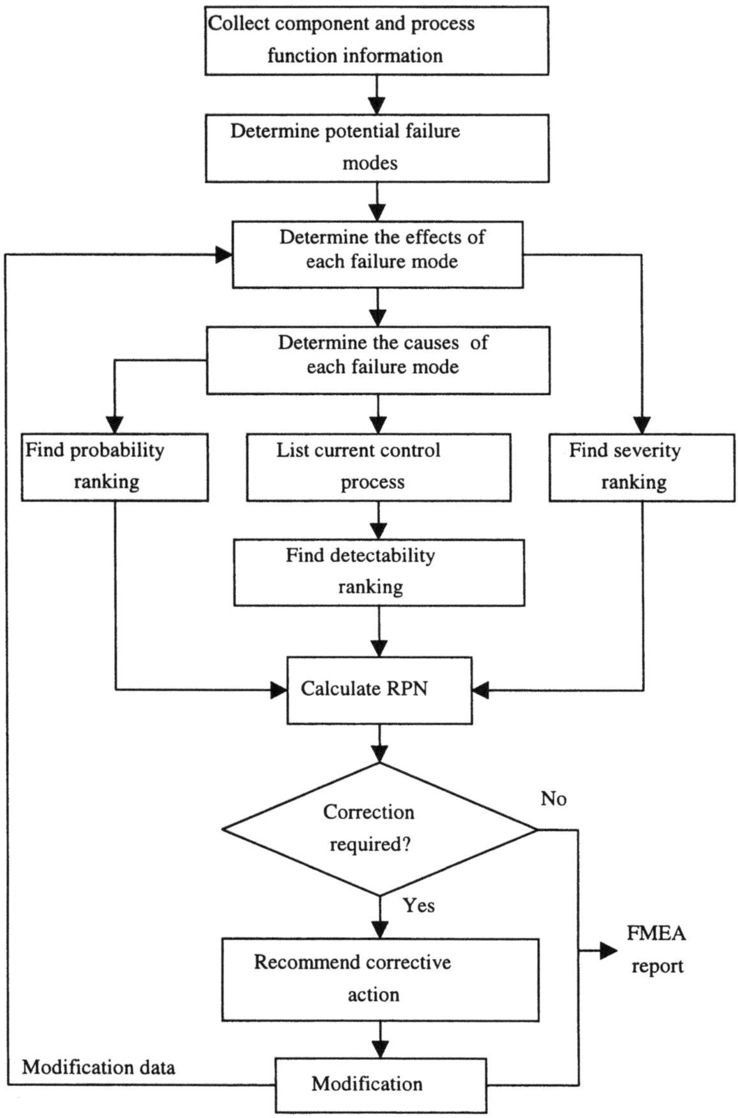

Figure 7.1 RPN calculation and FMEA process

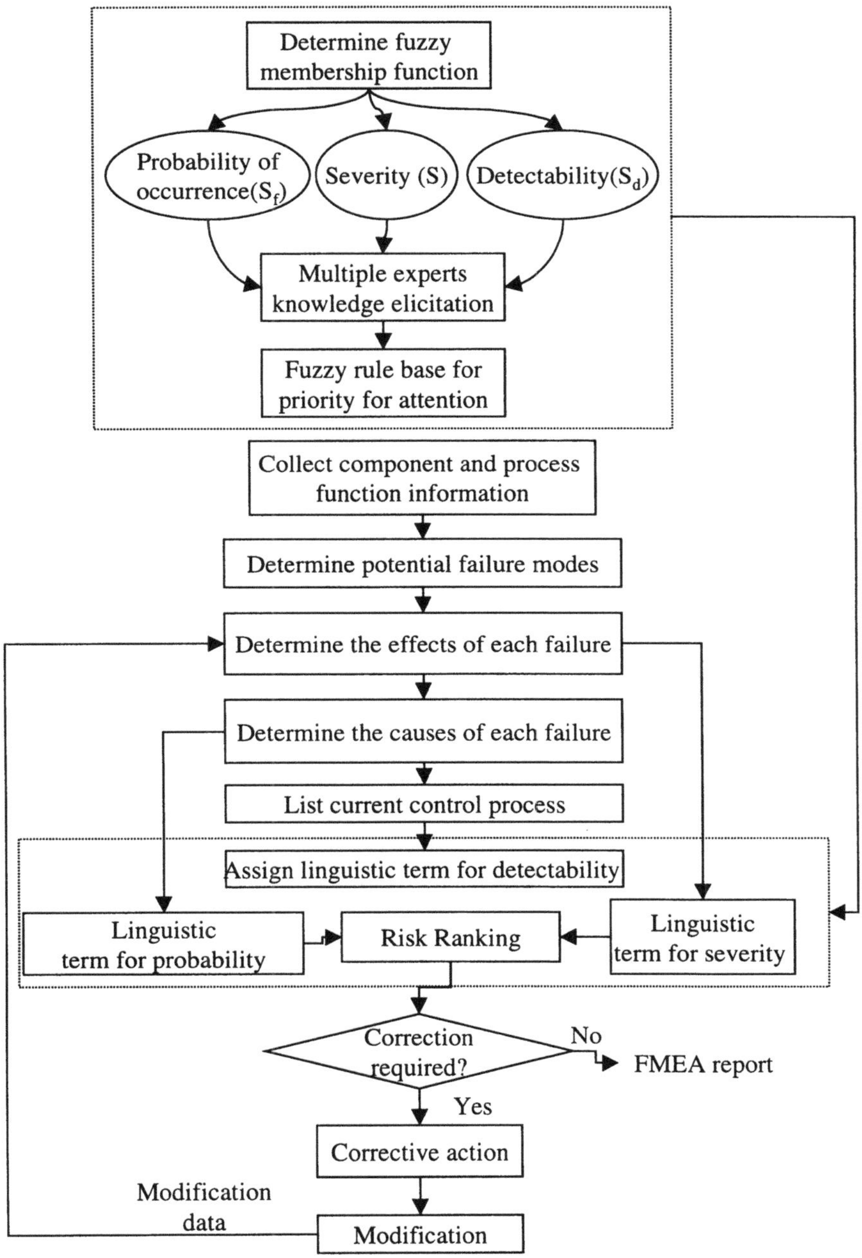

Figure 7.2 Flowchart of the described fuzzy rule base method

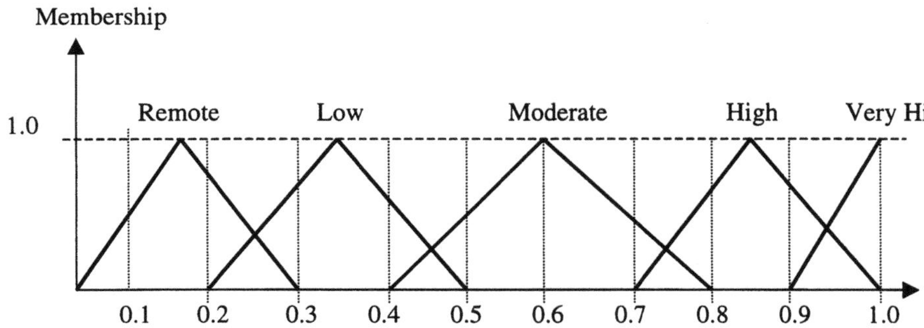

Figure 7.3 Graphical representation of the membership function for the linguistic terms

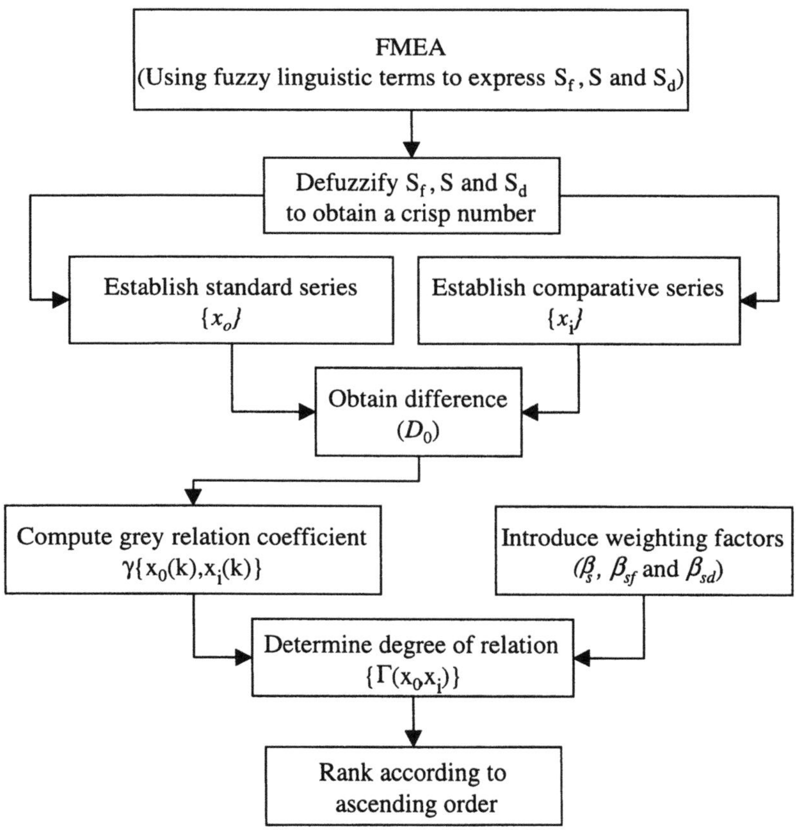

Figure 7.4 Flowchart of the grey theory method

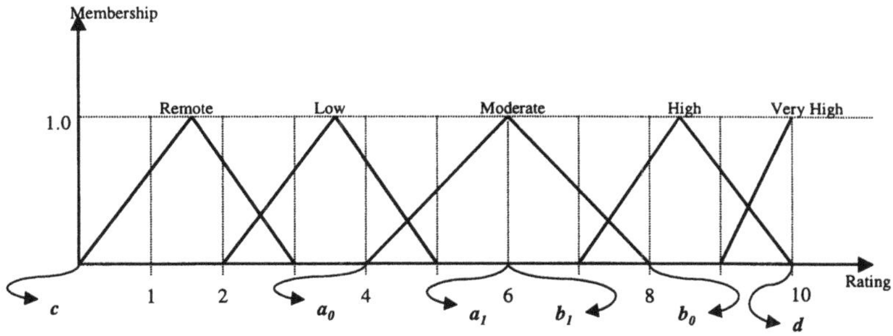

Figure 7.5 Defuzzification of the linguistic term *Moderate*

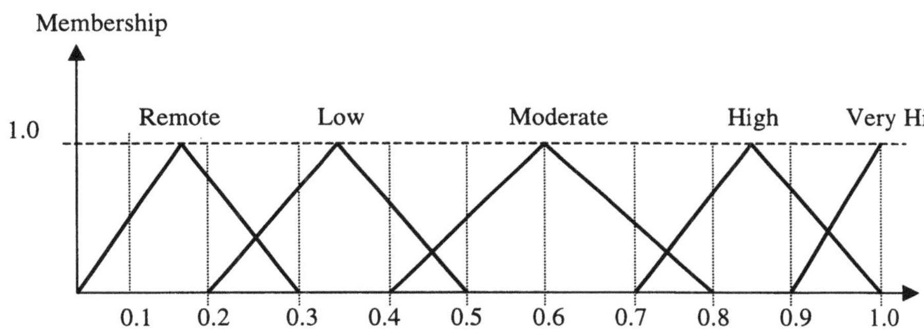

Figure 7.6 Membership function for the linguistic terms (generated by the experts)

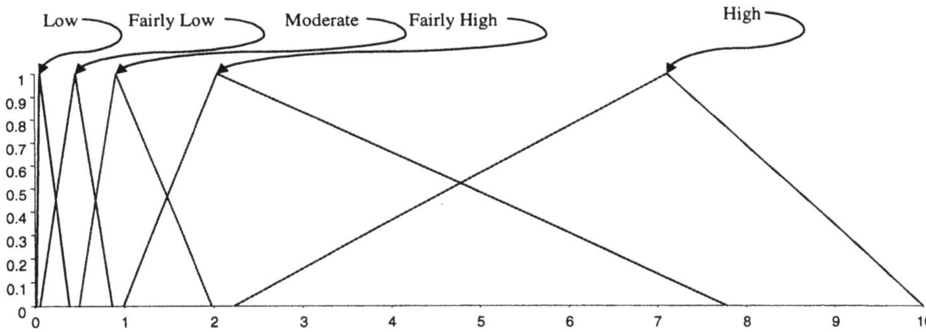

Figure 7.7 Membership function for the *priority for attention*

Chapter 8

Maintenance Modelling

Summary

The data analysis in Chapter 2 showed that more than 50% of accidents on fishing vessels involved machinery failure. Upon the investigation of several fishing vessels in the UK, it was found that maintenance activities on board these vessels were almost non-existent. This Chapter reviews different maintenance concepts in the first instance. This is followed by a summary of the advantages and disadvantages of these concepts. The current maintenance practice of fishing vessels is reviewed and a maintenance model is presented to reduce machinery failure on these vessels by means of implementing an inspection regime based on the delay-time concept. The described approach provides an alternative solution to the current maintenance practice to reduce cost incurred and downtime suffered by fishing vessels and/or control the level of safety criticality.

Keywords: Cost, delay time analysis, inspection, maintenance.

8.1 Introduction

Maintenance is defined as the combination of all technical and administrative actions, including supervision actions, intended to retain an entity in, or restore it to a state in which it can perform a required function. It involves planned and unplanned activities being carried out to ensure an acceptable state of operation. Selection of a maintenance strategy will depend on one or a combination of the following criteria: maximisation of reliability, minimisation of downtime and minimisation of total maintenance cost (Savic et al. (1995)).

The impact of the maintenance policy on total maintenance cost is hard to predict (Rischel and Christy (1996)). Any breakdown in machine operation results in disruption of production and leads to additional costs due to downtime, loss of production, decrease in productivity and quality and inefficient use of personnel, equipment and facilities (Ashayeri et al. (1996)).

Maintenance costs form a significant part of the overall operating costs in maritime operations. Maintenance also affects reliability and can thus have environmental and safety consequences. The International Management Code for the Safe Operation of Ships and for Pollution Prevention (ISM Code) addresses the management aspects. The importance of maintenance is demonstrated by the fact that it is the only shipboard activity to have one whole element assigned to it (i.e. ISM Code element 10) (IMO (1997)).

ISM Code element 10 focusing on maintenance of ship and equipment inter alia states that "The Company should establish procedures in its SMS (Safety Management System) to identify equipment and technical systems the sudden operational failure of which may result in hazardous situations. The SMS should provide for specific measures aimed at promoting the reliability of such equipment or systems". This is consistent with what Reliability Centred Maintenance (RCM) delivers. RCM focuses the maintenance resources only on those items that affect the system reliability, thereby making the maintenance programme cost effective in the long run (Mokashi et al. (2002)).

The recent Preventive Maintenance (PM) developments have seen to have a more generic framework with an aim of maximising the profitability of a working system, which is also demonstrated by another maintenance management approach "Total Productive Maintenance" (TPM) developed by Nakajima of the Japan Institute of Plant Maintenance (JIPM) (Nakajima (1997)). In such a philosophy of maximising profitability, different elements such as down time, cost, safety criticality may need to be studied together.

In PM, maintenance activities are performed before equipment failure. PM involves the repair, replacement and maintenance of equipment in order to avoid unexpected failure during use.

In the maritime industry, there are some specific problems with regard to maintenance, that need to be considered when developing a maintenance model. These problems include:

1. The high degree of isolation from repair and spares facilities.
2. The high cost of transport unit (i.e. the ship).
3. The high cost of a maritime system out of service.
4. Varying costs, availability and quality of labour and spares throughout the world.
5. Maritime personnel are operators as well as maintainers.
6. The frequency with which personnel join and leave ships, creating a need for continuity of ships maintenance plans.
7. Severe safety and insurance conditions, necessitating rigorous survey requirements.

Several of these problems are undeniably important to the fishing industry as will be discussed in Section 8.3.

8.2 Modern Maintenance Concepts

RCM sometimes referred to as Preventive Maintenance Optimisation (PMO) has become popular in recent years within several industries. The concept has been discussed and elaborated by several authors (Worledge (1993), Rausand (1998), Sherwin (1999)). RCM is a procedure for determining maintenance strategies based on reliability techniques and encompasses well-known analysis methods such as Failure Mode, Effects and Criticality Analysis (FMECA). RCM procedure takes into account the prime objectives of a maintenance programme:

1. Minimise costs.
2. Meet safety and environmental goals.
3. Meet operational goals.

The RCM process begins with a Failure Mode and Effects Analysis (FMEA), which identifies the critical plant failure modes in a systematic and structured manner. The process then requires the examination of each critical failure mode to determine the optimum maintenance policy to reduce the severity of each failure. The chosen maintenance strategy must take into account cost, safety, environmental and operational consequences. The effects of redundancy, spares costs, maintenance crew costs, equipment ageing and repair times must be taken into account along with many other parameters.

Classical RCM, as it was first developed, is expensive to implement since rigorous FMEA had to be developed. Classic RCM includes calculating the probability of failure for each piece of equipment (reliability calculations for each system) and it takes teams of engineers' months/years to complete, and requires a lot of historical data. As such it consumes a lot of time.

The streamlined RCM approach however, recognises the value of the personnel along with their experience and takes advantage of their extensive experience running the facility. By talking to the personnel on site, the equipment can be categorised and the initial phase of a RCM programme can be set up.

Streamlined RCM divides facility equipment into four major categories:

1. Reactive Maintenance.
2. Preventive Maintenance.
3. Predictive Maintenance.
4. Proactive Maintenance.

These four major categories summarise the available maintenance concepts in the industry. Each concept can be implemented as a stand-alone regime or it could be integrated with each other to produce a sound regime.

8.2.1 Reactive Maintenance

Reactive maintenance is referred to as many different names, such as breakdown maintenance, repair, fix-when-fail and run to failure maintenance. When applying this maintenance strategy, a piece of equipment receives maintenance (repair or replacement) only when the deterioration of the equipment's condition causes functional failure. The strategy of reactive maintenance assumes that failure is equally likely to occur in any part, component or system. Thus, this assumption precludes identifying a specific group of parts for possible repairs as being more necessary or desirable than others.

The major downside of reactive maintenance is unexpected and unscheduled equipment downtime. If the equipment fails and repair parts are not available, delays ensue while parts are ordered and delivered. When this is the sole type of maintenance practised, both labour and materials are used inefficiently. Labour resources are thrown at whatever breakdown is most pressing. A purely reactive maintenance programme ignores the many opportunities to influence equipment survivability. However, it can be effective if used selectively and performed as a conscious decision based on the results of an RCM analysis. Equipment that can be reactively maintained must be non-critical and will not pose any serious hazards or affect the operation of the system as a whole.

8.2.2 Preventive Maintenance

In PM, maintenance activities are performed before equipment failure. PM involves the repair, replacement and maintenance of equipment in order to avoid unexpected failure during use. PM with inspection intervals is a commonly used maintenance strategy (Ben-Daya and Hariga (1998), Lofsten (1999), Crocker (1999)). The objective of any PM programme is to minimise the total cost of inspection, repair and also equipment downtime. Two approaches have evolved from performing PM (Mann et al. (1999)). The traditional approach is based on the use of statistical and reliability analysis of equipment failure. The second approach involves the use of sensor-based monitoring of equipment condition in order to predict when a failure will occur. Under this condition-based PM, intervals between PM work are not fixed, but are carried out only "when needed".

Traditional PM is keyed to failure rates and times between failures. It assumes that these variables can be determined statistically, and that one can therefore replace a part that is "due for failure" shortly before it fails. The availability of statistical failure information tends to lead to fixed schedules for the overhaul of equipment or the replacement of parts subject to wear. PM is based on the assumption that the overhaul of equipment by disassembly and replacement of parts restores it to a "like-new" condition with no harmful side effects.

Failure rate or its reciprocal, Mean Time Between Failure (MTBF), is often used as a guide to establishing the interval at which the maintenance tasks should be carried out. The major weakness in using these measurements to establish task periodicity is that, failure rate data determines only the average failure rate. In reality, a failure is equally likely to occur at random times and with a frequency unrelated to the average failure rate. There has been considerable progress in recent years in developing PM models for particular equipment addressing this problem (Hariga (1994), Srikrishna et al. (1996), Luce (1999)). Other works include an attempt to model PM using Bayesian approach (Percy and Kobbacy (1996)) and the reduction of PM cost due to uncertainty (Cavalier and Knapp (1996)).

In summary, PM can be costly and ineffective when it is the sole type of maintenance practised.

8.2.3 Predictive Maintenance

Predictive maintenance or Condition Monitoring (CM) uses primarily non-intrusive testing techniques, visual inspections and performance data to assess equipment condition. It replaces arbitrarily timed maintenance tasks with maintenance scheduled only when warranted by equipment condition. Continuous analysis of equipment condition monitoring data allows planning and scheduling of maintenance or repairs in advance of catastrophic and functional failure.

The CM data collected is used in one of the following ways to determine the condition of the equipment and to identify the precursors of failure:

1. *Trend analysis* - Reviewing data to see if the equipment is on an obvious and immediate "downward slide" toward failure (Newell (1999)).
2. *Pattern recognition* - Looking at the data and realising the casual relationship between certain events and equipment failure (Parrondo et al. (1998)).
3. *Test against limits and ranges* - Setting alarm limits (based on professional intuition) and seeing if they are exceeded (Sherwin and Al-Najjar (1999)).
4. *Statistical process analysis* - If published failure data on a certain piece of equipment/component exists, comparing failure data collected on site with the published data to verify/disapprove that the published data can be used for the system analysed.

CM does not lend itself for all types of equipment or possible failure modes and therefore should not be the sole type of maintenance practised.

8.2.4 Proactive Maintenance

Proactive maintenance provides a logical culmination to the other types of maintenance described above. It improves maintenance through better design, installation, maintenance procedures, workmanship and scheduling.

Proactive maintenance is characterised by the following attributes:

1. Maintaining a feedback loop from maintenance to design engineers, in an attempt to ensure that design mistakes made in the past are not repeated in future designs.
2. Viewing maintenance and supporting functions from a life-cycle perspective. This perspective will often show that reducing maintenance activity to save money in the short term often costs more in the long term.
3. Constantly re-evaluating established maintenance procedures in an effort to improve them and ensure that they are being applied in the proper mix.

Proactive maintenance uses the following basic techniques to extend machinery life:

1. Proper installation and precision rebuild.
2. Failed-part analysis.
3. Root-cause failure analysis.
4. Rebuild verification.
5. Age exploration.
6. Recurrence control.

The major difference in proactive maintenance compared to other maintenance programmes is that it does not just treat the symptom but determines the root causes of repeated failures and addresses them.

8.2.5 Summary of Maintenance Techniques

Each of the maintenance concepts reviewed is associated with certain advantages and disadvantages. Hence, these concepts should be used in a right combination so as to ensure a sound and cost-effective maintenance regime. RCM attempts to integrate these techniques and its application has proven to be successful in the past (Goodfellow (2000), Fonseca and Knapp (2000), Hauge et al. (2000)). Table 8.1 summarises the advantages and disadvantages of the described maintenance concepts.

8.3 Current Maintenance Practice on Fishing Vessels

The current maintenance practice on fishing vessels varies according to the operating policies of the owner/operator. On most occasions, the crew does not carry out regular maintenance while at sea. As such, all maintenance work is completed while the vessel is at the discharging port. The time between discharge ports can be as long as 3 to 6 months, which allows for failures on the machinery to propagate and lead to a catastrophic breakdown.

The voyage duration of the vessel depends solely on the success of the catch. Hence, the vessel will stay at the fishing grounds as long as it is possible to maximise the catch. Should the vessel suffer any breakdown during this period, the vessel's crew will attempt to carry out emergency repairs. The amount of repair and replacement of damaged equipment is very limited, mainly due to the following reasons:

1. Limited amount of spares carried on board the vessel.
2. Limited number of tools available to carry out the repairs.
3. The competency of the crew to carry out complicated repairs.
4. Rough weather conditions (especially for small vessels).
5. Available manpower on board the vessel may not be sufficient to carry out major repairs.

Due to these reasons, only temporary repairs are carried out to enable the vessel to steam to the closest port, where more permanent repairs can be carried out. However, if temporary repairs are not sufficient to enable the vessel to move to the closest port, either a shore team is called out to the ship or the ship is towed back to the closest port by tugboats. Both these options are very costly especially when the vessel is stranded in the middle of the ocean.

During the discharging period at port, equipment requiring maintenance will be visited by personnel contracted by the ship owner. The time spent at port by the vessel will depend on the unloading time required. This could vary from a few days to a few weeks. Hence, the time available to carry out repairs is limited. In order to enable the best utilisation of the available time, a repair list is prepared by the ship's Chief Engineer while the vessel is at sea. This list is sent to the shore office (if one exists) to plan the maintenance activities at the next discharging. This list will be combined with a list created by the superintendent of the vessel - upon an inspection of the ship when it arrives at the discharging port. Large fishing companies that have a structured organisational hierarchy adopt this method. Skipper owned vessels, will depend on their contacts ashore to arrange for the repairs to be expedited.

There are several routine maintenances that are carried out regularly on board fishing vessels. These include:

- Filter cleaning.
- Fishing net mending.
- Oil changing.
- General cleaning and lubricating of machinery.
- De-rusting and painting.

These activities can be summarised as the bare minimum requirement of an engineering system.

It has been observed that many fishing vessels call into a floating dock once a year to carry out a complete inspection/repair/overhaul of equipment on board. These repairs and overhauls are normally carried out by yard workers or specially contracted personnel. These vessels also come in for dry-docking every 3 to 5 years (depending on the condition of the vessel) to carry out repairs on seawater valves, replacement of hull anodes, inspection of propeller, tail shaft and rudder and any other fitting which lies beneath the water line.

Considering the current status of maintenance practice on fishing vessels and the high number of accidents caused by the lack of maintenance activities, it is suggested that a maintenance regime be introduced. This regime should be practical (considering the limitations associated with fishing vessels) and effective. Taking into account the ability and competency of crew on board fishing vessels, it is recommended that an inspection regime be implemented in the first instance. This can be followed by an implementation of other maintenance concepts in the future, together with appropriate training for the crew.

This Chapter describes an approach to determine inspection intervals to complement regular maintenance planning. The purpose of inspection at intervals is to increase the up time of systems with comparatively high downtime costs. By regularly carrying out inspections on equipment, abnormalities can be identified and corrective action can be taken to prevent a catastrophic failure. However, carrying out regular inspection on a system that is continuously operating may result in higher operating cost due to downtime and the cost of inspection. A model using Delay Time Analysis (DTA) is described to estimate the expected downtime, cost and safety criticality for various inspection intervals. The optimal inspection period can be obtained depending upon the criteria chosen such that the downtime or cost be minimised or safety maximised.

8.4 Background of Delay-Time Analysis

The time to failure of equipment is a function of its maintenance concept, and to capture this interaction the conventional time to first failure of reliability theory requires enrichment. This may be achieved using the delay-time concept.

Considerable work has been carried out on the modelling of this concept to production plants (Christer and Walker (1984a), Christer et al. (1995), Christer et al. (1998)). Other works include the application to gearbox failure on buses (Leung and Kit-leung (1996)), preventive maintenance modelling for a vehicle fleet (Christer and Walker (1984b)) and application to concrete structures (Burley et al. (1989), Redmond et al. (1997)).

Before a component breaks down (assuming it is not a sudden failure), there will be telltale signs of reduced performance or abnormalities. The time between the first identification of abnormalities (initial point) and the actual failure time (failure point) will vary depending on the deterioration rate of the component. This time period is called the delay time or opportunity window to carry out maintenance or an inspection. The delay time is illustrated by means of a diagram as shown in Figure 8.1. The opportunity window is the period within which the defect could have been identified by inspection and corrective action taken before it led to a failure. The delay time *h*, reflects the characteristic of the plant/system.

Identifying the opportunity window in a system is important to minimise the number of failures. As an example, consider Figure 8.2 where a system is operated with an inspection period of 6 months. Plotting the failures on the same time scale as the inspection activities, it can be seen that if the inspection period had been reduced from every 6 months (*A*) to every 3 months (*B*), the failures would not have happened, as it would have been detected during the inspection and necessary repairs would have been carried out.

Following the argument of Christer and Walker (Christer and Walker (1984c)), a fault arising within a period (0,T) has a delay time *h*, the occurrence probability of this event being f(h)Δh where f(h) is the probability density function of the delay time. A fault will be repaired as a breakdown repair if the fault arises in the period (0, T-h), otherwise an inspection repair as seen in Figure 8.3.

Summing up all possible values of *h*, the probability of a defect arising as a breakdown failure *b(T)* can be expressed as:

$$b(T) = \int_0^T \frac{(T-h)}{T} f(h)\,dh \tag{8.1}$$

b(T) is independent of the arrival rate of the defect per unit time (k_f) but dependent on h. A delay time can only be estimated or identified when the defect has occurred and led to a breakdown failure. Hence if *b(T)* is the probability of a defect arising as a breakdown failure, and a breakdown failure can exist when a defect has arisen, then it is fair to say that *b(T)* is a conditional probability (keeping in mind that this expression excludes sudden failure, i.e. no opportunity window).

An estimation of the probability distribution function can be achieved in several ways as will be discussed in Section 8.5.1.

8.5 Model Development

The flowchart in Figure 8.4 illustrates the approach to delay-time analysis of fishing vessels (Pillay (2001), Pillay et al. (2001a, b)). The described approach is an integration of three models, that is, the downtime estimation model, cost estimation model and safety criticality estimation model. These models require failure data and a probability distribution function of the delay time. The data is then used in a mathematical formula to generate various values for the inspection period T for the corresponding expected downtime D(T), expected cost C(T) and expected safety criticality S(T). Each model developed will produce an optimal inspection period such that downtime, cost or safety criticality is minimised. A best compromise is then achieved by plotting D(T), C(T) and S(T) against the inspection time T.

8.5.1 Expected Downtime Model

After studying the operating practice, the existing maintenance and failure data, the system can be modelled using the following assumptions (Pillay (2001), Pillay et al. (2001a, b)):

- Inspections take place at regular time intervals of T hours and each requires a constant time.
- Downtime owing to inspection = d
- Average downtime for breakdown repair = d_b
- Arrival rate of defects per unit time = k_f
- Inspection period = T
- Failures are repaired immediately with downtime $d_b << T$
- Inspections are perfect in that any defect present will be identified.
- Defects identified will be repaired within the inspection period.
- The time of origin of faults is uniformly distributed over the time between inspections.
- The delay time is independent of its time of origin.

It could be argued that some of the above assumptions may not be practical. For example, inspections/repairs could never be carried out perfectly. However, such assumptions which have been widely used by many safety/reliability researchers are made mainly for demonstrating the described method with ease.

As a consequence of the above assumptions, the model of b(T) given in Equation (8.1) can be simplified as:

$$b(T) = \frac{1}{T}\int_0^T (T-h)f(h)\,dh \qquad (8.2)$$

Consequently, the expected downtime per unit time function D(T) is given by Equation (8.3) below:

$$D(T) = \left\{\frac{d + k_f T b(T) d_b}{T + d}\right\} \qquad (8.3)$$

The product $k_f T$ will give the expected number of defects within the time horizon considered. This is normally based on the historic data gathered for the equipment or system.

Substituting b(T) from Equation (8.2) gives:

$$D(T) = \left\{\frac{d + k_f T\left[\frac{1}{T}\int_0^T (T-h)f(h)\,dh\right] d_b}{T + d}\right\} \qquad (8.4)$$

8.5.1.1 Delay Time Parameter Estimation

Delay time distribution can be predominantly estimated using subjective or objective methods. Several models have been developed for these two approaches (Baker and Wang (1992, 1993), Wang (1997)). The objective models generally require a large amount of data complemented with survey questionnaires, which should reflect the operations of the analysed system over a considerable period of time. These requirements, however, are difficult to fulfil when considering operating systems on board fishing vessels. The subjective models would be more suitable for the intended application, however, they are complex, resource intensive and time consuming. As such, for demonstration purposes, different known distribution functions are experimented to determine the distribution function that produces the best results. As it will be demonstrated later, the research indicates that a truncated standard normal distribution and a Weibull distribution are the most appropriate for dealing with failure data of fishing vessel systems. The truncated standard normal distribution is then used to determine the optimum inspection period for the expected cost and safety criticality model.

When the probability distribution function of the delay time, *f(h)*, follows a normal distribution, i.e.

$$f(h) = \frac{1}{\sqrt{2\pi\alpha}} e^{-(h-\mu)^2/2\sigma^2} \qquad (8.5)$$

where μ = mean and σ^2 = standard deviation of h.

Care is necessary when using the normal distribution. Since $h \geq 0$, there is always a positive chance with the normal distribution that the observation is negative. Hence, a truncated standard normal distribution would be more appropriate. When *f(h)* follows a standard normal distribution truncated at 0 with $\mu = 0$ and $\sigma^2 = 1$, then:

$$f(h) = \frac{2}{\sqrt{2\pi}} e^{-h^2/2} \qquad (8.6)$$

In the above, a truncated standard normal distribution with $\mu = 0$ and $\sigma^2 = 1$ is used for simplification purposes. The distribution is normalised in such a way that its density function integrates to one. If sufficient amount of failure data is available, then variances for a truncated standard normal distribution or even other types of distributions can be specified to produce the best results. However, determination of such variances is out of the scope of this Chapter.

Substituting Equation (8.6) into Equation (8.4) gives:

$$D(T) = \left\{ \frac{d + k_f T \left[\frac{1}{T} \int_0^T (T - h)(\frac{2}{\sqrt{2\pi}} e^{-h^2/2}) \, dh \right] d_b}{T + d} \right\} \tag{8.7}$$

Equation (8.7) will give the estimated downtime per unit time of the equipment. A practical way of expressing this downtime is by means of its availability within a specified time period. The availability of the system, A, is calculated using Equation (8.8):

$$A = \frac{TOT - TDT}{TOT} \tag{8.8}$$

where:

TOT = total operating time

TDT = total downtime

The total downtime can be estimated using Equation (8.9) below:

$$TDT = \left(\frac{TOT}{T^*} \right) \left[d + k_f T^* b(T^*) d_b \right] \tag{8.9}$$

where T^* = optimum inspection period (when downtime is minimised).

The optimal inspection period T can be obtained graphically by plotting the expected downtime D(T) against the inspection period T. The optimal period will be such that D(T) is minimised or alternatively, such that the availability is maximised. The result obtained from the described method only reflects the availability of the component analysed and does not account for any redundancy features incorporated within the system.

8.5.2 Expected Cost Model

This model estimates the expected cost per unit time of maintaining the equipment on an inspection regime of period T. The probability of a defect arising as a breakdown failure is given in Equation (8.1) as b(T). As an inspection repair cost applies to all components even if the component is in good condition, the probability of fault arising as an inspection repair is 1 - b(T).

There are three cost elements which need to be considered in this modelling phase. These three elements are (Pillay (2001), Pillay et al. (2001a, b)):

1. Cost of a breakdown.
2. Cost of an inspection repair.
3. Cost of an inspection.

Using the same assumptions and notations described in Section 8.5.1, Equation (8.4) is modified to include the various costs involved in an inspection maintenance regime to give:

$$C(T) = \frac{[k_f T\{Cost_B b(T) + Cost_{IR}[1 - b(T)]\} + Cost_i]}{(T + d)} \quad (8.10)$$

where:

C(T) = the expected cost per unit time of maintaining the equipment on an inspection system of period *T*.

$Cost_B$ = breakdown repair cost.

$Cost_{IR}$ = inspection repair cost.

$Cost_i$ = inspection cost.

The above terms are described in detail later. When the probability distribution function of delay time *f(h)* follows a truncated standard normal distribution as shown in Equation (8.6) and substituting this into Equation (8.10) to obtain an expression for the expected cost, C(T), will give:

$$C(T) = \frac{[k_f T\left\{Cost_B\left[\frac{1}{T}\int_0^T (T-h)(\frac{2}{\sqrt{2\pi}}e^{-h^2/2})dh\right] + Cost_{IR}\left\langle 1 - \left[\frac{1}{T}\int_0^T (T-h)(\frac{2}{\sqrt{2\pi}}e^{-h^2/2})dh\right]\right\rangle\right\} + Cost_i]}{(T + d)} \quad (8.11)$$

8.5.2.1 Breakdown Repair Cost

When considering the cost associated with the breakdown of machinery, all failure modes and consequences need to be known. This can be achieved with the use of an FMEA. The process of carrying out an FMEA can be found in Chapter 3 and a modified FMEA specific for fishing vessels is presented in Chapter 7. Using the results from this analysis, each consequence is then quantified in monetary terms. The breakdown repair cost includes costs associated with the effects of a failure and also costs associated with the corrective action taken to restore the equipment back to its working condition. This can be represented by Equation (8.12) below:

$$Cost_B = Cost_k^{effect} + Cost_k^{c} \quad (8.12)$$

$Cost_k^{effect}$ is the cost associated with the effects of an equipment failure and $Cost_k^{c}$ is the cost associated with the corrective action carried out on the failed equipment. The various factors considered in predicting the costs associated with the effects of a failure are given in Equation (8.13) and where necessary they can be further elaborated. The various costs involved in carrying out corrective action are given in Equation (8.19) and will be explained later.

Predicting Costs Associated with the Effects of a Failure ($Cost_k^{effect}$)

The cost associated with the effects of an equipment failure, $Cost_k^{effect}$, is given by:

$$Cost_k^{effect} = \Sigma^M_{m=1} (Cost_m^{q}\theta_{mk}\varepsilon_k q_k \delta_k^{PF} + Cost_m^{\omega}\theta_{mk}\varepsilon_k \omega_k \delta_k^{PF}) \quad (8.13)$$

where:

$Cost_m^{q}$ = cost rate for effect *m*.

$Cost_m^{\omega}$ = cost per occurrence for effect *m*.

θ_{mk} = redundancy factor for failure k and effect m.

ε_k = operating time factor for failure k.

q_k = mean probability of failure k.

δ_k^{PF} = P-F factor for failure k.

ω_k = mean frequency of failure k.

M = number of effects.

The cost rate or cost per hour indicates the estimated cost per unit time due to the occurrence of the effect. The cost per occurrence indicates the fixed cost incurred every time the effect takes place.

The redundancy factor indicates whether a cause will produce the assigned effect on its own or whether other concurrent failures will need to occur in order for the effect to take place. A redundancy factor often needs to be determined if the effect is a hazardous one as there will almost certainly be protective systems in place to mitigate against failures, which would lead to a hazard. If the cause will produce the assigned effect without other concurrent failures taking place then the default value of 1 should be assigned to the redundancy factor. If the cause will only produce the assigned effect when other concurrent failures occur then a factor between 0 and 1 should be applied. The redundancy factor represents the probability that the failure cause will produce the assigned effect. For example, consider the analysis of the failure cause, "valve stuck closed" in a hydraulic winch system. This failure might lead to a hazardous event unless the system was shut down until the repair could be conducted. The protection system provided to protect against the hazardous event might consist of sensors and alarms and require the intervention of automatic shut-down systems and operator actions. If the protection system were to fail then the hazardous effect would occur, the unavailability redundancy factor should be set to the estimated probability that the protection system would not work on demand. Therefore, if the probability of failure of the protection system is estimated at 0.0001, the redundancy factor should be set to 0.0001.

The operating time factor indicates the fraction of the system lifetime or sampling period for which the specified failure effects are applicable. If the failure mode will always result in the specified effects then this factor should be set to 1. If the system operates in different phases, and the effects of failure are only applicable during certain phases then this value should indicate the ratio of applicable phase time lengths to the total lifetime:

$$\varepsilon_k = \tau_A/\tau_L \tag{8.14}$$

where:

ε_k = operating time factor.

τ_A = sum of applicable phase time lengths.

τ_L = system lifetime/sampling period.

The potential failure (P-F) interval indicates the time period before an actual failure during which potential failures are revealed. If the P-F interval is set to zero, failures will only be revealed if they have already occurred. Inspections of items with P-F intervals of zero are only effective for hidden failures. If potential failures can be identified before they occur (P-F interval > 0) then it may be worth inspecting items at regular intervals. The P-F factor is used to model the effects of non-zero P-F intervals for inspection tasks and alarm monitoring. For inspection tasks, the P-F factor is given by:

$$\delta_{PF} = 1 - i_{PF} / (\tau_i + mttr) \text{ for } i_{PF} / (\tau_i + mttr) < 1 \tag{8.15}$$

$$\delta_{PF} = 0 \text{ for } i_{PF} / (\tau_i + mttr) \geq 1 \tag{8.16}$$

where:

i_{PF} = P-F interval for the inspection task.

τ_i = inspection interval.

$mttr$ = corrective outage duration (including logistic delay).

For condition alarms with non-zero P-F intervals the P-F factor is given by:

$$\delta_{PF} = 1 - i_{PF} / mttr \text{ for } i_{PF} / mttr < 1 \tag{8.17}$$

$$\delta_{PF} = 0 \text{ for } i_{PF} / mttr \geq 1 \tag{8.18}$$

In all other cases the P-F factor is set to 1.

Predicting Costs Associated with Corrective Action ($Cost_k^c$)

The cost associated with the corrective action carried out on the failed equipment, $Cost_k^c$, is given by:

$$Cost_k^c = Cost_k^{opc}\omega_k + Cost_k^{cre,q}\omega_k mttr_c + Cost_k^{cre,\omega,c}\omega_k + \Sigma^N_{n=1}Cost_{nk}^{spa}U_{nk}^c\omega_k \tag{8.19}$$

where:

$Cost_k^{opc}$ = operational cost for corrective maintenance for failure k.

$Cost_k^{cre,q}$ = cost rate for crew.

$mttr_c$ = corrective task duration.

$Cost_k^{cre,\omega,c}$ = corrective call-out cost for crew.

ω_k = mean frequency of failure k.

$Cost_{nk}^{spa}$ = corrective spare n unit cost.

U_{nk}^c = no of spares used of type n during one corrective task.

N = no of types of spares that need to be replaced.

The operational cost parameter indicates any costs associated with the maintenance task other than the maintenance crew cost. This parameter is used to indicate any operational costs incurred by taking items off-line during maintenance.

The cost rate or cost per unit time defines the cost when the maintenance crew is performing scheduled or non-scheduled maintenance or inspection tasks. The corrective call-out cost represents any fixed costs associated with the call-out of the maintenance crew for corrective repairs. The scheduled call-out cost represents any fixed costs associated with each scheduled maintenance or inspection action.

8.5.2.2 Inspection Repair Cost

The inspection repair cost will include all the expenses incurred to carry out the inspection and corrective action taken (if necessary). This will include the cost of maintenance engineers, spares consumed and loss of operational time. The expected cost for corrective action under inspection repair is less compared to breakdown repair (from the experience of maintenance engineers and ship owners/operators). This is due to the number of components that have to be overhauled/changed when a breakdown occurs, probably attributed to the "knock-on effect" of a component/machinery failure. Hence, the inspection repair cost is given by Equation (8.20) and the value of $Cost_k^c$ in this equation will be less than the value of $Cost_k^c$ in Equation (8.19).

$$Cost_{IR} = Cost_k^i + Cost_k^c \qquad (8.20)$$

where $Cost_k^i$ is the cost associated with inspection tasks and $Cost_k^c$ is the cost associated with corrective action.

Predicting Costs Associated with Inspection Tasks ($Cost_k^i$)

The cost associated with inspections carried out on the equipment, $Cost_k^i$, is given by:

$$Cost_k^i = Cost_i^{op,g} + Cost_k^{cre,q} mtti + Cost_k^{cre,\omega,s} \qquad (8.21)$$

where:

$Cost_i^{op,g}$ = operational cost for task group i (includes inspection task for failure k)

$Cost_k^{cre,q}$ = cost rate for crew.

$mtti$ = inspection duration.

$Cost_k^{cre,\omega,s}$ = scheduled call-out cost for crew.

The inspection duration indicates the mean time taken to inspect the item. This time is only used to calculate the maintenance crew costs. A task group is used to group together different maintenance tasks, which are to be performed at the same time. Performing an inspection task on a group of items at the same time can often be more cost effective than inspecting the items at different intervals. The values of the cost rate for crew and scheduled call-out cost for crew should be the same as the values used in Equation (8.19).

Predicting $Cost_i$

$Cost_i$ represents the cost involved for the vessel crew to carry out the inspection.

8.5.3 Expected Safety Criticality Model

This model estimates the safety criticality per unit time of the equipment when it is inspected with a periodicity of T (Pillay (2001), Pillay et al. (2001a, b)). If b(t) is the probability of a defect arising as a breakdown failure k then, Cr_k^{safety} is the safety criticality of the said failure and Cr_k^{oper} is the operational safety criticality when the defect does not arise and/or is not a breakdown failure. The estimation of Cr_k^{safety} and Cr_k^{oper} is given by Equations (8.22) and (23) respectively.

$$Cr_k^{safety} = \sum_{m=1}^{M} S_m^{safety} \theta_{mk} \varepsilon_k \delta_k^{PF} \omega_k \qquad (8.22)$$

where:

Cr_k^{safety} = safety criticality associated with failure k.

S_m^{safety} = safety severity for the mth effect for failure k.

θ_{mk} = redundancy factor for failure k and effect m.

ε_k = operating time factor for failure k.

δ_k^{PF} = P-F factor for failure k.

ω_k = mean frequency of failure k.

$$Cr_k^{oper} = \sum_{m=1}^{M} S_m^{oper} \theta_{mk} \varepsilon_k \delta_k^{PF} \omega_k \tag{8.23}$$

where:

Cr_k^{oper} = operational safety criticality associated with failure k.

S_m^{oper} = operational safety severity for the mth effect for failure k.

θ_{mk} = redundancy factor for failure k and effect m.

ε_k = operating time factor for failure k.

δ_k^{PF} = P-F factor for failure k.

ω_k = mean frequency of failure k.

M = number of effects.

The safety criticality and operational safety criticality of a failure can be identified by performing an FMEA study on the system. The values of these two parameters can be estimated subjectively using a scale of 0 to10 (0 being least critical and 10 being most critical). The values are assigned based on the probability of occurrence and severity, and are considered for four categories (personnel, environment, equipment and catch). All the other variables in Equations (8.22) and (23) will have the same values as defined in Equation (8.13) of Section 8.5.2.

Maintaining the assumptions and notations presented in Section 8.5.1, the expected safety criticality is given by Equation (8.24).

$$S(T) = \frac{k_f T Cr_k^{safety} b(T) + Cr_k^{oper}[1 - b(T)]}{T + d} \tag{8.24}$$

where:

S(T) is the expected safety criticality per unit time.

Cr_k^{safety} and Cr_k^{oper} are given by Equations (8.22) and (8.23), respectively.

8.6 An Example

The application of the delay time concept to determine the optimum inspection interval is demonstrated using a main hydraulic winch operating system on a fishing vessel (Pillay et al. (2001a, b)). This vessel is a 1266 GRT (Gross Tonnage), deep-sea trawler with an L.O.A (Length overall) of 60 meters. The winches are used to deploy the nets and haul the catch on to the ship. The supporting winches, that is, the Gilson winch and tipping winches are not considered in this example. The schematic diagram in Figure 8.5 shows the layout of the main hydraulic piping system and the associated components within the system. The main pumps provide the hydraulic power to the port and starboard winches as well as the net drum motor. The 1010 pumps are used to control the tension and balance the loads on the main winches.

The fishing vessel used in this test case has a voyage profile as illustrated in the bar chart in Figure 8.6. The voyage duration of the vessel depends solely on the success of the catch and the duration at port depends on the discharging time and the amount of work to be carried out on the ship as discussed in Section 8.3. As an example of an analysis at the component level, the actual maintenance periods and failures of a brake seal for a winch are shown in Figure 8.7. This particular vessel operates on a yearly inspection/maintenance regime. This entails that once a year, a thorough check of the vessel is performed. Any components that are identified to require maintenance or replacement (during this inspection) are either overhauled or replaced accordingly to bring the equipment back to "as good as new".

8.6.1 Modelling Process

It can been seen from Figure 8.7 that many of the failures go unnoticed and the actual failures occur between the inspection/maintenance periods. For this example, only on two occasions (between voyages 3 and 4 and voyages 10 and 11), the initial failures were detected for the brake seal and the necessary action was taken.

The following information was gathered for this particular system, which included a combination of logged records and reports complemented by expert judgements (where no data was available):

Inspection downtime (d) = 15 minutes = 0.01041 days

Downtime for breakdown repair (d_b) = 4.5 days

Total operating hours of winch (for 25 voyages) = 1344 hrs = 56 days

Arrival rate of defects (k_f) = 0.535 per day [30 failures for 25 voyages]

The actual process of carrying out the inspection itself would take about 45 minutes for this particular system. Most of the inspection can be carried out when the hydraulic system is not operating, this includes visual inspection, off-load and function testing. Hence, the downtime caused by inspection would be much lower than 45 minutes. From the experience, only 15 minutes is required to carry out an on load pressure test for such a system. Therefore, the inspection downtime d is set to be 15 minutes or 0.01041 days.

The downtime for breakdown repair takes into account any logistic delays that may occur while waiting for spares to be sent from shore suppliers. Most fishing vessels carry minimum amount of spares on board. Hence, should a breakdown occur at sea on the hydraulic system, the ship might be operationally crippled for a period of time. From the experience, this period could be a few of hours or days, depending on the position of the vessel at the time of breakdown.

Substituting the values obtained for the hydraulic system into Equation (8.7) gives the following equation:

$$D(T) = \left\{ \frac{0.01041 + (0.535T)\left[\frac{1}{T}\int_0^T (T-h)\left(\frac{2}{\sqrt{2\pi}}e^{-h^2/2}\right)dh\right]4.5}{T + 0.01041} \right\} \tag{8.25}$$

Using a computing software such as *Derive, MatLab or Studyworks* to solve Equation (8.25), a graph of D(T) against T can be plotted as shown in Figure 8.8.

From the graph in Figure 8.8, the optimal inspection period, T (such that the expected downtime is minimised), is determined to be 0.216 days or 5.18 operating hours. This inspection frequency will cause an expected minimum downtime of 0.0853 days or 3.04 hours per unit time. To express this result more clearly for a certain period of operating time, the availability of the equipment is calculated using Equations (8.8) and (8.9) for various inspection intervals. The total operating time is taken to be 56 days for a period of 25 voyages. The result of this analysis is shown in Figure 8.9. From the graph, the maximum attainable availability is 91.1% with a corresponding inspection interval of 0.216 days or 5.18 operating hours.

For this particular study, two other different probability distribution functions of delay-time were experimented with the Weibull distribution and the exponential distribution.

Equation (8.4) was altered according to the type of distribution used. For the Weibull distribution where:

$$f(h) = \frac{\alpha}{\beta} h^{\alpha-1} e^{-(h/\beta)^{\alpha}} \tag{8.26}$$

and substituting Equation (8.26) into Equation (8.4) gives the following;

$$D(T) = \left\{ \frac{0.01041 + (0.535T)\left[\frac{1}{T}\int_0^T (T-h)(\frac{\alpha}{\beta} h^{\alpha-1} e^{-(h/\beta)^{\alpha}})\,dh\right]4.5}{T + 0.01041} \right\} \tag{8.27}$$

Different values of α and β were substituted to plot the change in the D(T) versus T curve. The results are as shown in Figure 8.10. From these curves, it was determined that the optimum inspection period is between 0.3 to 0.8 days (7.2 to 19.2 operating hours).

Using the exponential distribution for the delay time, where:

$$f(h) = \lambda e^{-\lambda h} \tag{8.28}$$

and substituting Equation (8.28) into Equation (8.4) to obtain an expression for the downtime will give:

$$D(T) = \left\{ \frac{0.01041 + (0.535T)\left[\frac{1}{T}\int_0^T (T-h)(\lambda e^{-\lambda h})\,dh\right]4.5}{T + 0.01041} \right\} \tag{8.29}$$

Different values of λ (failure rate) were substituted into Equation (8.29) to produce the graph in Figure 8.11. Although different values of λ were experimented with (a range from MTBF = 40 to MTBF = 900) the curve remained the same.

The results obtained using an exponential distribution is not very useful as it does not reflect a curve that increases in D(T) as the inspection period increases. From these results, the most suited distribution was found to be the Weibull and the truncated standard normal distributions. These two distributions gave clear indications of the optimum inspection period. The values of α and β in the Weibull distribution can be estimated by a collection of test data or by using available failure data of the equipment, and since the failure data available is associated with a high degree of uncertainty, this distribution is not used here. As such for the purpose of demonstrating the delay time concept for fishing vessels, the truncated standard normal distribution is used for the expected cost and safety criticality models.

The data collected from the hydraulic system for the cost estimation is as follows:

Cost Associated with Inspection Task ($Cost_k^i$)

From the historical data, it is found that contract workers carry out inspection tasks as per PMS (Preventive Maintenance Schedule) every 365 days when machinery is not operating/at port. However, should the inspection be carried out on board the vessel by the vessel crew, the values for $Cost_i^{opc,g}$ and $Cost_k^{cre,q}$ are 0. The only possible cost could be a call out cost for crew to carry out special inspection activities such as, the calibration of pressure control valves on the hydraulic system. The inspection cost from Equation (8.21) is calculated to be:

$$Cost_k^i = Cost_k^{cre,\omega,s} = £100$$

Cost Associated with Corrective Action ($Cost_k^c$)

From the historical data, it is known that contract workers normally carry out corrective action at port upon inspection. However, if the corrective maintenance was carried out on board the vessel upon inspection, the values for $Cost_k^{op}$ and $Cost_k^{cre,q}$ would be equal to 0. The data used for this test case considers repairs carried out on the clutch seal and break seal of the hydraulic winch. The following parameters were quantified:

$$Cost_k^{cre,\omega,c} = £100$$

$$\omega_k = 2.5$$

$$\mathrm{Cost}_{\mathrm{bseal}}^{\mathrm{spa}} = £30$$

$$\mathrm{Cost}_{\mathrm{cseal}}^{\mathrm{spa}} = £30$$

$$\mathrm{U}_{\mathrm{bseal}} = 1$$

$$\mathrm{U}_{\mathrm{cseal}} = 1$$

In the above, the call-out cost is assumed to be very low, i.e. £100. This is because it was the company's own technical superintendent that actually went out to the vessel and calibrated the pressure control valves for the winch.

Substituting the above values into Equation (8.19) gives,

$$Cost_k^c = 100(2.5) + 30(2.5)(1) + 30(2.5)(1) = £400$$

The predicted cost associated with inspection repair from Equation (8.20) is calculated to give,

$$Cost_{IR} = 100 + 400 = £500$$

Cost Associated with the Effect of Equipment Failure ($Cost_k^{effect}$)

The failure of the winch has an effect on the personnel, environment, equipment and catch (Pillay et al. (2001a, b)). The cost rate ($Cost_m^q$) and cost per occurrence ($Cost_m^{\omega}$) on each of these categories are given in Table 8.2. Since much of the information was lacking, expert judgement and subjective reasoning were used to obtain reasonable estimates of the effects of the hydraulic winch failure.

The other parameters were quantified as follows:

$$\theta_{mk} = 1$$

$$\varepsilon_k = 1$$

$\omega_k = 2.5$

$\delta_k^{PF} = 1$

$q_k = 0.02$

Using these values, the sum of $Cost_k^{effect}$ is calculated from Equation (8.13) to be £25,000.

As described previously, $Cost_i$ represents the cost involved for the vessel crew to carry out the inspection. As this is part of their job description in this example, it does not represents any additional cost to the operating company, as such it is assumed to be 0.

These values are substituted into Equation (8.11) to give the profile of the expected cost C(T) against the inspection period *T*. The results of the analysis are presented in Figure 8.12. From the graph, the optimal inspection period for this system is determined to be 0.302 days or 7.24 operating hours and the expected cost at this interval is estimated to be £881.

To analyse the effect of the change in the cost elements that were difficult to quantify, a sensitivity analysis is performed on the optimal inspection period by altering the inspection repair cost ($Cost_{IR}$) and the inspection cost ($Cost_k^i$). The following five cases were considered:

Case 1: $Cost_{IR}$ and $Cost_k^i$ increased by 10%

Case 2: $Cost_{IR}$ and $Cost_k^i$ increased by 5%

Case 3: $Cost_{IR}$ and $Cost_k^i$ unchanged

Case 4: $Cost_{IR}$ and $Cost_k^i$ decreased by 5%

Case 5: $Cost_{IR}$ and $Cost_k^i$ decreased by 10%

The result of this analysis is shown graphically in Figure 8.13 and the expected cost and optimal inspection period for each case are given in Table 8.3. From the sensitivity analysis, it can be seen that the optimal inspection period is around 7 to 8 operating hours. The variation in *T* is observed to be small when inspection repair cost and inspection cost are varied.

The data collected for the safety criticality estimation is based on expert judgement and is shown in Table 8.4. The failure was evaluated for its safety and operational criticality for the four different categories on a scale of 0 to 10. The estimation of the safety severity parameter, (S_m^{safety}), is assumed for the worst case scenario. It is also assumed that if the failure does not lead to a catastrophic breakdown, the operational safety severity (S_m^{oper}) will be minimal.

The values of Cr_k^{safety} and Cr_k^{oper} in Equations (8.22) and (8.23) were evaluated assuming that θ_{mk}, ε_k, and $\delta_k^{PF} = 1$ and $\omega_k = 2.5$, to give:

$$Cr_k^{safety} = 25 + 25 + 25 + 25 = 100$$

$$Cr_k^{oper} = 2.5 + 2.5 + 2.5 + 2.5 = 10$$

These values are then substituted into Equation (8.24) to give the profile of the expected safety criticality S(T) against the inspection period *T*. The results of the analysis are presented in Figure 8.14. This graph indicates that the optimal inspection period when the safety criticality is at its minimum is 0.72 days or 17.28 operating hours. This inspection interval is much higher compared to when the cost or downtime is minimised.

The next step in the analysis is to determine the best compromise between the three inspection intervals obtained. There are several methods that can be used to determine the best compromise, these include multiple criteria decision making, minimax principle optimisation and the Bayesian approach optimisation (Almeida and Bohoris (1995)). As these methods require tedious mathematical computation, which is not required here, a simple graphical method is used to determine the best compromise inspection interval.

8.6.2 Optimisation Results

The example used to demonstrate the described approach generated three different optimal inspection periods. The inspection period is estimated to be 5.18 operating hours when the downtime is minimised, 7.24 operating hours when the cost is minimised and 17.28 operating hours when the safety criticality is minimised. As the change in the safety criticality is small for a large change in the inspection interval, this criterion is not as critical as the cost and downtime criteria. An such, in the first instance, the expected cost C(T) is plotted against the expected downtime D(T) as shown in Figure 8.15. The curve generated can be used to determine the best compromise between the cost and downtime criteria.

Points 1 and 2 on the graph show the best downtime and cost achievable respectively for the system with an inspection interval of T^*. Should point 1 be selected, when downtime is minimised the operating cost per unit time is £911. This is almost 3.4% higher than the minimum possible operating cost. However, if point 2 is selected, when the cost is minimised, the downtime suffered will be 0.18 hours per unit time. This translates to a reduction in availability of 0.34% from the maximum availability attainable by the equipment.

It is worth mentioning that in some cases, it could be relatively difficult to quantify cost factors with confidence. For example, for a fishing vessel, there are several elements that need to be considered when representing downtime as a loss value in monetary terms, such as wages, loss of consumption of perishables, etc. Therefore decisions on the inspection strategy may not be made solely based on C(T). In such cases, D(T) based decision on the inspection strategy may be more reasonable. If uncertainties in both D(T) and C(T) are considered, then decisions on the inspection strategy can be made based on a combination of D(T) and C(T).

The ideal inspection time is located at point 3 where both the cost and downtime are simultaneously minimised. However, such an operating condition does not exist for the system that was modelled. Therefore, the best compromise is identified at point 4, which is nearest to the ideal point. If the cost and downtime are of equal importance, the best compromise point can be obtained using minimax approach (Sen and Yang (1993)). From the analysis, the best compromise (point 4) is when the inspection period is 6.24 operating hours, cost is £886, the expected downtime is 2.06 hours per unit time and the availability of the equipment is 91%. Considering the inspection time interval obtained (6.24 operating hours), it would entail that an inspection is to be carried out after every two fishing operations (assuming that the main winches run on an average of 3 hours per operation).

The graph in Figure 8.16 gives a clearer indication of the three criteria modelled for the winch system. This graph plots the expected cost C(T), expected downtime D(T) and expected safety criticality S(T) against the inspection period *T*. The shaded area shown on the graph represents the approximate operating hours of the winch system (per fishing operation), which ranges from 3 to 6 hours. For convenience, the inspection can be carried out during this period as the penalty is within reasonable limits.

8.7 Conclusion

Inspections carried out during the operation phase of machinery will reveal any failures that have already been initiated at an earlier time. Upon identifying the "abnormal" condition, necessary action can be taken to arrest the problem before it propagates to become a failure. This Chapter demonstrates the delay-time concept with the use of data gathered from a fishing vessel. Assumptions and expert judgements were made where the data was incomplete. Since there was no record of the delay-time for failures, the probability distribution function of the delay-time could not be ascertained mathematically. As such, known distribution functions such as the standard normal, exponential and Weibull were used to demonstrate the concept.

The example of the brake seal failure with the current maintenance policy of every 365 days (when the vessel is at port), showed that almost 66% of the failures went unnoticed. This would entail high repair costs coupled with high operational costs due to the downtime suffered. With the integration of the delay-time concept within the current maintenance policy, the percentage of failures going unnoticed is expected to be as low as 5 to 10% (Pillay (2001), Pillay et al. (2001a)).

Although the procedure to determine the optimal inspection time is complex, it can be easily incorporated into a user friendly computer interface, which would require owners/operators to input information about the failure of the equipment. Hence, it could be easily adapted to any vessel within the maritime community. The described approach would appeal to owners and operators who are running their vessels at high maintenance costs. Fishing vessels are constantly subjected to rough operating conditions as these vessels operate under various constrains such as size of the vessel, equipment on board, competency of crew and weather conditions. Owners of such vessels would be enthusiastic to incorporate an inspection regime on their fleet, as this would entail a more cost efficient ship, which further translates into income for the company. The described approach does not require any condition monitoring equipment to be installed, hence it would not be expensive for the owners/operators to implement such an approach.

The inspection regime can be integrated into the existing maintenance procedures in order to minimise the operating cost and downtime suffered. The effectiveness of the described approach can be improved if sufficient data is available in order to generate a true probability distribution function for the delay time. Currently there is no procedure in place for testing the hydraulic equipment for operation before the start of a fishing operation. As such, having an inspection regime before every other operation could be very useful to minimise unforeseen accidents/incidents caused by equipment failure. Any inspection regime implemented on board a fishing vessel would enable gathering of useful information about the system, such as the time of actual failure and the time of initial failure (the time when the equipment starts to show signs of abnormalities). This information will enable a better prediction of the delay time interval and distribution to be conducted, hence, enhancing the accuracy of the model.

The final decision of the optimal inspection period will depend heavily on the needs and operating culture of the owner/operator of the vessel. The implementation of such a regime on fishing vessels will be influenced by the operating circumstances of the equipment and other factors such as availability of expertise, position of vessel and sea conditions. However, should the conditions for implementation be favourable, delay-time analysis can be used to optimise the system's inspection maintenance scheme.

8.8 References (Chapter 8)

1. Almeida A.T. and Bohoris G.A., (1995) "Decision Theory in Maintenance Decision Making", Journal of Quality in Maintenance Engineering, Vol. 1, No. 1, 1995, pp. 34-45.

2. Ashayeri J., Teelan A. and Selen W., (1996) "A Production and Maintenance Model for the Process Industry", International Journal of Production Research, Vol. 34, No. 12, pp. 3311-3326.
3. Baker R.D. and Wang W., (1992) "Estimating the Delay Time Distribution of Faults in Repairable Machinery for Failure Data", IMA Journal of Mathematical Application in Business and Industry, Vol. 3, pp. 259-282.
4. Baker R.D. and Wang W., (1993) "Developing and Testing the Delay Time Model", Journal of the Operational Research Society, Vol. 44, pp. 361-374.
5. Ben-Daya M. and Hariga M., (1998) "A Maintenance Inspection Model: Optimal and Heuristic Solution", International Journal of Quality and Reliability Management, Vol. 15, No. 5, pp. 481-488.
6. Burley E., Christer A.H., Ridgen S.R., (1989) "Inspection Practice for Concrete Structures in the UK", Proceeding of the British Cement Association Conference: Life of concrete structures, United Kingdom.
7. Cavalier M.P. and Knapp G.M., (1996) "Reducing Preventive Maintenance Cost Error Caused by Uncertainty", Journal of Quality in Maintenance Engineering, Vol. 2, No. 3, pp. 21-36.
8. Christer A.H. and Walker W.M., (1984a) "Reducing Production Downtime Using Delay-Time Analysis", Journal of the Operational Research Society, Vol. 35, pp. 499-512.
9. Christer A.H. and Walker W.M., (1984b) "An Operational Research Approach to Planned Maintenance: Modelling PM for a Vehicle Fleet", Journal of the Operational Research Society, Vol. 35, pp. 401-406.
10. Christer A.H. and Walker W.M., (1984c) "Delay-Time Models of Industrial Inspection Maintenance Problems", Journal of the Operational Research Society, Vol. 35, pp. 401-406.
11. Christer A.H., Wang W., Baker R. and Sharp J., (1995) "Modelling Maintenance Practice of Production Plant Using Delay-Time Concept", IMA Journal of Mathematical Application in Business and Industry, Vol. 6, pp. 67-84.
12. Christer A.H., Wang W., Sharp J. and Baker R., (1998) "A Case Study of Modelling Preventive Maintenance of Production Plant Using Subjective Data", Journal of the Operational Research Society, Vol. 49, pp. 210-219.
13. Crocker J., (1999) "Effectiveness of Maintenance", Journal of Quality in Maintenance Engineering, Vol. 5, No. 4, pp. 307-313.
14. Fonseca D.J. and Knapp G.M., (2000) "An Expert System for Reliability Centred Maintenance in the Chemical Industry", Expert-Systems-with-Applications, Vol. 19, No. 1, pp. 45-57.
15. Goodfellow J.W., (2000) "Applying Reliability Centred Maintenance (RCM) to Overhead Electric Utility Distribution Systems", Power Engineering Society Summer Meeting, Vol. 1, pp. 566-569.
16. Hariga M., (1994) "A Deterministic Maintenance Scheduling Problem for a Group of Non-Identical Machines", International Journal of Operations and Production Management, Vol. 14, No. 7, pp. 27-36.
17. Hauge B.S., Stevens A.M., Loomis R.J.Jr. and Ghose A., (2000) "Reliability-Centred Maintenance on the Space Shuttle Program", Proceedings of the Annual Reliability and Maintainability. Symposium, IEEE, Piscataway, NJ, USA, pp. 311-316.
18. IMO, International Safety Management (ISM) Code/Guidelines on implementation of the ISM Code, London, 1997.
19. Leung F. and Kit-Leung M., (1996) "Using Delay-Time Analysis to Study the Maintenance Problem of Gearboxes", International Journal of Operations and Production Management, Vol. 16, No. 12, pp. 98-105.

20. Lofsten H., (1999) "Management of Industrial Maintenance - Economic Evaluation of Maintenance Policies", International Journal of Operations and Production Management, Vol. 19, No. 7, pp. 716-737.
21. Luce S., (1999) "Choice Criteria in Conditional Preventive Maintenance", Mechanical Systems and Signal Processing, Vol. 13, No. 1, pp. 163-168.
22. Mann L., Saxeña A. and Knapp G.M., (1995) "Statistical-based or Condition-based Preventive Maintenance?", Journal of Quality in Maintenance Engineering, Vol. 1, No. 1, pp. 49-59.
23. Mokashi A.J., Wang J. and Verma A.K., (2002) "A Study of Reliability Centred Maintenance in Ship Operations", Marine Policy, Vol. 26, No. 5, pp. 325-335.
24. Nakajima S., (1997) "Introduction to TPM", 2nd ed. Productivity Press (India) Private Limited, originally published as TPM Nyumon by Japan Institute of Plant Maintenance, Japan.
25. Newell G.E., (1999) "Oil Analysis: Cost-effective Machine Condition Monitoring", Industrial Lubrication and Tribology, Vol. 51, No. 3, pp. 119-124.
26. Parrondo J.L., Velarde S. and Santolaria C., (1998) "Development of a Predictive Maintenance System for a Centrifugal Pump", Journal of Quality in Maintenance Engineering, Vol. 4, No. 3, 1998, pp. 198-211.
27. Percy D.F. and Kobbacy K.A.H., (1996) "Preventive Maintenance Modelling, a Bayesian Perspective", Journal of Quality in Maintenance Engineering, Vol. 2, No. 1, pp. 15-24.
28. Pillay A., (2001) "Formal Safety Assessment of Fishing Vessels", PhD Thesis, School of Engineering, Liverpool John Moores University, UK.
29. Pillay A., Wang J., and Wall A., (2001a) "Optimal Inspection Period for Fishing Vessel Equipment: a Cost and Down Time Model Using Delay Time Analysis", Marine Technology, SNAME, Vol. 38, No. 2, pp. 122-129.
30. Pillay A., Wang J., Wall, A. and Ruxton T., (2001b) "An Maintenance Study of Fishing Vessel Equipment Using Delay Time Analysis", Journal of Quality in Maintenance Engineering, Vol. 7, No. 2, 118-125.
31. Rausand M., (1998) "Reliability Centred Maintenance", Reliability-Engineering & System Safety, Vol. 60, No. 2, pp. 121-132.
32. Redmond D.F., Christer A.H., Ridgen S.R. and Burley E., (1997) "OR Modelling of the Deterioration and Maintenance of Concrete Structures", European Journal of Operational Research, Vol. 99, pp. 619-631.
33. Rischel T.D. and Christy D.P., (1996) "Incorporating Maintenance Activities into Production Planning: Integration at the Master Schedule Versus Material Requirements Level", International Journal of Production Research, Vol. 34, No. 2, pp. 421-446.
34. Savic D.A., Godfrey W.A. and Knezevic J., (1995) "Optimal Opportunistic Maintenance Policy Using Genetic Algorithm, 1: Formulation", Journal of Quality in Maintenance Engineering, Vol. 1, No. 2, pp. 34-49.
35. Sen P. and Yang J.B., (1993) "A Multiple Criteria Decision Support Environment for Engineering Design", Proceedings of 9th International Conference on Engineering Design, Hague, The Netherlands, August.
36. Sherwin D.J., (1999) "A Constructive Critique of Reliability Centered Maintenance", Proceedings of the Annual Reliability and Maintainability, Symposium, IEEE, Piscataway, NJ, USA, pp. 238-244.
37. Sherwin D.J. and Al-Najjar B., (1999) "Practical Models for Condition Monitoring Inspection Intervals", Journal of Quality in Maintenance Engineering, Vol. 5, No. 3, pp. 203-220.

38. Srikrishna S., Yadava G.S. and Rao P.N., (1996) "Reliability-Centred Maintenance Applied to Power Plant Auxiliaries", Journal of Quality in Maintenance Engineering, Vol. 2, No. 1, pp. 3-14.
39. Wang W., (1997) "Subjective Estimation of Delay-time Distribution in Maintenance Modelling", European Journal of Operational Research, Vol. 99, pp. 151-529.
40. Worledge D.H., (1993) "Industry Advances in Reliability Centred Maintenance", Proceedings of the European Safety and Reliability Conference (ESREL'93), Amsterdam, Netherlands, pp. 267-277.

Table 8.1 Advantages and Disadvantages of Maintenance Concepts

Maintenance	*Advantages*	*Disadvantages*
Reactive	• Cost effective for small, non-critical equipment.	• Possible costly downtime. • Possible damage to associated equipment. • High cost for medium/high priority equipment.
Preventive	• Provides first line of defence.	• Often wasteful. • Does not prevent certain failure. • Can introduce problems. • Requires large parts inventory.
Predictive	• Reduces inventory cost. • Reduces downtime. • Reduces damage to associated equipment. • Reduces unnecessary parts replacement.	• When implemented alone, does not address root causes of problems. • CM equipment are costly.
Proactive	• Addresses root causes of problems. • Reduces maintenance costs beyond predictive levels. • Extends equipment life.	• Cost.

Table 8.2 Cost rate and Cost per Occurrence Estimation for a Failure

Effect of failure on	$Cost_m^q$	$Cost_m^{\varpi}$
Personnel	£100/hr	£4000
Environment	£100/hr	£2000
Equipment	£100/hr	£1000
Catch	£100/hr	£3000

Table 8.3 Optimal Inspection Period based on the Sensitivity Analysis for Various Cases

Case	*Expected cost, C(T)*	*Optimal insp. period, T (operating hours)*
Case 1 (+10%)	*£938*	***7.92***
Case 2 (+5%)	*£909*	***7.27***
Case 3 (Unchanged)	*£881*	***7.24***
Case 4 (-5%)	*£852*	***7.32***
Case 5 (-10%)	*£822*	***6.94***

Table 8.4 Values of S_m^{safety} and S_m^{oper}

Effect of failure on	S^{safety}_m	S^{oper}_m	Cr_m^{safety}	Cr_m^{oper}
Personnel	10	1	25	2.5
Environment	10	1	25	2.5
Equipment	10	1	25	2.5
Catch	10	1	25	2.5

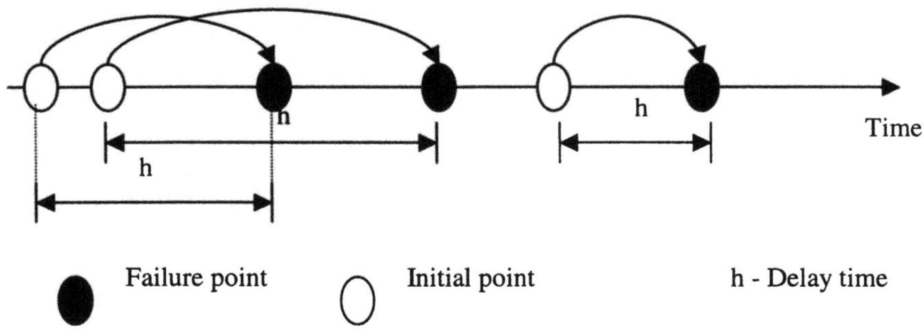

Figure 8.1 Delay time

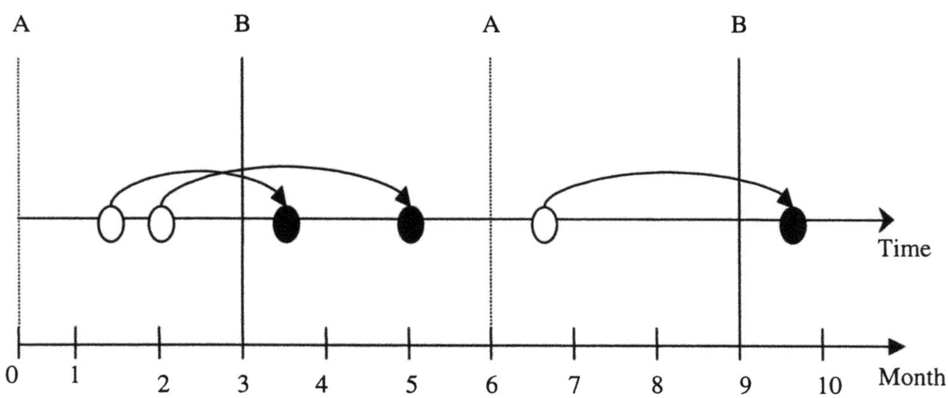

Figure 8.2 Inspection every 6 months (A) and 3 months (B)

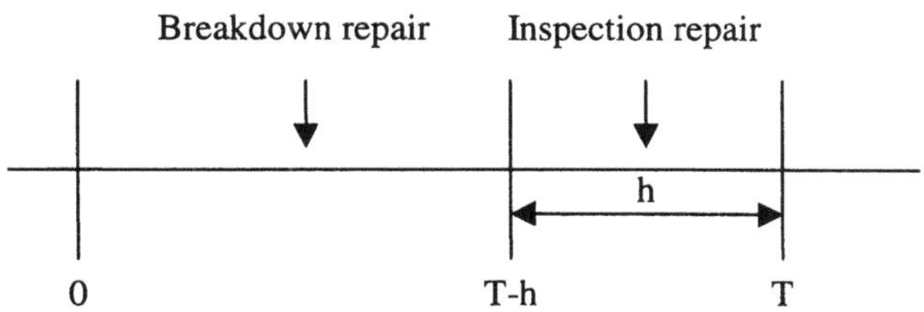

Figure 8.3 Breakdown and inspection repair

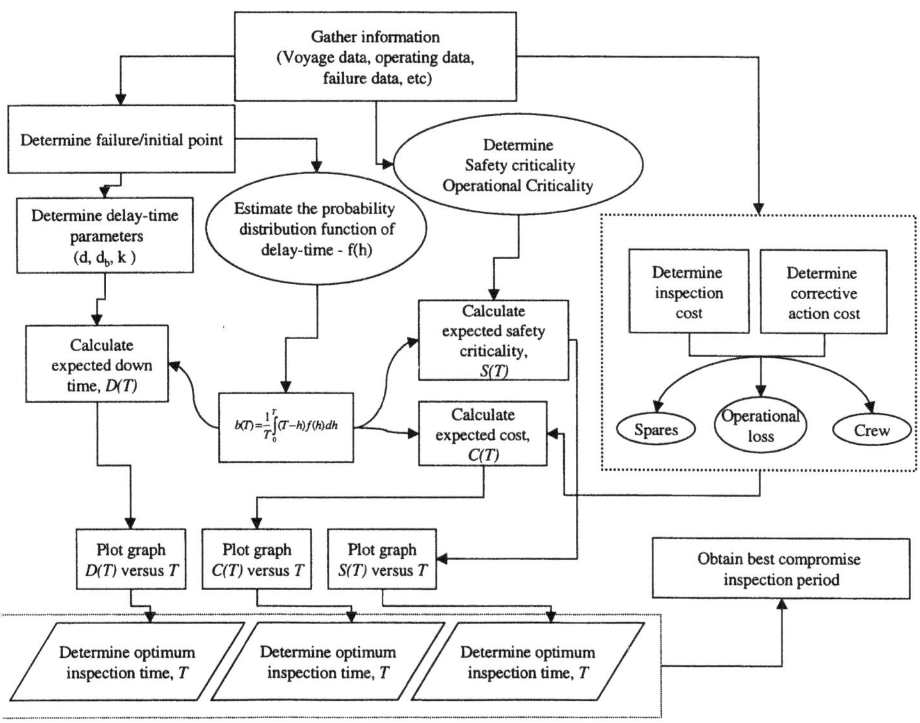

Figure 8.4 The approach flowchart

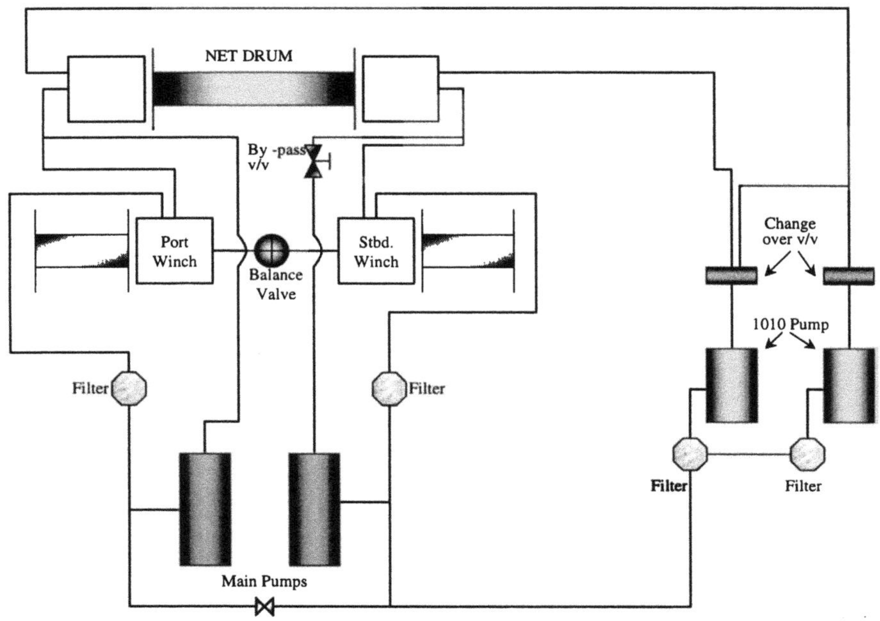

Figure 8.5. Hydraulic winch operating system of a fishing vessel

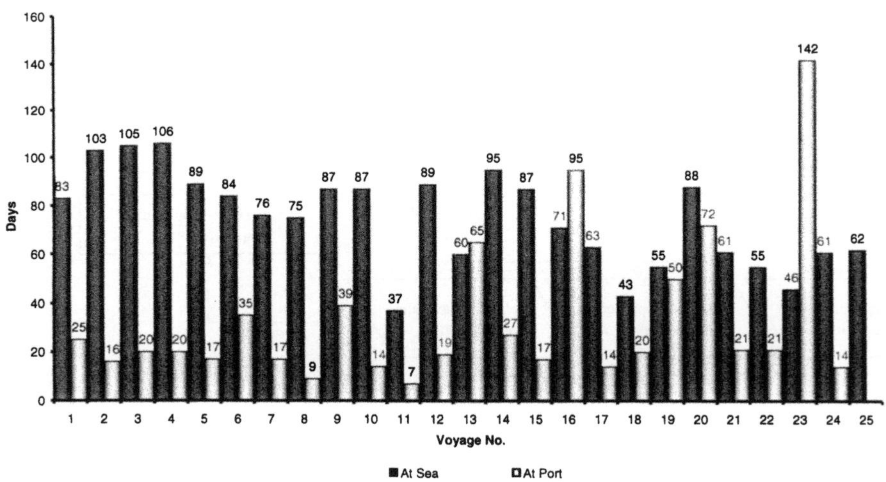

Figure 8.6 Voyage profile

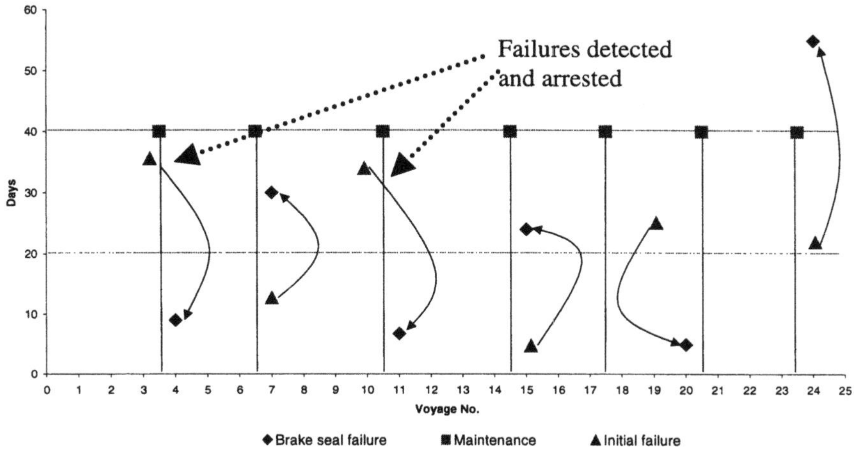

Figure 8.7 Initial point and failure point of brake seal

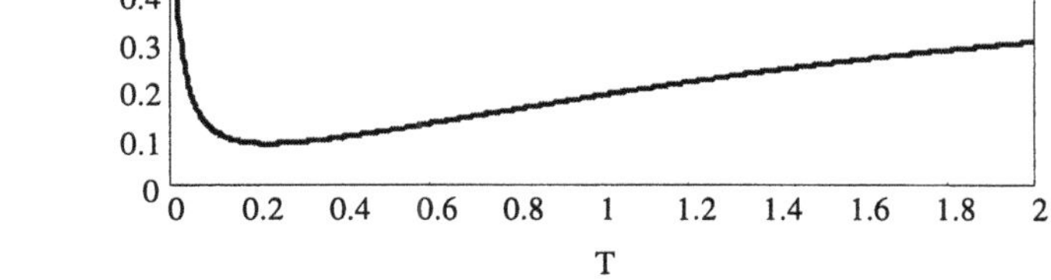

Figure 8.8 Optimal inspection period based on minimum D(T) for a truncated standard normal distribution of the delay time

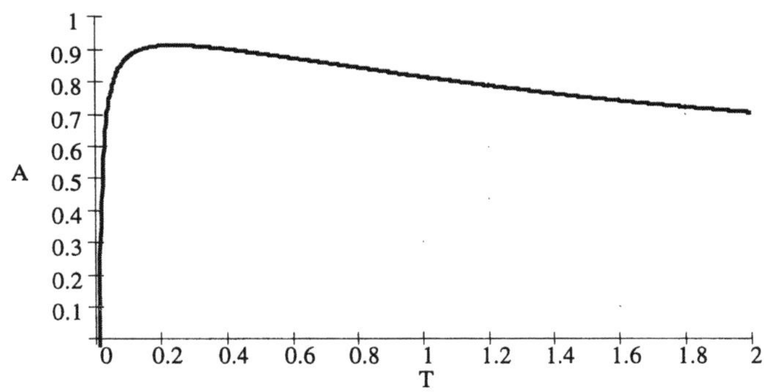

Figure 8.9 Optimal inspection period based on maximum availability for a truncated standard normal distribution of the delay time

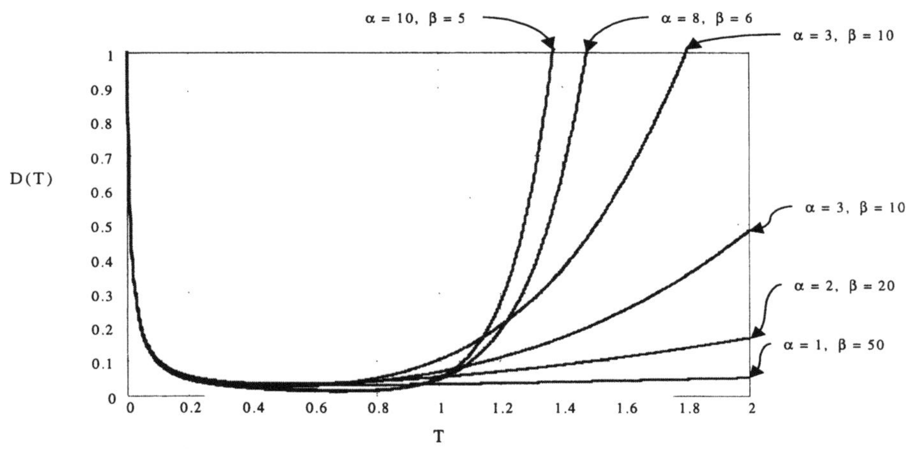

Figure 8.10 Optimal inspection period based on minimum D(T) for a Weibull distribution of the delay time

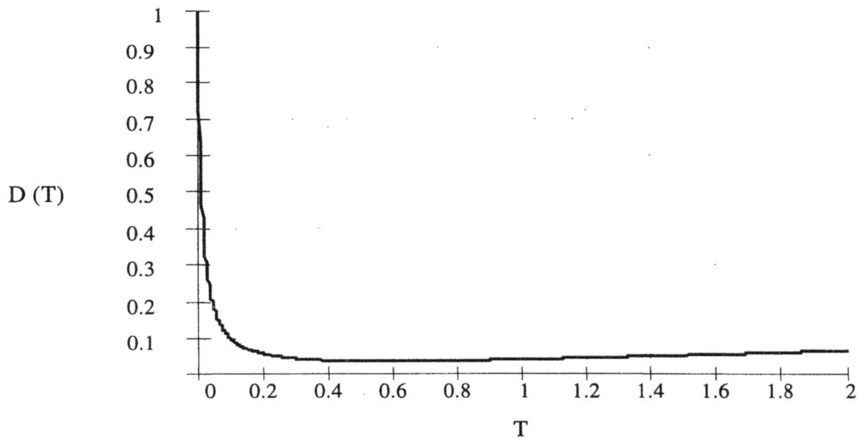

Figure 8.11 Optimal inspection period based on minimum D(T) for an exponential distribution of the delay time

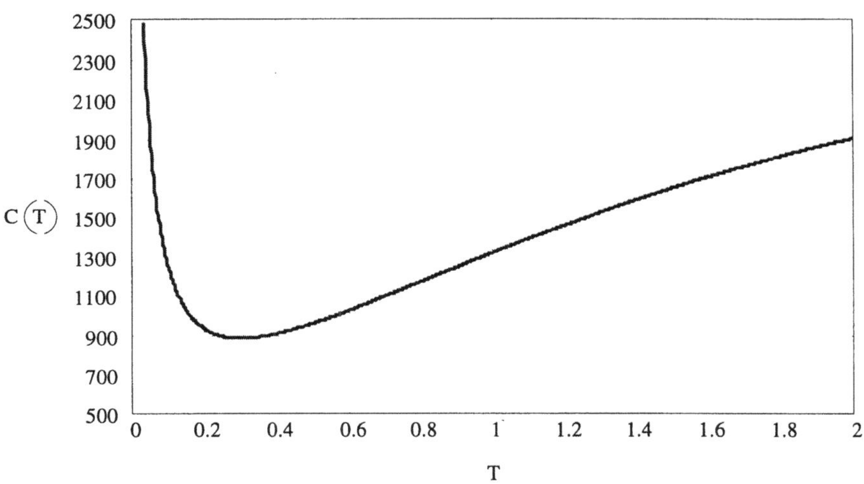

Figure 8.12 Optimal inspection period based on C(T)

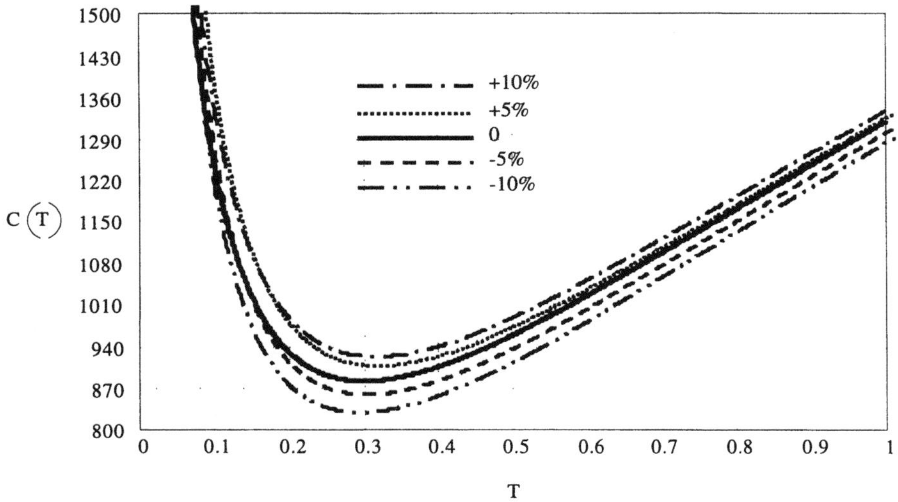

Figure 8.13 Sensitivity analysis for optimal inspection period based on C(T)

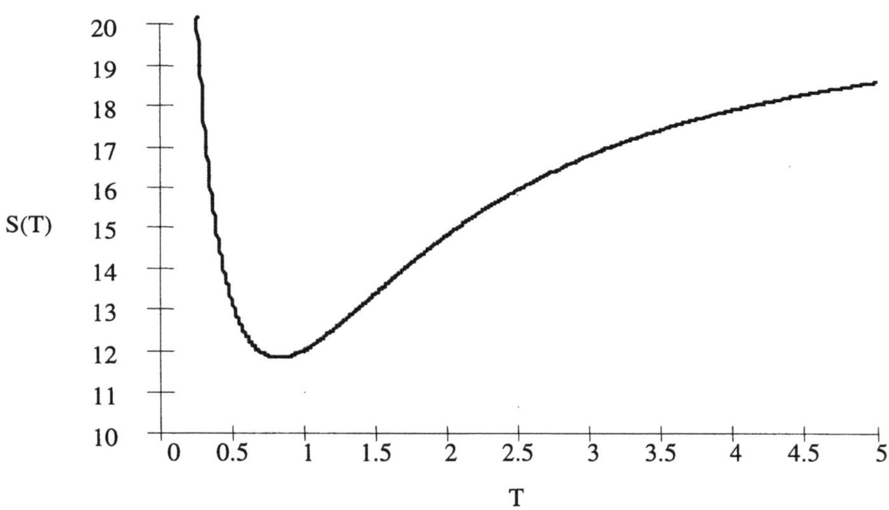

Figure 8.14 Optimal inspection period based on S(T)

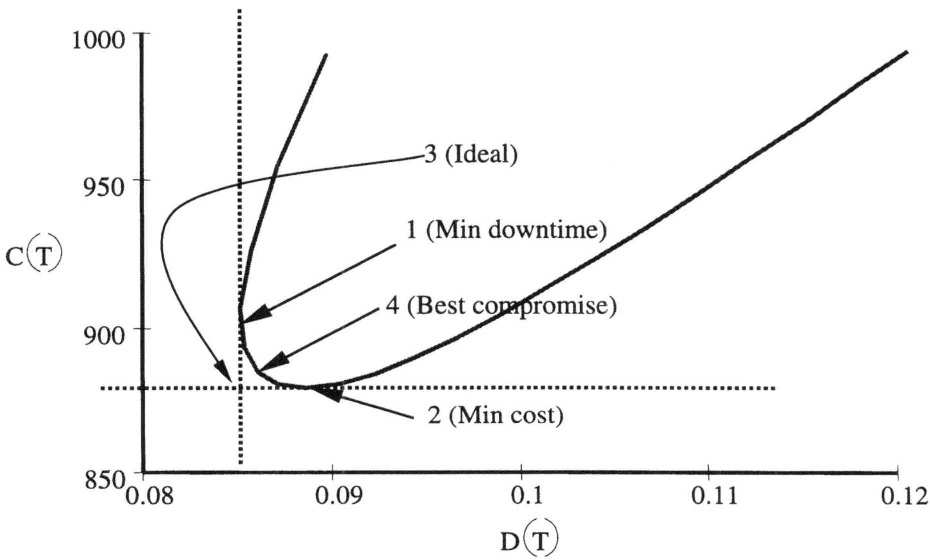

Figure 8.15 Expected cost C(T) against expected downtime D(T)

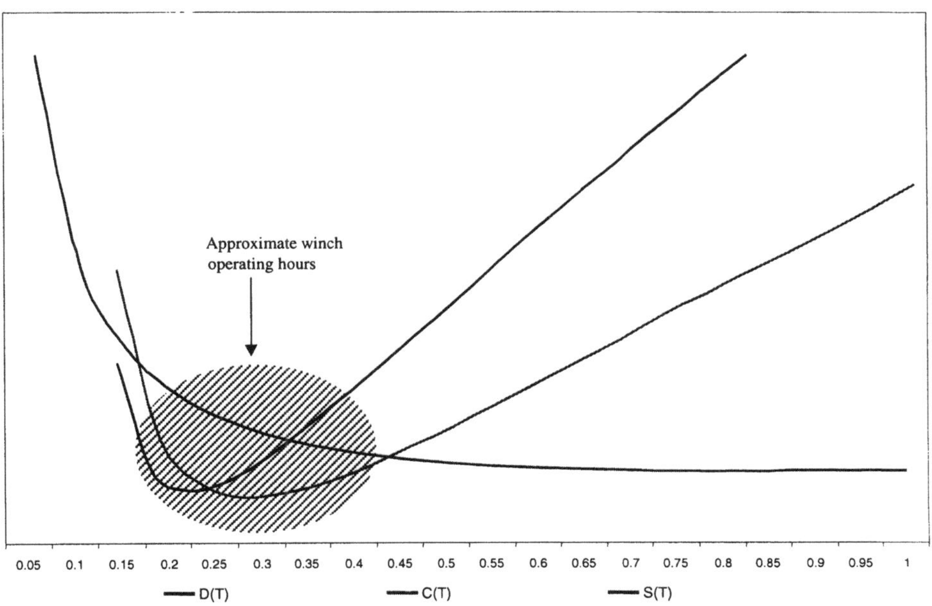

Figure 8.16 D(T), C(T) and S(T) against T

Chapter 9

Human Error Assessment and Decision Making Using Analytical Hierarchy Processing

Summary

A brief review of common human error assessment methods is presented highlighting the requirements and steps of each method. This is followed by an introduction to the Analytical Hierarchy Processing (AHP) method to aid decision-making. An approach to integrate human error assessment and decision-making using the AHP method is described. The aim of this approach is to reduce the occurrence probability and severity of human error during the operational phase of a fishing vessel. It utilises AHP theory to rank the impacts of human error and further integrates the available control options (to minimise these errors) within the analysis. The result obtained from the analysis reflects the most favoured control option that will address all the possible human errors within the system to a satisfactory level. A test case, which considers the shooting operation of a beam trawler, is used to demonstrate the described approach. Each step involved in the shooting operation is assessed for its vulnerability to human error with respect to the equipment being operated and this captures the operator-machine interaction.

Keywords: AHP, decision making, human error, human error assessment

9.1 Introduction

The cost of shipping casualties is normally expressed in terms of insurance values. The report of the Institute of London Underwriters (ILU) for 1995 stated that 95 ships were lost during the year (ILU (1996), ITSA (1996)). In 1996, the ILU recorded 1,190 lives lost at sea and the ship classification society Det Norske Veritas (DNV) estimated that accidents on board ships cost the industry around $US 10 billion a year (ILU (1996), IMO (1997)). It has been accepted that 80% of the accidents in the maritime industry is caused by human error. In the fishing industry, Lloyd's Register of World Fleet Statistics 1998 noted that the average age of the world fleet of fish catching vessels over 100 GRT was 20 years (ITWF (1999)). This could be a contributing factor to the high level of human error on these vessels. These older vessels lack automation and modern safety devices, hence the safe operation of the vessels is highly dependent on the competency of the crew on board.

Human error has played a critical role in the causes of many major marine accidents. The officers and crew of the *Herald of Free Enterprise* set to sea with their bow doors open (Sheen

(1987)). The crew and skipper of the Pescalanza and Sapphire did not close their watertight doors during heavy seas, which led to the sinking of the vessels by flooding (MAIB (2000)).

In these accidents, life, cargo and property had been lost due to the negligence and/or mistakes made by the operators of the system. Understanding errors and system failures are particularly important with respect to "high-consequence" systems. These are open systems whose behaviour has a significant effect not only on the system itself but also on the world outside the system. Hence, there is a need for an effective method to model the risks posed by human error in order to direct the limited resources to solutions that would reduce these risks.

9.2 Review of Human Error Assessment Methods

Engineers have developed a range of tools that can be used to represent and reason about the causes of major accidents (Leveson (1995)). For example, time-lines and fault trees have been recommended as analysis tools by a range of government and regulatory bodies. Unfortunately, these well-established techniques suffer from a number of limitations (Johnson (1998)). In particular, they cannot easily be used to represent and reason about the ways in which human errors and system failures interact during complex accidents (Hollnagel (1993)).

9.2.1 Methods for Quantification of Human Failures

Many methods for estimating human reliability were used in nuclear power plants. Such methods include confusion matrix (Fullwood and Hall (1988), Gertman and Blackman (1994)), expert estimation (Gertman and Blackman (1994)), Time Reliability Curve (TRC) (Dougherty and Fragola (1988), Moieni et al. (1994)), Maintenance Personnel Performance Simulation (MAPPS) (Fullwood and Hall (1988), Gertman and Blackman (1994)), Success Likelihood Index Method-Multi-Attribute Utility Decomposition (SLIM-MAUD) (Fullwood and Hall (1988), Gertman and Blackman (1994)), sociotechnical assessment of human reliability (Gertman and Blackman (1994)), Technique for Human Error Rate Prediction (THERP) (Dhillon (1986)), Sandia Recovery Model (SRM), INTENT (Gertman et al. (1992)) and Operator Reliability Calculation and Assessment (ORCA). Of the methods given, most deal with misdiagnosis or non-response errors and time dependent probability estimates. The most commonly used techniques are THERP, utilising generic Human Error Probabilities (HEP) from various industries, and SLIM-MAUD, using importance weightings from experts.

9.2.2 THERP

This method provides a mechanism for modelling as well as quantifying human error. It starts off with a task analysis that describes the tasks to be performed by the crew, maintainers or operators. Together with the task descriptions, Performance-Shaping Factors (PSF) such as stress and time available are collected to modify probabilities. The task analysis is then graphically represented in Human Reliability Assessment (HRA) event trees. The HEP for the activities of the task or the branches can be read and/or modified from the THERP tables as shown in (Gertman and Blackman (1994)). Details on the construction of HRA event trees and also the COGnitive EveNt Tree (COGENT) to represent cognitive activities and errors associated with human performance are also given in the book. Gertman and Blackman also provided a summary of the steps to THERP, which was adapted from the Nuclear Regulation-

NUREG/CR-1278 (Swain and Guttmann (1983)). THERP may suffer from the following limitations (Reason (1990), White (1995)):

1. It is difficult to represent the variability of human behaviour adequately.
2. The technique assumes that each task segment can be handled separately.
3. It is difficult to combine human and equipment reliability values.
4. It is difficult to identify inter-task dependencies.
5. The technique is not appropriate for continuous tasks.
6. The method does not determine motivation of the individual.
7. Analysts have the tendency to model only errors that appear in databases.

9.2.3 Accident Sequence Evaluation Programme (ASEP)

ASEP is a quicker version of THERP and is more conservative. It is a fine screening approach and can be complemented with THERP to warrant more detailed attention in the risk assessment. A more detailed discussion on ASEP can be found in (Swain (1987)).

9.2.4 SLIM-MAUD

The SLIM-MAUD method is centred on the assumption that the failure probability associated with task performance is based on a combination of PSFs that include the characteristics of the individual, the environment, and the task. It further assumes that experts can estimate these failure rates or provide anchor values to estimate them. A description on the steps to perform SLIM-MAUD was reported in (Gertman and Blackman (1994)) where two enhanced methods were developed. Dougherty and Fragola also provided the mathematics and an example for calculating SLIM-MAUD (Dougherty and Fragola (1988)). Davoudian et al provided an empirical evaluation of SLIM-MAUD and ranking to estimate HEPs, through the use of a simulated manufacturing environment under varying task conditions (Davoudian et al. (1994)).

9.2.5 Human Reliability Assessment (HRA)

HRA analyses the relationship between human behavioural tendencies and the work context to provide a better understanding in anticipating human errors, violations and severe system outcomes. This analysis requires a fundamental understanding of:

1. The way humans process information, including their capabilities and limitations at such processing (Wickens (1992)).
2. Human factors and ergonomics design consideration (Sanders and McCormick (1987)).
3. Skill, rule and knowledge based framework, which describes distinct levels of information processing at which workers perform (Rasmussen (1982, 1986)).
4. Psychosocial considerations that increase the likelihood of performing violations (CCPS (1994)).

The primary goals of HRA are to assess the risks attributable to human error and determine the ways of reducing system vulnerability due to human error impact. These goals are achieved by its three principal functions of identifying what errors can occur (human error identification), deciding how likely the errors are to occur (human error quantification), and, if appropriate, enhancing human reliability by reducing this error likelihood (human error reduction). The HRA process can be broken down into several steps as seen below:

Problem definition: This refers to deciding what human involvements are to be assessed (operators failing to deal with emergencies, operator's contribution to maintenance failures, etc.)

Task analysis: When the human aspect of the problem has been defined, task analysis can then define what human actions should occur in such events, as well as what equipment and other "interfaces" the operator should use. It may also identify what training (skills and knowledge) and procedures the operators will call upon.

Human Error Identification (HEI): Once the task analysis has been carried out, HEI then considers what can go wrong. The following types of errors are typically considered:

1. Error of omission - failing to carry out a required act.
2. Error of commission - failing to carry out a required act adequately; act performed without required precision, or with too much or too little force; act performed at wrong time; act performed in the wrong sequence.
3. Extraneous act - not required act performed instead of, or in addition to the required act.
4. Error-recovery opportunities - acts which can recover previous errors.

The HEI phase can identify many errors. Not all of these will be important for the study, as can be determined by reviewing their consequences on the system's performance. The ones that can contribute to a degraded system state, whether alone or in conjunction with other hardware/software failures or environmental events (or both together), must next be integrated into the risk analysis.

Representation: Having defined what the operator should do (via task analysis) and what can go wrong, the next step is to represent this information in a form which allows the quantitative evaluation of the human-error impact on the system to take place. It is usual for the human error impact to be seen in the context of other potential contributions to system risk. Human errors and recoveries are usually embedded within logical frameworks such as fault tree analysis and event tree analysis.

Human error quantification: Once the human error potential has been represented, the next step is to quantify the likelihood of the errors involved and then determine the overall effect of human error on the system safety and reliability. HEP is simply defined as "number of errors that occurred/number of opportunities for error".

Impact assessment: Once the errors have been quantified and represented in the risk assessment logic trees, the overall system risk level can be calculated. Then it can be determined whether or not the system has an acceptable level of risk. Impact assessments involve determining if the risk element is acceptable as well as which events (human, hardware, software or environmental - or any combination) contribute most to the level of risk. If the human error is a significant contributor to the overall risk at the system level, and if the system risk level is calculated to be too high, then the appropriate error will be targeted for reduction.

Error reduction analysis: Error reduction measures may be derived:

- According to the identified root causes of the error (from the error identification stage).
- From the defined factors that contribute to the HEP.

If error reduction is necessary to reduce the risk to an acceptable level, then following such error reduction measures, several iteration of impact assessments, error reduction and re-quantification may occur until satisfactory risk levels are achieved.

9.3 Human Error Probability

The analysis of many accidents has led to the appreciation that multiple equipment failures and process deviations combined with faulty human decisions and actions are often involved. Safety assessments, therefore, are not complete unless the interactions between equipment failures and human actions are considered. Since human behaviour is complex, and does not lend itself immediately to relatively straightforward reliability models, it is suggested that the following classifications of human interactions (that typically group all activities) need to be considered (Mahn et al. (1995)):

- Pre-initiator human interactions involving maintenance, testing, calibration, planning, etc.
- Initiators of accidents that involve operator awareness of potential accident initiators caused by errors in tests, or reconfiguration conditions involving control systems, protective logic, computer controlled functions and manual control.
- Post initiator interactions that involve procedure specified actions and recovery actions developed from training and experience.

These classifications of human interactions can be related to a simple error classification system consisting of three categories: (1) slips, (2) non-response, and (3) mistakes. This classification scheme can then be used to qualitatively incorporate human errors in accident scenarios. Table 9.1 provides generic human error probabilities for use in accident scenario assessment (Department of Energy (1996)).

The development of a generic set of human error probabilities is extremely difficult since there is a strong correlation on the actual person performing the task, the complexity of the task, the time required for task completion, and the training level of the person performing the task. Additionally, a worker may perform any specific task differently depending on the level of alertness due to fatigue or other factors.

A relatively simple model has been developed by Rasmussen to quantify human error rates based on the level of training (Rasmussen (1979, 1981)). This model divides the behaviour into three basic categories: skill-based, rule-based, and knowledge-based behaviours.

9.3.1 Skill-Based

Skill-based behaviours depend mostly on the operator's practice in performing the task. In short the operator can perform the task without ambiguity. A simplistic view is that skill-based errors are slips or lapses. These errors tend to be related to highly routine activities in familiar circumstances: omissions, repetitions, reversals, interference errors and double-capture slips. An example is incorrect use of foot pedal controls of fork-lift trucks. Some fork-lift trucks operate with three pedals (as a car), others have two pedals, reverse and forward. Removing a foot from either accelerator brings the vehicle to a halt. A common error is for the driver to press the backward accelerator in the belief (wrongly) that it is a brake pedal. Examples of slips and lapses include:

- Failing to disengage the gears before starting the engine (omission).
- Turning the ignition key to start the engine, when the engine is already running (repetition).
- Pressing the brake instead of the accelerator (reversal).

9.3.2 Rule-Based

Rule-based behaviour is at work when the operator does not have the same level of practice at performing the required task, but has a clear knowledge of the procedures. There may be some hesitation in recalling any procedure, the procedure may not be carried out in the proper sequence, or any step may not be performed precisely.

Rule-based errors are concerned with the misapplication or inappropriate use of problem solving rules. Individuals have a complex array of specific and general rules that they use to deal with everyday problems. Rules are of the type *if* <event> *then* <action>. Some simplistic examples relating to the operation of vehicles are:

- *if* <machine blockage> *then* <disengage power, switch off engine and investigate>
- *if* <pallet insecure> *then* <re-secure>
- *if* <towing a trailer on slopes> *then* <connect trailer brakes>

Sometimes the operator's rules are incomplete:

- *if* <emergency> *then* <apply handbrake, switch off engine, and dismount>

This is a perfectly good rule under most circumstances. However, with accidents involving contact with high voltage overhead lines, remaining in the cab provides protection against electrocution (principle of the Faraday Cage). A better additional rule would be:

- *if* <emergency involving electricity> *then* <stay in cab until supply isolated>.

The role of training in providing individuals with a set of safe rules is crucial.

9.3.3 Knowledge-Based

Knowledge-based action would include situations where the operator needs to contemplate the situation, interpret information or make a difficult decision. Also included in this grouping would be cases where a procedure is not well spelled out. In these cases the person performing the task must consider the actions to be taken and not act according to specific training.

Knowledge-based errors are concerned with performance in novel or new situations. Actions have to be planned "on-line" and the process is intellectually demanding. The problem solver will only resort to this type of activity when he has run out of rule-based solutions. An example of knowledge-based performance is that of first learning to operate a piece of machinery. The hydraulic controls of a winch provide a good example. Experimentation will help the operator to build a mental model of how the controls can be co-ordinated to achieve the desired movements. Eventually, the operator will adopt a set of rules derived from that mental model. With practice, the task will become skill-based. Training offers the opportunity to miss out the experimentation phase by guiding the trainee to a correct model of situations, based on the experiences of others.

Rasmussen provided per demand ranges and point estimates for these different categories (Rasmussen (1982)). These values are presented in Table 9.2. Swain and Guttmann suggested for screening purposes, the values of 0.05 and 1 are used for the rule-based and knowledge-based actions, respectively (Swain and Guttmann (1983)). However a value of 1 means 100% error rate for the knowledge-based action, a value that would appear to be unrealistically high.

One problem with the Rasmussen data is that it requires subjective analysis of the operator's training and capabilities. A set of human error rates was developed by Hunns for more specific tasks, not relying as much on the operator's capabilities and knowledge (Hunns (1982)). These data are presented in Table 9.3 and were based on extrapolation from human error rate databases. These data are similar to the rates of Rasmussen in Table 9.2 but provide some actual examples and do not require as much subjective analysis as the Rasmussen data.

The human error rates for some specific tasks have been provided by Dhillon and are presented in Table 9.4 (Dhillon (1986)). Dhillon points out that there are six basic categories of error sources that can eventually lead to an accident condition:

1. Operating errors
2. Assembly errors
3. Design errors
4. Inspection errors
5. Installation errors
6. Maintenance errors

Operating errors can be the result of:

1. Lack of proper procedures.
2. Task complexity and overload (of operator) conditions.
3. Poor personnel selection and training.
4. Operator carelessness and lack of interest.
5. Poor environmental conditions.
6. Departure from following correct operating procedures.

9.4 Analytical Hierarchy Processing

Analytical Hierarchy Processing (AHP) is a powerful and flexible decision making process to help set priorities and make the best decision when both qualitative and quantitative aspects of a decision need to be considered. By reducing complex decisions to a series of one-on-one comparisons, then synthesising the results, AHP not only helps decision-makers arrive at the best decision, but also provides a clear rationale that it is the best. Designed to reflect the way people actually think, AHP was developed more than 20 years ago by Dr. Thomas Saaty (Saaty (1980)), and continues to be the most highly regarded and widely used decision-making theory.

AHP is especially suitable for complex decisions, which involve the comparison of decision elements that are difficult to quantify. It is based on the assumption that when faced with a complex decision the natural human reaction is to cluster the decision elements according to their common characteristics. It involves building a hierarchy (ranking) of decision elements and then making comparisons between each possible pair in each cluster (as a matrix). This gives a weighting for each element within a cluster (or level of the hierarchy).

The AHP engages decision-makers in breaking down a decision into smaller parts, proceeding from the goal to criteria to sub-criteria down to the alternative courses of action. Decision-makers then make simple pair-wise comparison judgements throughout the hierarchy to arrive at overall priorities for the alternatives.

The literature survey on AHP indicates that the method has been effective to a wide range of applications. These include agricultural applications (Alho and Kangas (1997), Braunschweig (2000)), industrial engineering applications (Alidi (1996), Bhattarai and Fujiwara (1997)) and financial applications (Hachadorian (1987), Gerrits et al. (1994)). The application of AHP theory to ascertain business and financial risk has been relatively popular in the past (Jensen (1987a, b), Nezhad (1988), Simkin et al. (1990)). It has also found its place in risk and safety assessment of engineering systems (Shields and Silcock (1986), Saaty (1987), Hamalainen and Karjalainen (1989), Shields et al. (1990), Hamalainen and Karjalainen (1992), Frank (1995).

9.4.1 Principles and Background of AHP

When considering a group of activities (factors) for evaluation, the main objectives of this group are (Saaty (1990)):

1. To provide judgement on the relative importance of these activities.
2. To ensure that the judgements are quantified to an extent which also permits a quantitative interpretation of the judgement among these activities (factors).

The quantified judgements on pairs of activities C_i and C_j are represented by an n-by-n matrix.

$$A = (a_{ij}) \text{ where } i, j = 1, 2, \ldots, n. \tag{9.1}$$

The entries a_{ij} are defined by the following entry rules:

Rule 1. If $a_{ij} = \alpha$, then $a_{ji} = 1/\alpha$, $\alpha \neq 0$.

Rule 2. If C_i is judged to be of equal relative importance as C_j , then $a_{ij} = a_{ji} = 1$. Obviously $a_{ii} = 1$ for all i. Thus the matrix A has the following form:

$$A = \begin{bmatrix} 1 & a_{12} & \cdots\cdots & a_{1n} \\ 1/a_{12} & 1 & \cdots\cdots & a_{2n} \\ \cdot & \cdot & \cdots\cdots & \cdot \\ 1/a_{1n} & 1/a_{2n} & \cdots\cdots & 1 \end{bmatrix} \tag{9.2}$$

where each a_{ij} is the relative importance of activity i to activity j. Having recorded the quantified judgements of comparisons on pair (C_i, C_j) as numerical entry a_{ij} in the matrix A, what is left is to assign to the n contingencies C_1 , C_2 , C_3 , ..., C_n a set of numerical weights w_1, w_2, w_3, ..., w_n that should reflect the recorded judgements. The eigenvector of the comparison matrix provides the priority ordering (weight), and the eigenvalue is a measure of consistency. To find the priority vector or the weight of each factor included in the priority ranking analysis, the eigenvector corresponding to the maximum eigenvalue is to be determined from matrix analysis. One of the approximation methods to get the weight of each factor in the pair-wise comparison process is described below.

9.4.2 Weight Vector Calculation

In mathematical terms, the principal eigenvector is computed, and when normalised becomes the vector of priorities (weights). To reduce the excessive computing time needed to solve the problem exactly, and due to the results of complex numbers, a good estimate of that vector can be obtained by dividing the elements of each column in the comparison matrix by the sum of that column (i.e. normalise the column). The elements in each resulting row are added and the sum is divided by the number of the elements in the row. This is a process of averaging over the normalised columns. Mathematically, the equation for calculating w_1 is shown below:

$$w_1 = \frac{1}{n}\left[\left(\frac{a_{11}}{\sum_{i=1}^{n} a_{i1}}\right) + \left(\frac{a_{12}}{\sum_{i=1}^{n} a_{i2}}\right) + \cdots\cdots \left(\frac{a_{1n}}{\sum_{i=1}^{n} a_{in}}\right)\right] \tag{9.3}$$

In general, weights w_1, w_2, w_3, ..., w_n can be calculated using the following equation:

$$w_k = \frac{1}{n}\sum_{j=1}^{n}\left(\frac{a_{kj}}{\sum_{i=1}^{n} a_{ij}}\right) \; (k = 1, \;\ldots\ldots\; n) \tag{9.4}$$

where a_{ij} is the entry of row i and column j in a comparison matrix of order n.

9.4.3 Risk and AHP

Risks are by nature subjective, therefore, the AHP method may be suited for risk assessment in many situations. This technique allows subjective and objective factors to be considered in risk analysis and also provides a flexible and easily understood way to annualise subjective risk factors. The elements in each level are compared pair-wise with respect to their importance in making the decision under consideration. The verbal scale used in AHP enables the decision-maker to incorporate subjectivity, experience and knowledge in an intuitive and natural way.

After the comparison matrices have been created, the process moves on to the phase in which relative weights are derived for the various elements. The relative weights of the elements of each level with respect to an element in the adjacent upper level are computed as the components of the normalised eigenvector associated with the largest eigenvalue of their comparison matrix. The composite weights of the decision alternatives are then determined by aggregating the weights through the hierarchy. This is done by following a path from the top of the hierarchy to each alternative at the lowest level, and multiplying the weights along each segment of the path. The outcome of this aggregation is a normalised vector of the overall weights of the options. The mathematical basis for determining the weights has been established by Saaty (Saaty (1980)).

9.4.4 AHP for Human Error Assessment and Decision Making for Ship Operations

Several methods to quantify human error probability have been reviewed in Section 9.2. These methods suffer from the difficulty associated with any attempt to construct quantitative, predictive models of human behaviour. The qualitative methods on the other hand, require multi-disciplinary teams to carry out an analysis and this is regarded as being resource intensive. The more recent HRA methods have included cognitive aspects of decision making and the "time" dimension. However, it has not yet captured the fundamental nature of the interaction between actions and machine responses (Cacciabue et al. (1993)). These interactions lie in the mutual dynamic influence of the operator, the plant and the interfaces.

The use of AHP to evaluate human error on ship operations does not ignore small events or operations that are normally rationalised and eliminated as being not important in traditional methods. A chain of these small rationalisations results in a larger problem later. The AHP method looks at every event/operation and ranks them against each other to determine the importance of each event/operation over the other (without eliminating them from the analysis).

The use of the AHP method enables the solutions for each possible human error identified, to be integrated within the analysis. This is unlike the methods reviewed in Section 9.2, where the solutions to reduce the risk levels (posed by human errors) are evaluated in the first instance, and then a re-iteration of the whole analysis is performed (assuming the implementation of the solution) to confirm the risk reduction. An approach using the AHP method for human error assessment and decision making applied to ship operations is presented in Section 9.5.

9.5 Application of AHP to Vessel Operations

The flowchart in Figure 9.1 illustrates the steps involved in carrying out the application of AHP to vessel operations (Pillay (2001), Pillay and Wang (2001)). This approach can be executed in the following seven distinct steps:

1. Describe system - The system or operation under consideration is described in detail, highlighting all the equipment within the system that will be operated to achieve the desired objective of the defined operation.
2. Identify tasks to be carried out - Identify all tasks that are to be carried out to achieve the objective of the operation and present these tasks in the order that they should be carried out. This information can be represented by means of a flowchart. The order by which the tasks are carried out should reflect the normal safe working procedure of the system. To

enable effective use of this information in the AHP phase, all tasks are annotated according to the equipment that are operated.

3. Determine operator behaviour - For each of the tasks identified in Step 2, determine the required operator's behaviours. Three types of behaviours are considered namely, skill-based, rule-based or knowledge-based behaviour. These behaviours are discussed in Sections 9.3.
4. Determine the probability of occurrence - Using a generic database, determine the probability that a human error might occur while carrying out the task specified in Step 2. Use the information developed in Step 3 to assign the probability of occurrence of the human error.
5. Determine the severity of occurrence - The severity of a human error should take into account the consequences of the error on the system, operation, environment and operator. This can be quantified in monetary terms or downtime.
6. Determine Risk Control Options (RCOs) - Considering the system/operation under study, determine several options that could address the risks estimated (associated with each task defined in Step 2).
7. AHP analysis - Using the data gathered in Steps 2, 4, 5 and 6, carry out the AHP analysis to determine the most favourable RCO. This RCO will address all the risks associated with tasks where human errors could manifest.

Step 7 (AHP analysis) involves 4 distinct steps, which are described below:

(a) Set-up - Decision making criteria are generated, often by brainstorming or past experience. Hierarchical relationships are drawn between the criteria and are then represented in a matrix form.

(b) Weighting - The matrices are filled with the criteria comparisons. The comparisons allow calculation of the criteria-weighting vector.

(c) Ranking - The different RCOs are ranked on their ability to satisfy the various criteria.

(d) Evaluation - The final solution ratings are then calculated using the ratings determined in step (c) and the weighting vector calculated in step (b).

The first task is to decide on the problem statement. This statement becomes the goal of the hierarchy (Level One) and will be broken down into nested levels (Level Two). Level Two will comprise the different elements needed to be considered to achieve the goal set in the problem statement. The elements in Level Two are further broken-down to represent the various constituents that make up or belong to each specific element. The hierarchical structure is assumed to exist inherently in the problem considered and can be identified.

The hierarchy records the flow of detail from the problem statement (Goal) to broad issues (Level Two) and more specific levels (Level Three). While the concerns on a particular level are not equally important, they should be on the same order of magnitude. This feature in AHP allows decisions to be made involving different orders of magnitude criteria, by placing each criterion in its proper matrix in the objective hierarchy. Figure 9.2 shows an example of the hierarchy represented diagrammatically.

Once the hierarchy has been completed, matrices are constructed with the criteria labels on each axis. There will be one Level Two matrix and a number of associated matrices for the sub-elements of each element. For example, Figure 9.2 will have one Level Two matrix and three Level Three matrices. These Level Three matrices may be broken down in finer detail where applicable. The two axes of a matrix will contain the names of the elements on the level being considered. For example, the Level Two matrix in Figure 9.2 will have the form shown in Table 9.5. The elements below Level Two would also be represented in a matrix form. Table 9.6 shows an example for Element 1 (constituent *A*) matrix. The complete representation of Element 1 would comprise three matrices (as Element 1 has the constituents *A, B* and *C*).

As the described method does not use historical data (probability of occurrence in terms of hard numbers or severity in terms of number of deaths), the uncertainty in these parameters is captured by representing them in terms of preference/importance against each other. Hence, the analysis is targeted at improving the current situation by identifying the areas that need improving, rather than trying to quantify the occurrence likelihood/severity of an undesired event.

Upon generating the matrices for all the elements, it must now be filled with the comparisons of the relative importance of the elements on the two axes. The comparisons are used to calculate the weighting vector that will give the relative importance of all the elements. The entire weighting vector is calculated from comparisons made between just two elements at a time. Table 9.7 shows the scale (1 to 9) proposed by (Saaty (1980)) for indicating the relative importance between the elements.

Considering the example of the Level Two matrix in Table 9.5, and assuming that Element 1 is weakly more important than Element 2 and strongly more important than Element 3. Then, the matrix in Table 9.5 may be represented as seen in the matrix below:

$$\mathit{Level\ Two} = \begin{bmatrix} 1 & 3 & 7 \\ 1/3 & 1 & 7/3 \\ 1/7 & 3/7 & 1 \end{bmatrix}$$

In the matrix, Element 1 is of equal importance with respect to itself, so 1 is placed in the upper left-hand corner. A consistent matrix formulation allows the remainder of the matrix to be completed given the information in the top row. Since the relationship is known between Element 1 and Element 2, and Element 1 and Element 3, the relationship between Element 2 and Element 3 can be determined. In this case the matrix entry for Element 2 versus Element 3 would contain 7/3.

The weighting vector is then determined to give the percentage of the total weight applied to each element. The first column in the Level Two matrix, (1, 1/3, 1/7) is normalised so that the sum of the entries is 1.0. The weighting of Element 1 will be given as 1/(1+1/3+1/7) = 0.677 or 67.7%. Similarly Elements 2 and 3 can be calculated to be 22.6% and 9.78%. The normalised weighting vector for Elements 1, 2 and 3 is $[0.667\ 0.226\ 0.097]^T$. The sum of all three weightings is equal to 100%.

The comparison process is repeated for all the matrices to be used in the analysis. The weighting vectors of the lower matrices will be normalised so that their total weight will equal that of the previous level (Level Two). For example, for Element 1, sub-elements A_1, A_2, A_3, B_1, B_2, C_1, C_2 and C_3 will be given a total weight of 67.7%. All sub-elements are analysed in the same fashion to the lowest level possible and the results are normalised to reflect the weight of each sub-element in the hierarchy.

The next step is to generate the possible solutions to achieve the problem statement/goal. Each solution is compared against each of the lowest level sub-elements. The possible solutions are assumed to reduce the likelihood of human error occurring and/or the possible consequences. The evaluation represents the "effectiveness" of the solution in controlling the risks. These evaluations (of the solutions) are recorded with a user defined numerical scale, as appropriate for the sub-elements. For any given element, a normalised score is determined for each solution by taking the assigned score (which may have units) and dividing it by the sum of the assigned scores across all of the solutions. This fraction is then multiplied by the weighting coefficient for the element. This will give a normalised score for each solution based on the element considered. These normalised results are then summed for the different elements in the matrix, to arrive at a final rating for each solution. The result of this series of operations is a weighted rating for each solution. The highest rated solution will best meet the problem statement (goal).

9.6 An Example

The purpose of this analysis is to address the high level of human errors that occur during the fishing operation on board fishing vessels. As an example, the initial shooting operation of the fishing nets is considered.

9.6.1 Initial Shooting Operation

At the commencement of the voyage the beams are stowed port and starboard alongside and inboard of the bulwark rails. The cod ends are held by the Gilson wires up at the cod end lifting blocks with the netting hanging down to the chain mat that is beneath the beam. As soon as the vessel clears the harbour, the derricks are lowered to an angle of approximately 45 degrees. This reduces the top weight on the vessel, improving stability, but importantly, it is to prevent the derricks from moving past vertical and falling inboard as the vessel rolls.

On reaching the fishing grounds, the vessel stops. Working one side at a time, the derrick is sufficiently raised to lift the beam and chain mat up and over the rail. The derrick is then lowered back to 45 degrees on completion of the manoeuvre. While the cod ends are held by Gilson wire over the lifting block, the netting is paid overboard.

Attached between the inboard end of the beam and the cod end lifting becket is a heavy rope, referred to as a 'lazy decky'. This rope is pulled to swing the beam around to bring it normal to the vessel side. The vessel moves ahead slowly and the Gilson wires are lowered slightly, sufficient to allow the cod ends to be swung over the rail, but still with the Gilson hooks attached in the lifting becket. The weight is carried by the cod ends lifting blocks. The crew, on each side, then takes the 'lazy decky' and makes it fast on a bulwark rail pin such that the weight of cod end is carried by the pin. The remainder of the 'lazy decky' lies in a bight on the deck up to the point where it goes over the rail to hang in a bight between the vessel and the beam. Once the weight of the cod end has been transferred to the rail pins, the Gilson hooks are released, the derricks are lowered fully outboard, and the vessel is brought up to speed. When the crew in the wheelhouse, either the skipper or the mate, is satisfied that the vessel is running straight and true, he signals to the crewman to release the 'lazy decky' ropes from the rail pins. The cod ends then stream astern with the netting stretched out. Warp is then paid out, typically 200 fathom for sea depth of 40 fathom. Due to the double purchase around the block on the beam, 200 fathoms of warp is in effect a 100 fathom pay-out giving a typical warp/depth ratio of 2.5:1. The complete initial shooting operation can be represented diagrammatically as seen in Table 9.7.

9.6.2 Hierarchy Set-Up

The various tasks identified above are used to set up the hierarchy of elements. The goal of the analysis is determined to be the safe initial shooting operation. The elements in Level Two are set to be the probability of a human error occurring and the severity of human error. The sub-elements (Level Three) are determined by grouping the equipment that are operated i.e. derrick, vessel, lazy decky, net and Gilson. Each task carried out in relation to these equipment is considered within this level i.e. derrick 1, derrick 2, derrick 3, etc. The hierarchy can be represented diagrammatically as seen in Figure 9.4.

9.6.3 Level Two Matrix

The probability of occurrence and severity make up the two elements in Level Two as seen in Figure 9.4. These two elements are compared against each other to determine the weighting vector of each element. The comparison scale in Table 9.5 is used to determine the importance of the two elements. Considering the goal of the analysis, it is decided that both these elements are equally important to a safety assessment, hence, the Level Two matrix is determined as:

$$Level\,Two = \begin{bmatrix} 1 & 1 \\ 1 & 1 \end{bmatrix}, \text{ and the } Weighting\,Vector = \begin{bmatrix} 0.5 \\ 0.5 \end{bmatrix}$$

9.6.4 Human Error Probability Evaluation

First, the importance of each element (derrick, vessel, lazy decky, net and Gilson) is determined. Using the comparison scale in Table 9.7, the matrix below is obtained for the probability importance of each element.

$$\textbf{Probability} = \begin{bmatrix} 1.00 & 7.00 & 3.00 & 9.00 & 5.00 \\ 0.14 & 1.00 & 0.43 & 1.29 & 0.71 \\ 0.33 & 2.33 & 1.00 & 3.00 & 1.67 \\ 0.11 & 0.78 & 0.33 & 1.00 & 0.56 \\ 0.20 & 1.40 & 0.60 & 1.81 & 1.00 \end{bmatrix}$$

The weighting vector and normalised vector are determined by considering the weighting vector obtained in the Level Two matrix and are shown below:

$$Weighting\,Vector = \begin{bmatrix} 0.5595 \\ 0.0799 \\ 0.1865 \\ 0.0622 \\ 0.1119 \end{bmatrix} \text{ and } Normalised\,Vector = \begin{bmatrix} 0.2798 \\ 0.04 \\ 0.0933 \\ 0.0311 \\ 0.056 \end{bmatrix}$$

The probability of human error is considered for each of the task carried out by determining the type of human behaviour required to carry out the task successfully. Using the generic human error data by Rasmussen (Table 9.2), each task is assigned the operator behaviour and the generic error probability. This data is then used to compare each task against the others to determine the Level Three matrix. The various tasks identified in this example and the associated generic data are provided in Table 9.8.

The matrices for the probability of occurrence for each task are determined as follows:

$$Derrick = \begin{bmatrix} 1.00 & 1.00 & 1.00 & 0.11 \\ 1.00 & 1.00 & 1.00 & 0.11 \\ 1.00 & 1.00 & 1.00 & 0.11 \\ 9.00 & 9.00 & 9.00 & 1.00 \end{bmatrix}$$

$$Weighting\ Vector = \begin{bmatrix} 0.0833 \\ 0.0833 \\ 0.0833 \\ 0.7500 \end{bmatrix}, \quad Normalised\ Vector = \begin{bmatrix} 0.0233 \\ 0.0233 \\ 0.0233 \\ 0.2098 \end{bmatrix}$$

$$Vessel = \begin{bmatrix} 1.00 & 5.00 & 1.00 & 9.00 \\ 0.20 & 1.00 & 0.20 & 1.80 \\ 1.00 & 5.00 & 1.00 & 9.00 \\ 0.11 & 0.56 & 0.11 & 1.00 \end{bmatrix}$$

$$Weighting\ Vector = \begin{bmatrix} 0.4327 \\ 0.0865 \\ 0.4327 \\ 0.0481 \end{bmatrix}, \quad Normalised\ Vector = \begin{bmatrix} 0.0173 \\ 0.0035 \\ 0.0173 \\ 0.0019 \end{bmatrix}$$

$$Lazy\ Decky = \begin{bmatrix} 1.00 & 1.00 & 1.00 \\ 1.00 & 1.00 & 1.00 \\ 1.00 & 1.00 & 1.00 \end{bmatrix}$$

$$Weighting\ Vector = \begin{bmatrix} 0.3333 \\ 0.3333 \\ 0.3333 \end{bmatrix}, \quad Normalised\ Vector = \begin{bmatrix} 0.0311 \\ 0.0311 \\ 0.0311 \end{bmatrix}$$

$$Net = \begin{bmatrix} 1.00 & 1.00 \\ 1.00 & 1.00 \end{bmatrix}$$

$$Weighting\ Vector = \begin{bmatrix} 0.50 \\ 0.50 \end{bmatrix}, \quad Normalised\ Vector = \begin{bmatrix} 0.0155 \\ 0.1155 \end{bmatrix}$$

$$Gilson = \begin{bmatrix} 1.00 & 0.20 \\ 5.00 & 1.00 \end{bmatrix}$$

$$Weighting\ Vector = \begin{bmatrix} 0.1667 \\ 0.8333 \end{bmatrix},\ Normalised\ Vector = \begin{bmatrix} 0.0093 \\ 0.0466 \end{bmatrix}$$

9.6.5 Human Error Severity Evaluation

The importance of each element (derrick, vessel, lazy decky, net and Gilson) is determined using the comparison scale in Table 9.7. The matrix below is obtained for the severity importance of each element.

$$Severity = \begin{bmatrix} 1.00 & 7.00 & 3.00 & 9.00 & 5.00 \\ 0.14 & 1.00 & 0.43 & 1.29 & 0.71 \\ 0.33 & 2.33 & 1.00 & 3.00 & 1.67 \\ 0.11 & 0.78 & 0.33 & 1.00 & 0.56 \\ 0.20 & 1.40 & 0.60 & 1.81 & 1.00 \end{bmatrix}$$

The weighting vector and normalised vector are determined by considering the weighting vector obtained in the Level Two matrix and are shown as follows:

$$Weighting\ Vector = \begin{bmatrix} 0.5595 \\ 0.0799 \\ 0.1865 \\ 0.0622 \\ 0.1119 \end{bmatrix} \text{ and } Normalised\ Vector = \begin{bmatrix} 0.2798 \\ 0.0400 \\ 0.0933 \\ 0.0311 \\ 0.0560 \end{bmatrix}$$

The matrices for the severity of the consequences of human error for each task are determined as follows:

$$Derrick = \begin{bmatrix} 1.00 & 7.00 & 3.00 & 5.00 \\ 0.14 & 1.00 & 0.43 & 0.71 \\ 0.33 & 2.33 & 1.00 & 1.67 \\ 0.20 & 1.40 & 0.60 & 1.00 \end{bmatrix}$$

$$Weighting\ Vector = \begin{bmatrix} 0.5966 \\ 0.0852 \\ 0.1989 \\ 0.1193 \end{bmatrix},\ Normalised\ Vector = \begin{bmatrix} 0.1669 \\ 0.0238 \\ 0.0556 \\ 0.0334 \end{bmatrix}$$

$$Vessel = \begin{bmatrix} 1.00 & 0.20 & 0.33 & 0.11 \\ 5.00 & 1.00 & 1.67 & 0.56 \\ 3.00 & 0.60 & 1.00 & 0.33 \\ 9.00 & 1.80 & 3.00 & 1.00 \end{bmatrix}$$

$$Weighting\ Vector = \begin{bmatrix} 0.0556 \\ 0.2778 \\ 0.1667 \\ 0.5000 \end{bmatrix},\ Normalised\ Vector = \begin{bmatrix} 0.0022 \\ 0.0111 \\ 0.0067 \\ 0.0200 \end{bmatrix}$$

$$Lazy\ Decky = \begin{bmatrix} 1.00 & 0.11 & 0.33 \\ 9.00 & 1.00 & 3.00 \\ 3.00 & 0.33 & 1.00 \end{bmatrix}$$

$$Weighting\ Vector = \begin{bmatrix} 0.0769 \\ 0.6923 \\ 0.2308 \end{bmatrix},\ Normalised\ Vector = \begin{bmatrix} 0.0072 \\ 0.0646 \\ 0.0215 \end{bmatrix}$$

$$Net = \begin{bmatrix} 1.00 & 5.00 \\ 0.20 & 1.00 \end{bmatrix}$$

$$Weighting\ Vector = \begin{bmatrix} 0.8333 \\ 0.1667 \end{bmatrix},\ Normalised\ Vector = \begin{bmatrix} 0.0259 \\ 0.0052 \end{bmatrix}$$

$$Gilson = \begin{bmatrix} 1.00 & 0.11 \\ 9.00 & 1.00 \end{bmatrix}$$

$$Weighting\ Vector = \begin{bmatrix} 0.1000 \\ 0.9000 \end{bmatrix},\ Normalised\ Vector = \begin{bmatrix} 0.0056 \\ 0.0504 \end{bmatrix}$$

9.6.6 Risk Control Options (RCO)

Several viable Risk Control Options (RCO) are generated in order to reduce the level of risks posed by human errors during the initial shooting operation. These risk control options are evaluated for their effectiveness against each of the operator tasks identified. For this example, an arbitrary scale (1 to 10) is used to compare each RCO, 1 being not effective and 10 being most effective. When assigning a score on scale 1 to 10, several factors are considered, such as cost, ease of implementation, efficiency, time before solution becomes effective, etc. Six RCOs have been identified to reduce the probability and severity of human errors of the initial shooting operation. These RCOs include:

RCO 1 - Training of crew

RCO 2 - Redesign system

RCO 3 - Incorporate additional interlocks

RCO 4 - Change operating procedures

RCO 5 - Additional crewing

RCO 6 - Install warning devices (audio and visual alarms, indications, etc.)

The matrices for the effectiveness of each RCO in reducing the probability of occurrence are presented in the form as seen in Table 9.9. Similarly all tasks are compared with the different RCOs.

9.6.7 RCO Evaluation to Reduce Probability of Occurrence

$$Derrick = \begin{bmatrix} 6 & 1 & 7 & 3 & 4 & 9 \\ 6 & 2 & 8 & 5 & 1 & 7 \\ 6 & 1 & 7 & 3 & 4 & 9 \\ 6 & 6 & 9 & 7 & 1 & 9 \end{bmatrix}$$

$$Normalised\ results = \begin{bmatrix} 0.0047 & 0.0008 & 0.0054 & 0.0023 & 0.0031 & 0.0070 \\ 0.0048 & 0.0016 & 0.0064 & 0.0040 & 0.0008 & 0.0056 \\ 0.0047 & 0.0008 & 0.0054 & 0.0023 & 0.0031 & 0.0070 \\ 0.0331 & 0.0331 & 0.0497 & 0.0386 & 0.0055 & 0.0497 \end{bmatrix}$$

$$Vessel = \begin{bmatrix} 8 & 1 & 3 & 2 & 4 & 7 \\ 7 & 1 & 2 & 2 & 4 & 7 \\ 7 & 1 & 3 & 2 & 4 & 7 \\ 9 & 5 & 3 & 2 & 6 & 8 \end{bmatrix}$$

$$Normalised\ results = \begin{bmatrix} 0.0055 & 0.0007 & 0.0021 & 0.0014 & 0.0028 & 0.0048 \\ 0.0011 & 0.0002 & 0.0003 & 0.0003 & 0.0006 & 0.0011 \\ 0.0050 & 0.0007 & 0.0022 & 0.0014 & 0.0029 & 0.0050 \\ 0.0005 & 0.0003 & 0.0002 & 0.0001 & 0.0003 & 0.0005 \end{bmatrix}$$

$$Lazy\ Decky = \begin{bmatrix} 4 & 3 & 7 & 2 & 6 & 9 \\ 5 & 2 & 8 & 3 & 6 & 8 \\ 6 & 4 & 8 & 3 & 7 & 8 \end{bmatrix}$$

$$Normalised\ results = \begin{bmatrix} 0.0040 & 0.0030 & 0.0070 & 0.0020 & 0.0060 & 0.0090 \\ 0.0049 & 0.0019 & 0.0078 & 0.0029 & 0.0058 & 0.0078 \\ 0.0052 & 0.0035 & 0.0069 & 0.0026 & 0.0060 & 0.0069 \end{bmatrix}$$

$$Net = \begin{bmatrix} 2 & 2 & 8 & 3 & 4 & 10 \\ 3 & 3 & 7 & 3 & 4 & 10 \end{bmatrix}$$

$$Normalised\ results = \begin{bmatrix} 0.0011 & 0.0011 & 0.0043 & 0.0016 & 0.0021 & 0.0054 \\ 0.0016 & 0.0016 & 0.0036 & 0.0016 & 0.0021 & 0.0052 \end{bmatrix}$$

$$Gilson = \begin{bmatrix} 5 & 3 & 6 & 5 & 6 & 7 \\ 7 & 3 & 10 & 6 & 6 & 10 \end{bmatrix}$$

$$Normalised\ results = \begin{bmatrix} 0.0015 & 0.0009 & 0.0017 & 0.0015 & 0.0017 & 0.0020 \\ 0.0078 & 0.0033 & 0.0111 & 0.0067 & 0.0067 & 0.0111 \end{bmatrix}$$

9.6.8 RCO Evaluation to Reduce Severity of Possible Consequences

$$Derrick = \begin{bmatrix} 6 & 1 & 5 & 5 & 4 & 7 \\ 6 & 2 & 7 & 5 & 4 & 7 \\ 6 & 1 & 5 & 6 & 4 & 7 \\ 6 & 4 & 6 & 7 & 5 & 9 \end{bmatrix}$$

$$Normalised\ results = \begin{bmatrix} 0.0358 & 0.0060 & 0.0298 & 0.0298 & 0.0238 & 0.0417 \\ 0.0046 & 0.0015 & 0.0054 & 0.0038 & 0.0031 & 0.0054 \\ 0.0115 & 0.0019 & 0.0096 & 0.0115 & 0.0077 & 0.0134 \\ 0.0054 & 0.0036 & 0.0054 & 0.0063 & 0.0045 & 0.0081 \end{bmatrix}$$

$$Vessel = \begin{bmatrix} 5 & 1 & 7 & 5 & 4 & 7 \\ 5 & 1 & 7 & 5 & 4 & 7 \\ 5 & 1 & 7 & 5 & 4 & 7 \\ 9 & 5 & 8 & 7 & 6 & 8 \end{bmatrix}$$

$$Normalised\ results = \begin{bmatrix} 0.0004 & 0.0001 & 0.0005 & 0.0004 & 0.0003 & 0.0005 \\ 0.0019 & 0.0004 & 0.0027 & 0.0019 & 0.0015 & 0.0027 \\ 0.0011 & 0.0002 & 0.0016 & 0.0011 & 0.0009 & 0.0016 \\ 0.0042 & 0.0023 & 0.0037 & 0.0033 & 0.0028 & 0.0037 \end{bmatrix}$$

$$Lazy\ Decky = \begin{bmatrix} 4 & 3 & 7 & 6 & 6 & 7 \\ 5 & 2 & 8 & 6 & 6 & 8 \\ 6 & 4 & 8 & 7 & 7 & 8 \end{bmatrix}$$

$$Normalised\ results = \begin{bmatrix} 0.0009 & 0.0007 & 0.0015 & 0.0013 & 0.0013 & 0.0015 \\ 0.0092 & 0.0037 & 0.0148 & 0.0111 & 0.0111 & 0.0148 \\ 0.0032 & 0.0022 & 0.0043 & 0.0038 & 0.0038 & 0.0043 \end{bmatrix}$$

$$Net = \begin{bmatrix} 2 & 2 & 8 & 6 & 5 & 8 \\ 3 & 3 & 7 & 6 & 5 & 8 \end{bmatrix}$$

$$Normalised\ results = \begin{bmatrix} 0.0017 & 0.0017 & 0.0067 & 0.0050 & 0.0042 & 0.0067 \\ 0.0005 & 0.0005 & 0.0011 & 0.0010 & 0.0008 & 0.0013 \end{bmatrix}$$

$$Gilson = \begin{bmatrix} 5 & 3 & 6 & 5 & 6 & 7 \\ 7 & 3 & 7 & 6 & 6 & 8 \end{bmatrix}$$

$$Normalised\ results = \begin{bmatrix} 0.0009 & 0.0005 & 0.0010 & 0.0009 & 0.0010 & 0.0012 \\ 0.0095 & 0.0041 & 0.0095 & 0.0082 & 0.0082 & 0.0109 \end{bmatrix}$$

9.6.9 Summary of Results

The results obtained from Sections 9.6.7 and 9.6.8 are collated to determine the best RCO. Tables 9.10 and 9.11 show the summary of these results obtained in percentage. There are several tasks carried out using the derrick during the initial shooting operation and the "effectiveness" of RCO 1 is evaluated for each of these tasks. The total or overall effectiveness of RCO 1 dealing with tasks associated with operating the derrick is obtained by adding the individual "effectiveness" value for each task. For example, the score for RCO 1 in the category of derrick in Table 9.10 is 0.0047 + 0.0048 + 0.0047 + 0.0331 = 4.73%. This represents the effectiveness of RCO 1 to reduce the probability of human error occurring when the derrick is operated. In Table 9.11, the same principles are applied for the evaluation of the effectiveness of RCO 1 to reduce or mitigate the severity of human error occurring when the derrick is operated.

Each of these tables (Tables 9.10 and 9.11) represents 50% of the weight (as the RCO evaluation has been normalised) of the elements in Level Two of the hierarchy. The final ranking of the RCOs is achieved by adding the final ratings of these tables for the respective RCOs. Table 9.12 shows the final results obtained for this analysis. From this table, it can be determined that the best control option to reduce the probability of occurrence and the severity (of human error) during the initial shooting operation is RCO 6. The results entail that by installing various warning and indication devices onto/for the equipment used for the initial shooting operation on a fishing vessel, the level of human error can be reduced and safer operation can be achieved.

9.7 Conclusion

Human errors on fishing vessels have contributed to a great number of accidents in the past, as seen in Chapter 2. Almost 20% of all accidents on these vessels are caused by negligence/carelessness of the crew. As such, the ultimate aim for carrying out a human error assessment on fishing vessel is to determine the best method by which accidents caused by these errors can be reduced. This would entail decreasing the risk level by either reducing the probability of a human error occurring or the severity of the possible consequences.

This Chapter describes a method using AHP to achieve this aim. The approach integrates the risk control option within the human error assessment framework to determine the best option for the identified hazards. The advantages of using the described approach for fishing vessels include:

1. The use of a flexible modelling and measurement approach to evaluation.
2. The application of structure to facilitate decision making through the use of a model which imposes strict independence, ordinality, or homogeneity of preferences.
3. Allowing the decision-maker to arrive at consistent and objective evaluations.
4. The simplicity of the use of the model.

5. The confidence that all human errors identified are evaluated (without being omitted by rationalisation) in the decision making process.
6. The interaction between operator and machine is captured within the analysis.

In this Chapter, only human errors are considered in the analysis. However, this can be extended to include failures induced by other causes, such as machinery failure. Hence, it can be easily integrated into the Formal Safety Assessment (FSA) framework as discussed in Chapter 5. Step 4 of the FSA framework requires the evaluation of different risk control options. The AHP method presented here can be used for this purpose, and the results obtained from the analysis can be applied to Step 5 of an FSA.

9.8 References (Chapter 9)

1. Alho J.M. and Kangas J., (1997) "Analysing Uncertainties in Experts' Opinions of Forest Plan Performance", Forest Science, Vol. 43, No. 4, pp. 521-528.
2. Alidi A.S., (1996) "Use of the Analytic Hierarchy Process to Measure the Initial Viability of Industrial Projects", International Journal of Project Management, Vol. 14, No. 4, pp. 205-208.
3. Bhattarai S. and Fujiwara O., (1997) "Evaluation of Appropriate Scale of Hydropower Development for Nepal: An Analytic Hierarchy Process Approach", Proceeding of International Conference on Large Scale Water Resources Development in Developing Countries: New Dimensions of Problems and Prospects, October 20-24, Kathmandu, Nepal.
4. Braunschweig T., (2000) "Priority Setting in Agricultural Biotechnology Research: Supporting Public Decisions in Developing Countries with the Analytic Hierarchy Process", ISNAR Research Report No. 16, The Hague: International Service for National Agricultural Research.
5. Cacciabue P.C., Carpignano A. and Vivalda C., (1993) "A Dynamic Reliability Technique for Error Assessment in Man-Machine Systems", International Journal of Man-Machine Studies, Vol.38, pp. 403-428.
6. CCPS, (1994) "Guidelines for Preventing Human Error in Process Safety", American Institute of Chemical Engineers, Centre for Chemical Process Safety, New York.
7. Davoudian K., Wu J.S. and Apostolakis G., (1994) "The Work Process Analysis Model (WPAM)", Reliability Engineering & System Safety, Vol. 45, pp. 107-125.
8. Department of Energy, (1996) "Hazard and Barrier Analysis Guidance Document", EH-33, Office of Operating Experience and Feedback, Rev.0, November, USA.
9. Dhillon B.S., (1986) "Human Reliability with Human Factors", Pergamon Press Inc., New York.
10. Dougherty E.M. and Fragola J.R., (1988) "Human Reliability Analysis: a Systems Engineering Approach with Nuclear Power Plant Applications", John Wiley & Sons Inc., Chichester, United Kingdom.
11. Frank M.V., (1995) "Choosing among Safety Improvement Strategies: a Discussion with Example of Risk Assessment and Multi-Criteria Decision Approaches for NASA", Reliability Engineering & Systems Safety, Vol. 49, No. 3, pp. 311-324.
12. Fullwood R.R. and Hall R.E., (1988) "Probabilistic Risk Assessment in the Nuclear Power Industry", 1st Ed., Pergamon Press.

13. Gerrits M., Ridgley M.A. and Rijsberman F.R., (1994) "Risk Prioritisation and Site Selection for Sanitation in a Rhine Estuary", Third International Symposium on the Analytic Hierarchy Process, July 11-13, Washington DC, USA.
14. Gertman D.I. and Blackman H.S., (1994) "Human Reliability and Safety Analysis Data Handbook", 1st Ed. John Wiley & Sons Inc., New York, USA.
15. Gertman D.I., Blackman H.S., Haney L.N., Seidler K.S. and Hahn H.A., (1992) "INTENT: A Method for Estimating Human Error Probabilities for Decision Based Errors", Reliability Engineering & System Safety, Vol. 35, pp. 127-136.
16. Hachadorian G.E., (1987) "Hierarchical Determination of the Risk of Forced Divestment to Foreign Owned Enterprises in LDC's", Mathematical Modelling Vol. 8, pp. 163-166.
17. Hamalainen R.P. and Karjalainen R., (1992) "Decision Support for Risk Analysis in Energy Policy", European Journal of Operational Research, Vol. 56, No. 2, pp. 172-183.
18. Hamalainen R.P. and Karjalainen R., (1989) "Structuring the Risks of Energy Production", Proc. of the International Conference on Multicriteria Decision-Making, December 6-8, Bangkok, Thailand, pp. 677-689.
19. Hollnagel E., (1993) "The Phenotype of Erroneous Actions", International Journal of Man-Machine Studies, Vol.39, pp. 1-32.
20. Hunns D.M., (1982) "Discussions around a Human Factors Data-Base, an Interim Solution: the Method of Paired Comparisons, Section 1.6.3 of High Risk Safety Technology, A. E. Green, J. Wiley and Sons, Ltd., Chichester, United Kingdom.
21. ILU, (1996) "Extract from Database 1989-1996", The Institute of London Underwriters, London, United Kingdom.
22. IMO, (1997) "Optimum Maritime Safety Demands: a Focus on People", World Maritime Day 1997, International Maritime Organization, London, United Kingdom.
23. ITWF, (1999) "Troubled Waters: 1999 CSD Thematic Review (Oceans and Seas)", International Transport Workers Federation, March, ISBN 0-901969-22-2.
24. ITSA, (1996) "Serious Ship Accidents Rose in First Quarter of 1996", ITSA Report, International Transport Safety Association, Vol. 2, No. 4.
25. Jensen R.E., (1987a) "International Investment Risk Analysis: Extensions for Multinational Corporation Capital Budgeting Models", Mathematical Modelling, Vol. 9, No. 3, pp. 265-284.
26. Jensen R.E., (1987b) "A Dynamic Analytic Hierarchy Process Analysis of Capital Budgeting under Stochastic Inflation Rates, Risk Premiums, Liquidity Preferences: Theory", Advances in Financial Planning and Forecasting, Vol. 2, pp. 269-302.
27. Johnson C., (1998) "Representing the Impact of Time on Human Error and System Failure", Interacting with Computers, Vol. 11, pp. 53-86.
28. Leveson N., (1995) "System Safety and Computers", Addison Wesley, Reading, Massachusetts, USA.
29. Mahn J.A., Hannaman G.W. and Kryska P., (1995) "Qualitative Methods for Assessing Risk", SAND95-0320, May, Sandia National Laboratories, Albuquerque, Canada.
30. MAIB, (2000) "Safety Digest: Lessons from Marine Accident Reports", Marine Accident Investigation Branch, DETR, 1/2000, United Kingdom.
31. Moieni P., Spurgin A.J. and Singh A., (1994) "Advances in Human Reliability Analysis Methodology. Part I: Frameworks, Models and Data", Reliability Engineering & System Safety, Vol. 44, pp. 27-55.
32. Nezhad H.G., (1988) "Risk Assessment for International Investment", Reprints of the International Symposium on The Analytic Hierarchy Process, Sept. 6-9, Tianjin University, Tianjin, China, pp. 371-380.

33. Pillay A., (2001) "Formal Safety Assessment of Fishing Vessels", PhD Thesis, School of Engineering, Liverpool John Moores University, UK.
34. Pillay A. and Wang J., (2001), "Human Error Modelling in Fishing Vessel Safety Assessment", Proceeding of International Mechanical Engineering Congress and Explosion (ASME Conference), New York, November 11-16.
35. Rasmussen J., (1979) "On the Structure of Knowledge - a Morphology of Mental Models in a Man Machine Context", RIS-M-2192, RISO National Laboratory, Denmark.
36. Rasmussen J., (1981) "Models of Mental Strategies in Process Plant Diagnosis, Human Detection and Diagnosis of Systems Failures", J. Rasmussen and W. B. Rouse (Eds.), Plenum Press, New York, pp. 241-258.
37. Rasmussen J., (1982) "Human Errors: a Taxonomy for Describing Human Malfunction in Industrial Installations", Journal of Occupational Accidents, Vol. 4, pp. 311-333.
38. Rasmussen J., (1986) "Information Processing and Human-Machine Interaction", Amsterdam Press, North-Holland.
39. Reason J., (1990) "Human Error", Cambridge University Press, Cambridge, United Kingdom.
40. Saaty T.L., (1980) "The Analytic Hierarchy Process", McGraw Hill Company, New York, USA.
41. Saaty T.L., (1987) "Risk - Its Priority and Probability: the Analytic Hierarchy Process", Risk Analysis, Vol. 7, No. 2, pp. 159-172.
42. Sanders M.S. and McCormick, E.J., (1987) "Human Factors in Engineering Design", 6th Ed, McGraw-Hill, New York, USA.
43. Sheen N., (1987) "Formal Investigations into the Sinking of the MV Herald of Free Enterprise", UK Department of Transport, Report of Court 8074, Her Majesty's Stationery Office, United Kingdom.
44. Shields T.J. and Silcock G., (1986) "Application of the Hierarchical Approach to Fire Safety", Fire Safety Journal, Vol. 11, No. 3, pp. 235-242.
45. Shields T.J., Silcock G.W. and Donegan H.A., (1990) "Towards the Development of a Fire Safety Systems Evaluation for Public Assembly Buildings", Construction Management and Economics, Vol. 8, No. 2, pp. 147-158.
46. Simkin M.G., Bagranoff N.A. and Crawford P.J., (1990) "Using a Decision Support System to Manage Exchange Rate Risk", Oklahoma Business Bulletin, Vol. 58, No. 3, pp. 32-35.
47. Swain A.D. and Guttmann H.E., (1983) "Handbook of Human Reliability Analysis with Emphasis on Nuclear Power Plant Applications", NUREG/CR-1278, August, USA.
48. Swain A.D., (1987) "Accident Sequence Evaluation Programme Human Reliability Analysis Procedure", Report No. NUREG/CR-4722, USNRC, Washington D.C.
49. White D., (1995) "Application of System Thinking to Risk Management: a Review of the Literature", Journal of Management Decision, Vol. 33, No. 10, pp. 35-45.
50. Wickens C.D., (1992) "Engineering Psychology and Human Performances", 2nd Ed., Harper Collins, New York, USA.

Table 9.1 Generic Human Failure Probabilities

Human Error Probability	Description of human interaction and error	Example factors for a facility specific adjustment
3×10^{-3} to 3×10^{-4}	Pre-Initiator actions - test, maintenance, and calibrations leaving a component, or system with un-revealed fault. Include typical errors in maintenance that cause overall system unavailability (10^{-3})	No written procedure available, or newly defined action; verbal instructions, no checking for completed action, poor equipment/procedure identification label matching.
	Errors include: slips, non-responses, or mistakes leading to skipping a procedure, selecting an incorrect procedure, omitting a step in a procedure, improper communication, transposition of labelling, or misunderstanding task responsibility.	Use established, practised, written procedures, discussed in training, work progress verified with signed checklist, apply self-checking, use tag-out system to maintain configuration control, etc.
1×10^{-2} to 1×10^{-4}	Initiator actions - test, maintenance and calibration activities that trigger events. Include contribution of errors that cause initiating events - covered in initiating event frequencies (10^{-3})	Signals and instruments inappropriate for the action and procedure, lack of cues, or verbal instructions for interlocks, need for process knowledge, requires interpretation of indirect information, etc.
	Typical error modes include slips, non-responses and mistakes.	Indications permit easy transfer through procedures, discussed in training, practiced before hand, administrative control of tags, training involves understanding of the basic principles, and feedback of lessons learned from event precursors.
1 to 1×10^{-3}	Post-Initiator actions - response actions that are not successful in terminating or mitigating the event. Include recovery actions subsequent to initiating events: (1) following multiple failures and (2) directly following an initiating event.	Actions typically outside control room, involves more than one person, lack of a clear cue, knowledge of the process required, process knowledge substituted for emergency procedures, etc.
	Errors include slips, mistakes, and non-responses for control and mitigation actions following an initiating event.	Actions in a control room, include redundant cues, memorised and practised responses, clear man-machine interface, action priorities stressed in training which includes simulation of process dynamics, recoverability from errors, training on infield procedures and long time available for action.

Table 9.2 Error Rates of Rasmussen

	Per demand error rate range	Per demand error rate point estimate
Skill-based	5E-5 to 5E-3	1E-3
Rule-based	5E-4 to 5E-2	1E-2
Knowledge-based	5E-3 to 5E-1	1E-1

Table 9.3 Error Rates of Hunns

Classification of error type	Typical probability
Processes involving creative thinking, unfamiliar operations where time is short; high stress situations	0.1-1
Errors of omission where dependence is placed on situation cues or memory	1E-2
Errors of commission such as operating wrong button, reading wrong dial, etc.	1E-3
Errors in regularly performed, common-place tasks	1E-4
Extraordinary errors - of the type difficult to conceive how they could occur; stress-free, powerful cues militating for success	<1E-5

Table 9.4 Error Rates of Dhillon

Error	Rate per demand	Rate per plant-month
Reading a chart recorder	6E-3	
Reading an analogue meter	3E-3	
Reading graphs	1E-2	
Interpreting incorrectly an indicator	1E-3	
Turning a control in the wrong direction under high stress	0.5	
Using a checklist incorrectly	0.5	
Mating a connector	1E-2	
Choosing an incorrect panel control out of several similar controls	3E-3	
Reading a gauge incorrectly	5.0E-3	
Closing a valve improperly	1.8E-3	
Soldering connectors improperly	6.5E-3	
Actuating switch inappropriately	1.1E-3	
Failure to tighten nut and bolt	4.8E-3	
Failure to install nut and bolt	6E-4	
Improper adjustment of mechanical linkage	1.7E-2	
Procedural error in reading instructions	6.5E-2	
Connecting hose improperly	4.7E-3	
Failure to pursue proper procedure by an operator		0.040
Installation error		0.013
Misinterpretation or misunderstanding of requirements by the operator		0.0076
Inadvertent or improper equipment manipulation by the operator		0.071
Improper servicing or re-assembly by the maintenance personnel		0.015

Table 9.5 Example of Level Two Matrix

Level Two	Element 1	Element 2	Element 3
Element 1	EL_{11}	EL_{12}	EL_{13}
Element 2	EL_{21}	EL_{22}	EL_{23}
Element 3	EL_{31}	EL_{32}	EL_{33}

Table 9.6 Example of Level Three Matrix

Element 1	A_1	A_2	A_3
A_1	A_{11}	A_{12}	A_{13}
A_2	A_{21}	A_{22}	A_{23}
A_3	A_{31}	A_{32}	A_{33}

Table 9.7 Comparison Scale

1	Both elements of equal importance
3	Left weakly more important than top
5	Left moderately more important than top
7	Left strongly more important than top
9	Left absolutely more important than top

1/3	Top weakly more important than left
1/5	Top moderately more important than left
1/7	Top strongly more important than left
1/9	Top absolutely more important than left

Table 9.8 Identified Task and Generic Human Error Data

Task	Operator behaviour	Error Probability
Derrick 1	Skill base	5.00E-03
Derrick 2	Skill base	5.00E-03
Derrick 3	Skill base	5.00E-03
Derrick 4	Knowledge base	5.00E-01
Vessel 1	Knowledge base	5.00E-01
Vessel 2	Rule base	5.00E-02
Vessel 3	Knowledge base	5.00E-01
Vessel 4	Skill base	5.00E-03
L.D 1	Skill base	5.00E-03
L.D 2	Skill base	5.00E-03
L.D 3	Skill base	5.00E-03
Net 1	Skill base	5.00E-03
Net 2	Skill base	5.00E-03
Gilson 1	Rule base	5.00E-02
Gilson 2	Knowledge base	5.00E-01

Table 9.9 RCO Matrix

Derrick	RCO 1	RCO 2	RCO 3	RCO 4	RCO 5	RCO 6
Derrick 1						
Derrick 2						
Derrick 3						
Derrick 4						

Table 9.10 Summary of Results for Probability Element

	Derrick	Vessel	Lazy Decky	Net	Gilson	Total rating
RCO 1	*4.73%*	*1.22%*	*1.40%*	*0.26%*	*0.92%*	***8.53%***
RCO 2	*3.63%*	*0.19%*	*0.84%*	*0.26%*	*0.42%*	***5.34%***
RCO 3	*6.70%*	*0.47%*	*2.17%*	*0.79%*	*1.28%*	***11.42%***
RCO 4	*4.73%*	*0.32%*	*0.75%*	*0.32%*	*0.81%*	***6.94%***
RCO 5	*1.25%*	*0.66%*	*1.79%*	*0.42%*	*0.84%*	***4.97%***
RCO 6	*6.93%*	*1.14%*	*2.37%*	*1.05%*	*1.31%*	***12.81%***

Table 9.11 Summary of Results for Severity Element

	Derrick	Vessel	Lazy Decky	Net	Gilson	Total rating
RCO 1	*5.73%*	*0.76%*	*1.33%*	*0.22%*	*1.04%*	***9.08%***
RCO 2	*1.30%*	*0.30%*	*0.65%*	*0.22%*	*0.46%*	***2.93%***
RCO 3	*5.02%*	*0.85%*	*2.06%*	*0.78%*	*1.06%*	***9.77%***
RCO 4	*5.15%*	*0.67%*	*1.61%*	*0.60%*	*0.90%*	***8.93%***
RCO 5	*3.91%*	*0.55%*	*1.61%*	*0.50%*	*0.92%*	***7.50%***
RCO 6	*6.87%*	*0.85%*	*2.06%*	*0.80%*	*1.21%*	***11.79%***

Table 9.12 Final Ranking of RCO

	Derrick	Vessel	Lazy Decky	Net	Gilson	Total rating
RCO 1	*10.46%*	*1.98%*	*2.74%*	*0.48%*	*1.96%*	***17.61%***
RCO 2	*4.93%*	*0.49%*	*1.49%*	*0.48%*	*0.88%*	***8.27%***
RCO 3	*11.72%*	*1.33%*	*4.23%*	*1.57%*	*2.34%*	***21.19%***
RCO 4	*9.88%*	*0.99%*	*2.36%*	*0.91%*	*1.72%*	***15.87%***
RCO 5	*5.16%*	*1.21%*	*3.40%*	*0.92%*	*1.76%*	***12.46%***
RCO 6	*13.80%*	*1.99%*	*4.43%*	*1.85%*	*2.53%*	***24.60%***

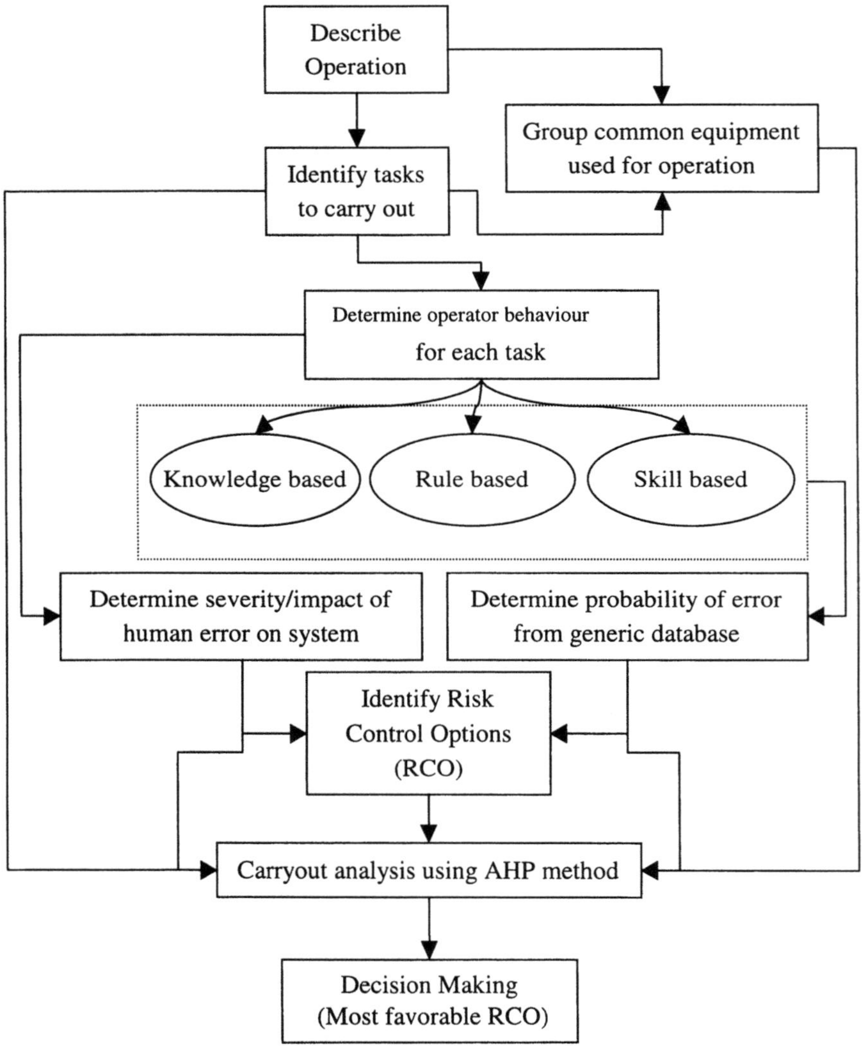

Figure 9.1 Flowchart of the approach

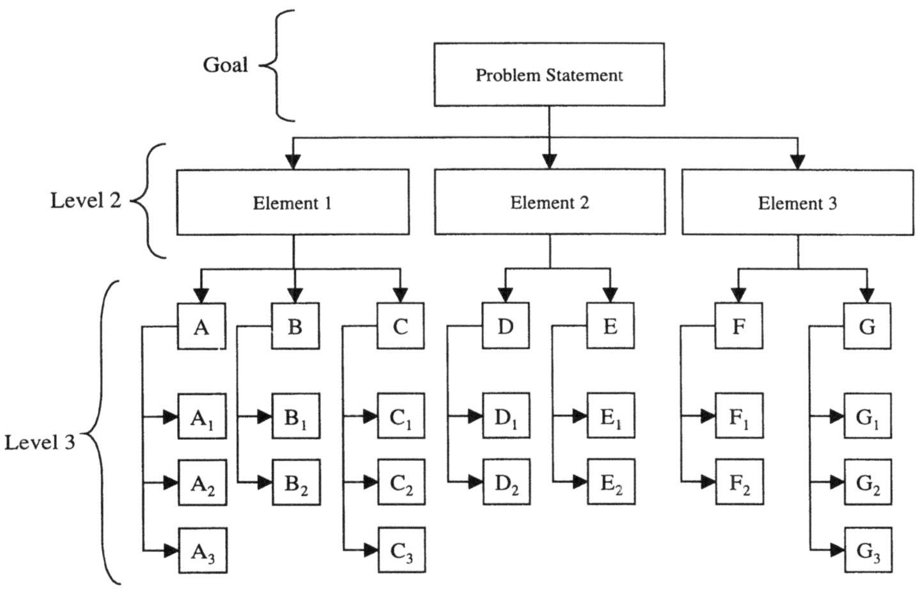

Figure 9.2 Example of hierarchy levels

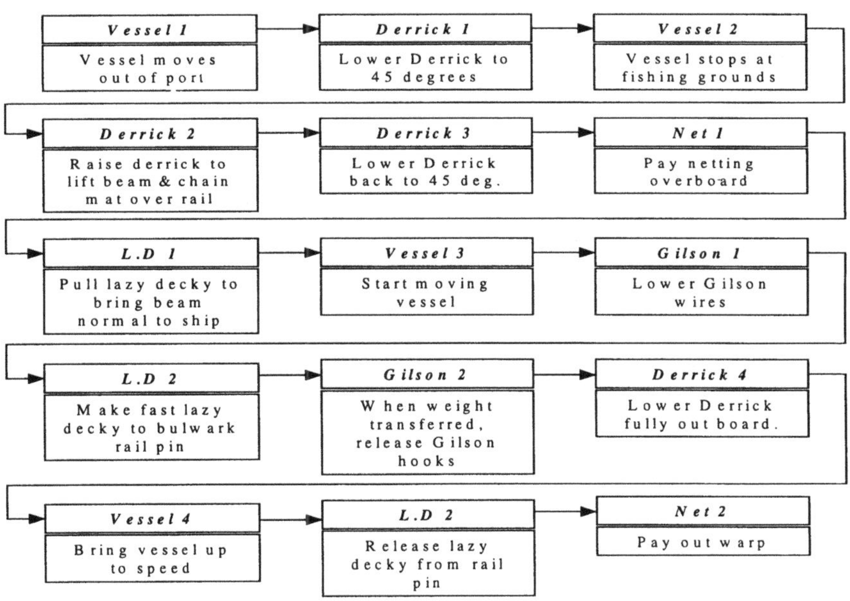

Figure 9.3 Diagrammatic representation of initial shooting operation

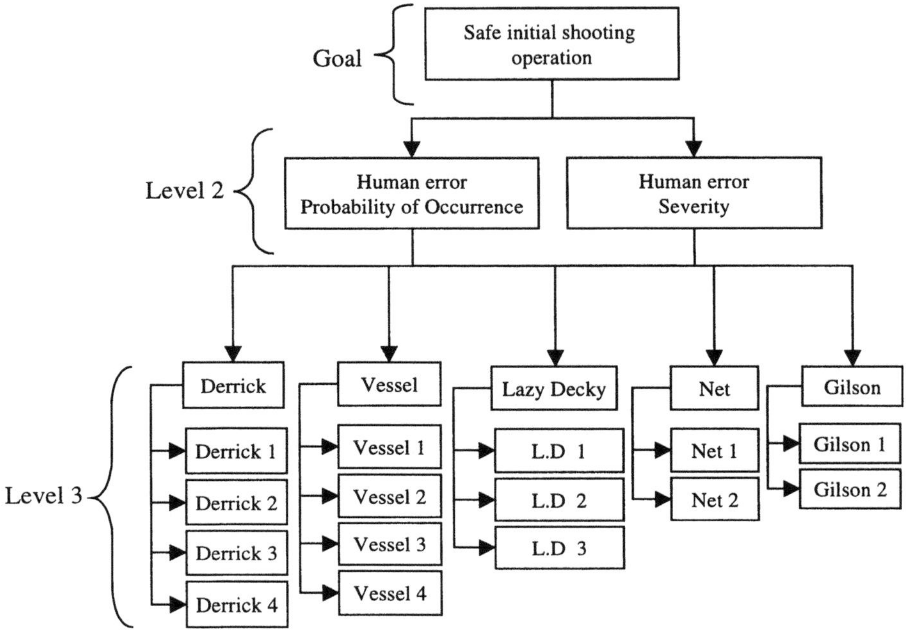

Figure 9.4 Initial shooting operation hierarchy levels

Chapter 10

Three Novel Risk Modelling and Decision Making Techniques

Summary

This Chapter presents three novel safety assessment and decision making approaches. They are (1) a safety based decision support system using artificial neural network techniques, (2) a safety optimisation framework using Taguchi concepts, and (3) a multiple criteria decision approach applied to safety and cost synthesis. Such approaches provide the safety analyst with more flexibility and may be more appropriate in situations where satisfactory results cannot be obtained using other methods.

A risk estimation framework incorporating artificial neural networks (ANNs) is described with two case studies demonstrating its use. Some suggestions are made for further research and development on ANN techniques in the context of maritime safety assessment. The possibility to pool records on notation fields such as system data, function information and casualty or defect data in an agreed and standardised database structure is discussed.

A safety optimisation framework using the Taguchi concepts is described with an example demonstrating its use. The Taguchi concepts are described and discussed. Orthogonal arrays are used to study multiple parameters simultaneously with a minimum of time and resources to produce an overall picture for more detailed safety based design and operational decision making. The signal-to-noise ratio is employed to measure quality, in this case, risk level. The outcomes produced using the described framework may provide the fundamental knowledge for safety analysts to make safety based design and operation decisions.

A new safety and cost modelling approach is described and demonstrated by an example. Three typical multiple criteria decision analysis methods for safety and cost synthesis are described. Their potential for use in safety based decision making is discussed.

Keywords: Analytical Hierarchy Processing, artificial neural networks, cost, decision making, decision support systems, marine safety assessment, multiple criteria decision analysis, risk assessment, Taguchi concept.

10.1 A Safety-Based Decision Support System Using Artificial Neural Network Techniques

10.1.1 Introduction

Artificial Neural Networks (ANNs) have been deemed successful in applications involving classification, identification, pattern recognition, time series forecasting and optimisation. ANNs are distributed information-processing systems composed of many simple computational elements interacting across weighted connections. It was inspired by the architecture of the human brain. The ability of ANNs to model a complex stochastic system could be utilised in risk prediction and decision-making research, especially in areas where multi-variate statistical analysis is carried out.

The paucity literature reported on applications of ANNs in marine and offshore safety engineering reflects that the concept of ANNs is still an extremely raw technique to this area. Published research literature providing a step by step explanation of input data identification through network architecture design and output analysis is somewhat sparse. Buxton et al. (1997) applied the techniques of ANNs to statistics of losses of bulk carriers due to fire to determine whether it is of potential value as a predictor of overall risk. More recently, some initial findings based on a feasibility study of using ANN techniques in offshore and maritime safety-based decision support system has been reported (Sii et al. (2000), Wang et al. (2001)).

Programmed computing involves utilising an algorithm and/or a set of rules for solving the problem and then correctly coding these decisions in software. However, programmed computing can only be applied in cases that can be described by known procedures or set of rules. Neuralcomputing is one of the first alternatives to programmed computing. Neuralconputing provides a new approach to information processing in which algorithm or rule development is not required. The primary information processing structure in neuralcomputing is an ANN (Hecht-Nielsen (1990)).

An ANN is depicted in Figure 10.1. The nodes of the graph are commonly called processing elements. The arcs of the graph are called connections. An adjustable value called weight is associated with each connected pair of processing elements. The weight, w_{ji}, represents the strength of the connection. The processing elements are organised into layers with full or random connections between successive layers. Nodes in the input layer receive input, and nodes in the output layer provide output. Nodes in the middle layers receive signals from the input nodes and pass signals to output nodes. The value entering a processing element is typically the sum of each incoming value multiplied by its respective connection weight. This is often referred to as internal activation or a summation function. The internal activation is then transformed by a non-linear function, which determines the strength of the output connection. The transformed signal will be transmitted to other nodes in the next connected layer which in turn may produce the input to one or more processing elements in subsequent layers. Because the output of the middle nodes is not directly observable, the middle layers can be thought of as hidden. Each processing element may have any number of incoming or outgoing connections but the output signal y_j from node j must all be the same.

ANNs build models based on historical data. The connection weights and threshold values developed by the model are then applied to a new data set. This process is analogous to fitting a regression model based on past data and then utilising the data for prediction. Both techniques require the identification and categorisation of both the input and the output. The major difference that exists is that the regression model requires specification of an exact functional model. Although the number of processing elements and layers in the ANNs determine the complexity of the relationships that the network can capture, this is not as stringent as the development of a specific functional form.

Regression analysis and neural network modelling also require the estimation or training of the model. In both cases, it is common to validate the resulting model against data not used during estimation or training. However, in the case of regression analysis, it is usually possible to evaluate the statistical significance of the estimated parameters in terms of confidence limits, but ANNs are unable to do that.

Like regression, although the most popular firing criterion for ANNs is minimisation of the squared errors, individual values rather than their sums are estimated. ANNs are applicable in any situation where there is an unknown relationship between a set of input factors and an outcome for which a representative set of historical examples of this unknown mapping is available. The objective of building an ANN model is to find a formula or program that facilitates predicting the outcome from the input factors.

The advantages of ANNs can contribute to risk modelling, especially in situations where conventional methods could not be used with confidence to describe the relationship between the input and output variables or there is an inconsistency in input-output relationships (Sii (2001)). An inconsistency in input-output relationship here refers to situations when conventional mathematical models fail to be applied to delineate the input-output relationship due to lack of precise knowledge, or information/data with a high level of fuzziness or ambiguity, or differences (if not contradictory) of opinions about that relationship among the risk analysts. Under such circumstances ANNs may be more appropriate to be used to elicit the true input-output relationship.

Different types of neural networks, such as the Multi-Layer Perceptron (MLP), the radial basis function networks (RBF) and B-spline (Haykin (1999)) networks, etc. can be used to model a system for risk assessment. In the development of an ANN model, the success depends upon a clear understanding of the actual problem, as the selection of network inputs, number of hidden layers and number of neurons in each layer, the non-linear transfer function, and the training algorithm should be based on the features of the problem to be modelled. As ANNs learn by examples, defining and preparing the training data set is also important. The training data must sample every possibility of the problem under all possible working conditions. The data sets including the input training set and the desired output should be as orthogonal as possible, that is, the variables contained in the data sets should be independent with no correlation. Once the problem description and data for the training sets are produced, the rest of the development of the ANN will simply fall into place. ANN testing is performed with a set of test data that is different from the training data used.

10.1.2 A Risk Estimation Framework

A risk estimation framework incorporating ANNs is depicted in a flowchart as seen in Figure 10.2. The framework comprises the following steps (Sii (2001), Wang et al. (2001)):

Step 1: Collect Data. Collect data sets, number series or system information that have a relationship or influence to a system failure from relevant sources such as classification societies, ship owners, flag states, insurance companies and experts.

Step 2: Prepare Data: Define and prepare training input sets and decide how to handle the gathered information for presentation to the ANNs. Determine the range of data and set minimum and maximum values to these levels.

Step 3: Extract test data set: In order to be able to test the trained network, it is common to set aside some of the data for testing (cross validation). Usually, the total data set is divided into two, one for training and the other for testing.

Step 4: Train the network: Select suitable network architecture by setting the number of network inputs equal to the number of input variables and the number of network outputs equal to that of problem output. Select the number of hidden layers and the number of neurons in each hidden layer.

Step 5: Test the ANN: Apply the test data sets to the trained ANN model to test its performance.

Step 6: Evaluate the ANN model: If the estimation generated by the model lies within acceptable accuracy then, proceed to the next step. Otherwise repeat Steps 2 to 6 all over again until the estimation produced falls within the acceptable accuracy. For various applications, accuracy level requirements would be different and are judged subjectively by the user.

Step 7: Use the ANN model to carry out risk prediction: Feed new casualty data to the ANN model, to perform risk estimation.

Step 8: The risk estimation or prediction generated by the ANN model can be applied to safety based design and operation support system as a source of expert input.

The general guidance for building the framework is briefly described as follows:

1. There could be multiple inputs and multiple outputs.
2. Experience indicates that one hidden layer would be enough to deal with majority of risk modelling problems. Using more than one hidden layer will increase the computational load but may achieve faster learning or better generalisation.
3. A Signoid transfer function is usually used while other types of transfer functions may also be applicable.
4. A fast back propagation training algorithm can be used which is available in the Neural Network Toolbox in MATLAB.
5. Techniques incorporating momentum can be used to decrease back-propagation's sensitivity to small details in the error surface. This will help the ANN avoid getting stuck in shallow minima which could prevent the ANN from finding a lower error solution.
6. Adaptive learning rate helps to decrease training time and improve reliability of the back-propagation.

10.1.3 Two Examples

10.1.3.1 Case Study 1

An ANN was developed as an aid in understanding the relationship among the different parameters or features of a generic type of vessel, such as vessel's size, age, degree of machinery redundancy, external factors, etc. Once this relationship is understood, it could be used in predicting vessel failure. A fast back-propagation ANN in the Neural Network toolbox in MATLAB, developed by the Math works Inc., is chosen as the software package. MATLAB Neural Network uses a sigmoid function as the activation function, therefore all data must be scaled between 0 and 1.0 (Sii (2001), Wang et al. (2001)).

Configuration

The configuration of the ANN is shown in Figure 10.3. There are two nodes on the input layer corresponding to the amplitudes for the vessel's size (the dead weight (dwt)) and age of typical bulk carriers. After performing a series of experiments on the effects of number of hidden neurons on training epoch, ten nodes are selected on the hidden layer to allow for the non-linearity of the problem. The output layer has one node corresponding to the amplitude for hull failure rate. Table 10.1 outlines the major neural network characteristics.

Training the ANN

For this network, a training pair is a set of known input and output values. To train the network, an output value is computed from the known input values and the random weights. This computed output is then compared to the known output. A change in the weight is computed and propagated back through the ANN. The modified weights are then used with the known inputs to compute another output value. This process continues until the sum-squared error (sse) difference between the known output and the computed output converges to some given tolerance, arbitrarily defined to be 0.02 for this problem (sum-squared error of 0.02 is commonly used for normalised data in optimisation (Matlab (1994)).

For this initial investigation, it was decided to try and model the problem with ten training pairs. These ten training pairs to be used in training the ANN were created by interpreting (arbitrarily chosen) from LR (Lloyds Register of Shipping) defect data for bulk carriers (Buxton et al. (1997)). The set of scaled training pairs is shown in Table 10.2.

Results

The experimental results on the effects of number of hidden nodes on training epoch or training time are shown in graphical form in Figure 10.4. It is obvious that as the number of hidden nodes increase the training epochs or training time decreases to achieve the same model accuracy, and it reached the smallest number of training epochs or the shortest training time when 10 hidden nodes were used. Further increase in the number of hidden nodes will gradually increase the training epoch. Hence, in this case 10 hidden nodes were selected. This experimental findings agree well with the choosing criteria for the number of hidden neurons as suggested by (Nelson and Illingworth (1990)). According to their postulation, in most cases, except for imaging, one uses four or five hidden layer neurons to one input neuron.

Once trained, the ANN was applied to predict five different test cases. The computed outputs are shown in Table 10.3, together with the actual output from LR defect database and the comparison made between them. It can be seen from Table 10.3 that the ANN model does not predict 100% accurately the failure rate in all the 5 test cases. The error is between 0% and 11.9%. Though the ANN was trained with limited number of training pairs, the computed outputs were considered to be quite optimistic.

10.1.3.2 Case Study 2

The objective of this case study is to develop an ANN model for predicting the possibility of failure or defect of a given vessel. It is based on a hypothetical vessel's design features and the ship owner's management quality.

The vessel's design features include:

- Fire-fighting capability.
- Navigation equipment level.
- Redundancy of machinery.

The ship owner management quality include:

- Quality of ship owner management.
- Quality of operation.

<u>Configuration</u>

An MLP network is chosen as shown in Figure 10.5. In this case, after performing several experiments on the optimal number of hidden neurons to be used for ANN training and learning; 20 hidden neurons are selected for the first hidden layer (In Figure 10.5, four nodes and two small circles with dotted lines that are not explicitly shown, are used to represent the actual 20 nodes for hidden layer). There are five nodes (neurons) on the input layer corresponding to the quality of ship owner's management, quality of operation, fire-fighting capacity, navigation equipment level and machinery redundancy. Twelve nodes (neurons) are chosen on the hidden layer to allow for the non-linearity of the problem. The output layer has one node (neuron) corresponding to the possibility of vessel failure.

The major neural network characteristics are outlined in Table 10.4.

The data used for ANN model training and learning is organised as shown in Table 10.5, which lists data that are within the following guidelines (Sii (2001), Wang et al. (2001)):

- IF either one or more of the factors from ship owner management quality or vessel's design features is classified as '*Very Low*', THEN the possibility of vessel failure is predicted to be '*Very High*';
- IF either one or more of the factors from ship owner management or vessel's design features is classified as '*Low*', THEN the possibility of vessel failure is predicted to be '*High*';
- The rest of the predicted possibilities of vessel failure will be computed according to the average scale values of the five factors listed below:

Level	*Scale values*
Very High	0.9, 1.0
High	0.7, 0.8
Average	0.4, 0.5, 0.6
Low	0.2, 0.3
Very Low	0.0, 0.1

In this particular case study, techniques incorporating momentum and adaptive learning rate are used to increase the speed and reliability of the back-propagation. This helps the network avoid getting stuck in shallow minima which would prevent the network from finding a lower error solution. Training time can also be decreased by the use of an adaptive learning rate which attempts to keep the learning step size as large as possible while keeping learning stable. The learning rate is made responsive to the complexity of the local error surface.

Training the ANN

It was decided to model the problem with twenty-five training sets created hypothetically, as shown in Table 10.5. After 282 epochs of training, for learning rate 0.01 (learning rate is a training parameter that controls the size of weight and bias changes during learning), it reached the pre-defined error goal 0.02.

Results

There is no clear and straightforward solution to the selection of number of hidden neurons. After performing several experiments on the optimal number of hidden neurons to be used for ANN training and learning, the experimental findings are depicted in Figure 10.6, indicating that as the number of hidden nodes increases the training epoch or training time becomes smaller or shorter. The shortest training time is reached when 12 hidden nodes were used.

The trained ANN model was applied to predict 10 different test cases. The predicted outputs were shown in Table 10.6. The predicted results were found to be good as they follow the pre-defined hypothetical criteria closely. For example, the possibility of vessel failure is predicted to be *High* IF ship owner management quality is *Low*; operation quality is *Very High*; fire-fighting capability is *Very High*; navigation equipment level is *Average*; and machinery redundancy is *Very high*.

10.1.4 Discussions

It has been demonstrated by the case studies that ANNs have the following characteristics:

- The capability of learning a set of non-linear patterns.
- ANNs are able to generalise and interpolate accurately within the range of the training data.
- A crucial point for a risk predicting or forecasting model, is the set of consistent, sufficiently independent variables (features) used to train and test the ANNs.
- There is no need to know the type of regression function.
- ANNs are easier to be applied, especially using the existing software packages.
- ANNs are powerful tools and a complement to statistical techniques when data are multi-variate with high degree of interdependence between factors, when the data are incomplete or "noisy", or when many hypotheses are to be pursued and high computational rates are required. With their unique features, they can lead to a powerful decision-making, predicting and forecasting tool.

10.1.5 Conclusion

ANN techniques would provide new insights into assessing and predicting the risks posed by ships with different characteristics. This will permit more rational comparison between alternative ship design and operational features. It is worth noting that ANNs can be a potential tool for risk assessment.

10.2 Taguchi Concepts and Their Applications in Maritime Safety Assessment

10.2.1 Introduction

Taguchi methods of robust experimental design have traditionally been employed in manufacturing settings (Roy (1990)). Literature review indicates that there appears to be virtually no study that uses Taguchi concepts to optimise safety in any discipline of engineering applications (Sii et al. (2001)). Beyond the original intention of Taguchi to apply his methods to manufacturing settings, there are other reasons perceived why Taguchi methods have not been employed at all in safety-based decision making studies. Firstly, the safety of a system is very difficult to measure precisely or quantitatively. This induces problems in the application of Taguchi methods as they actually depend heavily on the accurate measurement of variation of 'quantified' parameters of a process. Secondly, the outcome of a safety-based decision making problem is inherently much more inconsistent in quality than its manufacturing counterpart. This is primarily due to the fact that the safety performance of a system depends largely on the behaviour of the human involvement. High variation in quality makes it difficult to make bona fide judgements about the system performance since Taguchi methods rely on only a small part of the total information pertaining to variations. Finally, safety-related problems, generally speaking, have more 'noise' factors associated with them compared with their manufacturing counterparts. Despite these attributes of a safety- related problem, by appropriately identifying a "quantitative" measure of safety, Taguchi's concepts of robust designs can be employed successfully to optimise safety-related decision making problems (Sii et al. (2001)). This may provide an alternative tool for safety analysts, designers, regulatory bodies and managers to conduct decision making with confidence in situations where other methods cannot be effectively applied.

Engineering safety involves broadly three dimensions of management, engineering and operation, underpinned by the human factors of behaviour, decision and error. The goal for marine and offshore operations can be stated as follows: 'to be competitive in meeting the client's specifications with solutions that are cost-effective at an acceptable level of safety' (Kuo (1998)).

In the context of commercial operations "competitiveness" means "level of profitability", however, in non-commercial activities "effectiveness" would be more appropriate as it has to take into account the specific objective of the activity concerned. The real challenge is that the success in achieving the goal in any project is to meet all four sets of criteria simultaneously, that is, safety, competitiveness, specification and cost-effectiveness.

The requirement of meeting any one of these sets of criteria on its own is relatively straightforward if the others do not have to be taken into consideration. Typical examples of this would be:

- To adopt the latest technology for the production process without due regard to cost.
- To achieve a very high level of safety without taking into account the need for the operation to be competitive.

In engineering terms this is often referred to as a special "multiple-level-multiple-variable optimisation" problem. 'Multiple-level' means that each of the parameters such as specification, comprises requirements, and is with varying degrees of complexity. "Multiple-variable" implies that there is more than one variable or factor involved. "Optimisation" aims to find the best solution to the problem, and in this case the most competitive solution is being sought. Existing optimisation techniques can be used to solve the problem in which the relationships within each parameter and between each other are known and expressible in mathematical terms. However, when some of the relations are qualitative, such as those relating to human factors, the solution to optimisation problems can be extremely difficult to deal with.

The performance of complex engineering systems and quality of products or processes generally depend on many factors. Taguchi separates these factors into two main groups: control factors and noise factors (Ross (1988)). Control factors are those which are set by the designers or manufacturers; noise factors are those over which the designers or manufacturers have no direct control but which vary in the environment of the system or product (Phadke (1989)).

A great deal of engineering effort is consumed in conducting experiments to obtain information needed to guide decisions related to a particular artefact. It would further complicate the situation once safety is integrated into design, especially in the initial concept design stage. This is due to the typical problems associated with a lack of reliable safety data or a high level of uncertainty in safety data. This is particularly true when dealing with the high level of novelty in design and optimisation of marine and offshore safety within both technical and economic constraints.

Taguchi's quality engineering and robust design may offer a useful method to evaluate the performance of a system when uncertainty is present and to measure the quality in the design (Sii et al. (2001)).

10.2.2 The Taguchi Methods and Robust Design

Driven by the need to compete on cost and performance, many quality-conscious organisations are increasingly focusing on the optimisation of product design. This reflects the realisation that quality cannot be achieved economically through inspection. Designing in quality is cheaper than trying to inspect and re-engineer it after a product hits the production floor (Gunter (1987)). Thus, new philosophy, technology and advanced statistical tools must be employed to design high quality products at low cost.

Products have characteristics that describe their performance relative to customer requirements or expectations (Ross (1988)). The quality of a product/process is measured in terms of these characteristics. Typically, the quality is also measured throughout its life-cycle. The ideal quality a customer can expect is that every product delivers the target performance each time the product is used under all intended operating conditions and throughout its intended life and that there will be no harmful side effects (Phadke (1989)). The quality of a product is measured in terms of the total loss to society due to functional variation and harmful side effects (Taguchi (1986)). The ideal quality loss is zero.

Since the late 1950s Dr Taguchi has introduced several new statistical tools and concepts of quality improvement that depend heavily on the statistical theory for design of experiments (Gunter (1987), Phadke (1989), Wille (1990)). These methods of design optimisation developed by Taguchi are referred to as robust design (Phadke (1989)). The robust design method provides a systematic and efficient approach for finding the near optimum combination of design parameters so that the product is functional, exhibits a high level of performance, and is robust to noise factors (Bendell (1988), Phadke (1989)).

The challenge for a designer to design products with high quality is obviously driven by the need to compete on price and performance. Quality-conscious designers are increasingly aware of the need to improve products and processes (Roy (1990)). Delivering a high-quality product at low cost is an interdisciplinary problem involving engineering, economics, statistics, and management (Phadke (1989)). In the cost of a product, one must consider the operating cost, the manufacturing cost, and the cost of new product development. A high-quality product has low costs in all three categories. Robust design is a systematic method for keeping the producer's cost low while delivering a high-quality product and keeping the operating cost low. Taguchi espoused an excellent philosophy for quality control in manufacturing industries (Roy (1990)). His philosophy is founded on three very simple and fundamental concepts. These concepts are stated in (Roy (1990)) as follows:

- Quality should be designed into the product and not inspected into it.
- Quality is best achieved by minimising the deviation from the target. The product should be designed in such a way that it is immune to uncontrollable environmental factors (noise factors).
- The cost of quality should be measured as a function of deviation from the standard and the losses should be measured system-wide.

A leading indicator of quality is required by which one can evaluate the effect of changing a particular design parameter on the product's performance. This indicator is called signal-to-noise ratio. It isolates the sensitivity of the system's performance to noise factors and converts a set of observations into a simple number.

A product under investigation may exhibit a distribution which has a mean value that differs from the target value. The first step towards improving quality is to achieve a distribution as close to the target as possible. Efficiency experimentation is required to find dependable information with minimum time and resources about the design parameters (Phadke (1989)). Taguchi designs experiments using orthogonal arrays which make the design of experiments easy and consistent. The power of orthogonal arrays is their ability to evaluate several factors with a minimum number of experiments.

10.2.3 The Design Process

The early design phase of a product or process has the greatest impact on life cycle cost and quality (Kackar (1985), Phadke (1989), Taguchi et al. (1989)). Therefore, significant cost savings and improvements in quality can be realised by optimising product designs. The three major steps in designing a quality product are: system design, parameter design, and tolerance design (Bendell (1988), Phadke (1989), Taguchi (1986), Taguchi et al. (1989)).

System design is a process of applying scientific and engineering knowledge to produce a basic functional prototype design (Kackar (1985)). The prototype model defines the configuration and attributes of the product undergoing analysis or development. The initial design may be functional but it may be far from optimal in terms of quality and cost.

The next step, parameter design, is an investigation conducted to identify the settings of design parameters that optimise the performance characteristic and reduce the sensitivity of engineering designs to the source of variation (noise) (Kackar (1985)). Parameter design requires some form of experimentation for the evaluation of the effect of noise factors on the performance characteristic of the product defined by a given set of values for the design parameters. This experimentation aims to select the optimum levels for the controllable design parameters using the robust design method.

Experimenting one design variable at a time or by trial and error until a first feasible design is found, is a common approach to design optimisation (Bendell (1988), Phadke (1989)). However, this approach can lead to either a very long and expensive time span for completing the design or a premature termination of the design process due to budget and schedule pressures. The result in most cases is a product design which may be far from optimal. As an example, if the designer is studying 13 design parameters at three levels, varying one factor at a time would require studying 1,594,323 experimental configuration (3^{13}). This is a full factorial approach where all possible combinations of parameter values are tried. Obviously, the time and cost involved in conducting such a detailed study during the advanced design is prohibitive.

In contrast, Taguchi's robust design method provides the designer with a systematic and efficient approach for conducting experimentation to determine near optimum settings of design parameters for performance and cost (Bendell (1988), Kackar (1985), Logothetis and Salmon (1988), Phadke (1989), Meisl (1990)). The robust design method uses orthogonal arrays (OA) to study the parameter space, usually containing a large number of decision variables, with a small number of experiments. Taguchi's orthogonal arrays provide a method for selecting an intelligent subset of the parameter space. Using orthogonal arrays significantly reduces the number of experimental configurations. A typical tabulation is shown in Table 10.7 (Bendell (1988), Phadke (1989), Taguchi and Konishi (1987)). In this array, the columns are mutually orthogonal, that is, for any pair of columns, all combinations of factor levels occur, and they occur at an equal number of times. There are four factors 1, 2, 3 and 4, each at three levels. This is called an L_9 design, where 9 indicates the nine rows, configurations or prototypes to be tested, with test characteristics defined by the row of the table. The top row of the array shows the four different levels (i.e. alternative settings). The other rows represent different combinations of control factor levels. This set-up of nine level combinations satisfies the information need just as good as a full factorial experiment in which all 3^4 = 81 level combinations are tested.

The number of columns of an OA represents the maximum number of factors that can be studied using that array. Note that this design reduces 81 (3^4) configurations to 9. Some of the commonly used orthogonal arrays are shown in Table 10.8 (Bendell (1988)). As Table 10.8 shows, there are greater savings in testing for the larger arrays.

Using an L_9 OA means that nine experiments are carried out in search of the 81 control factors combinations which gives the near optimal mean, and also the near minimum variation away from this mean. To achieve this, the robust design method uses a statistical measure of performance called signal-to-noise (S/N) ratio (Phadke (1989)). Borrowed from electrical control theory, the S/N ratio developed by Dr Taguchi is a performance measure to choose control levels that best cope with noise (Bendell (1988), Bryne and Taguchi (1986), Phadke (1989)). The S/N ratio takes both the mean and the variability into account. In its simplest form, the S/N ratio is the ratio of the mean (signal) to the standard deviation (noise). The S/N equation depends on the criterion for the quality characteristic to be optimised. While there are many different possible S/N ratios, three of them are considered standard and are generally applicable in following situations (ASII (1989), Bryne and Taguchi (1986), Phadke (1989)):

- Biggest-is-best quality characteristic (strength, yield).
- Smallest-is-best quality characteristic (contamination).

- Nominal-is-best quality characteristic (dimension).

Whatever the type of quality or cost characteristic, the transformations are such that the S/N ratio is always interpreted in the same way, the larger the S/N ratio the better (Bryne and Taguchi (1986)).

The third step, tolerance design, is the process of determining tolerances around the nominal settings identified in the parameter design process (Kackar (1985)). Tolerance design is required if robust design cannot produce the required performance without special components or high process accuracy (Bendell (1988)). It involves tightening of tolerances on parameters where their variability could have a large negative effect on the final system. Typically tightening tolerances leads to higher cost (Phadke (1989)).

10.2.4 Background of Taguchi Concepts

The fundamental principle of robust design is to improve the quality of a product by minimising the effects of the causes of variation without eliminating those causes. Efficient experimentation is necessary to find dependable information about design parameters. The information should be obtained with minimum time and resources. Estimated effects of parameters must be valid even when other parameters are changed. Employing the signal-to-noise ratio to measure quality and orthogonal arrays to study many parameters simultaneously are the keys to high quality and robust design.

Since variation in product performance is similar to quality loss, analysis of variance (ANOVA) will be carried out to interpret experimental data and factor effects. ANOVA is a statistically based decision tool for detecting differences in average performance of groups of items tested (Ross (1988), Roy (1990)).

Phadke, following Taguchi, measures the quality of a product in terms of the total loss to society due to functional variation and harmful side effects. Under ideal conditions, the loss would be zero, that is, the greater the loss, the lower the quality (Phadke (1989)). In the following, how this quality loss can be quantified, factors that influence this loss, how this quality loss can be avoided are discussed.

10.2.4.1 The Taguchi Quality Loss Function

Quality is often measured in terms of the fraction of the total number of units that are defective. This is referred to as fraction defective. However, this implies that all units which are within the tolerances of the requirements are equally good. In reality, a product that is exactly on target gives the best performance. As the product's response deviates from the target its quality becomes progressively worse. Therefore, one should not be focusing on meeting the tolerances but on meeting the target.

The quality loss is crucial in Taguchi's theory. It is based on the assumption that when a functional characteristic y deviates from the specified target value m, the customer and the society in general experiences an economical loss due to poorer product quality. This economic loss is expressed as the loss function $L(y)$. Based on this, Taguchi defines the quality loss for not being on target by means of the quadratic quality loss function (Phadke (1989), Taguchi (1986)):

$$L(y) = k(y - m)^2 \tag{9.1}$$

where y is the quality characteristic of a product/process, k is a constant called the quality loss coefficient and m is the target value for y.

When the functional characteristic deviates from the target, the corresponding quality loss increases. Furthermore, when the performance of the product is outside the tolerance, the product is considered defective. A convenient way to determine the constant k is to determine first the functional limits for the value of y. Let $m \pm \Delta_0$ be the safety range for a vessel. Suppose the cost (loss) of losing or repairing the vessel is A_0 when the vessel goes beyond the safety range. By substitution into Equation (9.1), the following can be obtained:

$$k = \frac{A_0}{\Delta_0^2} \tag{9.2}$$

With the substitution of Equation (9.2) into Equation (9.1) it is able to calculate the quality loss for a given value of y. More on the determination of k can be found in (Phadke (1989)).

10.2.4.2 Signal-to-Noise Ratio (S/N Ratio)

Taguchi has developed a signal to noise ratio in order to provide a way of measuring the robustness of a product. In other words, he has used the signal-to-noise ratio as a predictor of quality loss after making certain simple adjustments to the system's function (Taguchi (1986), Phadke (1989)). This ratio isolates the sensitivity of the system's function to noise factors and converts a set of observations into a single number. It is used as the objective function to be maximised in robust design (Phadke (1989)).

The ratio takes into account the mean and the variance of the test results, and is as a rule of thumb always maximised. This leads to several specialised S/N ratios, depending on the nature of the comparison variable. There are three basic S/N ratios, but according to (Fowlkes and Creveling (1995)), the variety of S/N ratios is limitless. The three possible categories of quality characteristics or most widely used S/N ratios are:

- Smallest-is-better: $\eta = -10\log\left(\frac{1}{n}\sum_{i=1}^{n} y_i^2\right)$ e.g. seeking the minimum light weight/dead weight ratio.
- Nominal-is-best: $\eta = 10\log\left(\frac{\mu^2}{\sigma^2}\right)$ e.g. maintaining cell guide tolerances.
- Larger-is-better: $\eta = -10\log\left(\frac{1}{n}\sum_{i=1}^{n}\frac{1}{y_i^2}\right)$ e.g. seeking the maximum profit or the highest efficiency.

where:

η represents the S/N ratio

μ and σ^2 are the mean value and the variance of the variables

y_i is the comparison variable in experiment i for a certain combination of control factor levels

n is the number of experiments performed for that combination.

The conversion to the signal-to-noise ratio can be viewed as a scale transformation for convenience of better manipulation. It offers an objective way to look at two characteristics, namely, variation and mean value. Analysis using the signal-to-noise ratio has two main advantages:

- It provides a guideline to the selection of the optimum level based on least variation around the target and also on the average value closest to the target.
- It offers objective comparison of two sets of experimental data with respect to variation around the target and the deviation of the average from the target value.

For the robust design of a product, the following two steps are required:

- Maximise the signal-to-noise ratio η. During this step, the levels of the control factors to maximise η are selected while ignoring the mean.
- Adjust the mean on target. For this step, a control factor is used to bring the mean on target without changing η.

Further information on quality loss and signal-to-noise ratios can be found in texts written in (Phadke (1989), Ross (1988), Roy (1990), Suh (1990), Taguchi (1986)). They all provide detailed discussions on how to apply statistical methods and Taguchi's approach in the selection of design parameters for satisfying functional requirements.

10.2.4.3 Life Cycle Quality Loss

For a ship owner, it may be of interest to study how life cycle considerations fit in the theory of Taguchi. Let $y_1, y_2, \ldots y_n$ be n representative measurements of the quality characteristic y taken through the life cycle of a ship, and assume that y shall be as close to a specified target value m as possible. Then the average quality loss Q caused by this product may be expressed as:

$$Q = \frac{1}{n}[L(y_1) + L(y_2) + \ldots\ldots + L(y_n)] = \frac{k}{n}\left[(y_1 - m)^2 + (y_2 - m)^2 + \ldots\ldots + (y_n - m)^2\right]$$

$$= k\left[(\mu - m)^2 + \frac{n-1}{n}\sigma^2\right]$$

where $\mu = \frac{1}{n}\sum_{i=1}^{n} y_i\,(mean) \quad \sigma^2 = \frac{1}{n-1}\sum_{i=1}^{n}(y_i - \mu)^2\ (variance)$

When n is large (there are many measurements during the product life), the expression can be simplified as $Q = k\left[(\mu - m)^2 + \sigma^2\right]$. This simplified expression shows that the average quality loss depends on the following two terms:

- The deviation of quality characteristic y relative to the target value m.
- The mean variance of y relative to the observed mean value of y.

It is usually easy to reduce or eliminate the first term; reducing the variance of a product is generally more difficult and expensive. A systematic approach to optimise the performance is Taguchi's two step optimisation process.

10.2.4.4 Taguchi's Two Step Optimisation Process

Taguchi's two step optimisation process focuses on the product's performance on the target. It consists of the following two steps:

- The first step of the process is to reduce the variability of the product performance by selecting parameter values for minimum variability.
- The second step is to select and adjust parameters with strong influence on the mean and weak influence on the variability to put the performance on the target by identifying the influences of the different design parameters on the mean and variance.

10.2.4.5 Orthogonal Arrays

In Taguchi's theory, design parameters are set based on studies of the behaviour of the concept under different operating conditions. The parameters are set in such a way that the sensitivity of the concept performance with respect to uncontrollable factors is minimised. The sensitivity is analysed by the use of experiments, and through analysis of the information need and the use of orthogonal arrays – the core of the Taguchi's experimental design technique, the experiment efficiency is optimised. The experiments may be either analytical or physical, and the results are analysed using an appropriate comparison variable and a so-called signal-to-noise ratio.

The term orthogonal refers to the balance of the various combinations of factors so that no single factor is given more or less weight in the experiment than other factors. Orthogonality also refers to the fact that the effect of each factor can be mathematically assessed independent of the effects of the other factors (Fowlkes and Creveling (1995)). In orthogonal arrays, the columns are mutually orthogonal. Most books dealing with the Taguchi theory provide standardised orthogonal arrays. In a more advanced parameter design set-up, control factors with varying number of levels (usually 2, 3 and 4) can be performed simultaneously. It is also possible to study the interaction between the control factors in an experiment.

10.2.4.6 Degree of Freedom

Degree of freedom is a concept that is useful to determine how much information can be derived from an experiment in a matrix representation. The degree of freedom of a matrix experiment is one less than the combinations of levels in the experiment (i.e. number of rows in the orthogonal array): $DOFexp = \#combinations - 1$.

The degree of freedom needed to describe a factor effect (i.e. a factor's contribution to the result) is one less than the number of levels (values) tested for that factor: $DOF_f = \#levels - 1$.

The problem of solving a set of simultaneous equations for a set of unknowns is a good mathematical analogy for the experiment. The number of equations is analogous to the degree of freedom of a matrix experiment. The number of unknowns is analogous to the total degree of freedom of the factorial effects: $(Total\ DOF)_f = (\#factors)(DOF_f)$.

The *DOF* is used to select an appropriate orthogonal array for the experiment, i.e. for the testing of the parameter combinations. As a general rule, the selected standardised orthogonal array must have at least the same degree of freedom as the experiment. In addition, the number of rows must be at least one more than the $(Total\ DOF)_f$. One reason for the rationality of the Taguchi experiments is therefore that they do not produce more information than is needed.

10.2.4.7 Control factors

The control factors are values that can be specified freely by the designer. It is designers' responsibility to determine the best values of these parameters. Each control factor can take multiple values, called levels. Their settings or levels are selected to minimise the sensitivity of the product's response to all noise factors.

10.2.4.8 Noise factors

Noise factors are treated like the control factors in terms of *DOF* calculation and selection of orthogonal arrays, but might be more often represented by two-level parameters reflecting a probable operating interval. An example of this may be fuel price, where one may set an extreme high and expected price as the operating interval. The distinction between controllable and uncontrollable factors is very often an economical question, and in the extreme case with unlimited resources available, all factors may be controllable.

10.2.4.9 ANOVA Terms and Notations

The analysis of variance (ANOVA) computes parameters such as degree of freedom, sums of squares, mean squares, etc. and organises them in a standard tabular format. These parameters and their interrelationships are defined as shown below using the following notation:

V = Mean squares (variance)

S = Sum of squares

S' = Pure sum of squares

f = Degree of freedom

e = Error (experimental)

F = Variance ratio

P = Percent contribution

T = Total (of results)

N = Number of experiments

$C.F.$ = Correction factor

n = Total degrees of freedom

Variance

The variance of each factor is determined by the sum of the square of each trial sum result involving the factor, divided by the degree of freedom of the factor. Thus:

$V_A = S_A/f_A$ (for factor A)

$V_B = S_B/f_B$ (for factor B)

$V_e = S_e/f_e$ (for error terms)

Variance Ratio

The variance ratio is the variance of the factor divided by the error variance.

$F_A = V_A/V_e$

$F_B = V_B/V_e$

$F_e = V_e/V_e = 1$

Pure Sum of Squares

The pure sum of squares is:

$S'_A = S_A - f_A \times V_e$

$S'_B = S_B - f_B \times V_e$

$S'_e = S_e + (f_A + f_B) \times V_e$

Percent Contribution

The percent contribution of each factor is the ratio of the factor sum to the total, expressed in percentage.

$P_A = S_A \times 100/ S_T$

$P_B = S_B \times 100/ S_T$

$P_e = S_e \times 100/ S_T$

where S_T is total sum of square, obtained by: $S_T = \left(Y_1^2 + Y_2^2 + ...Y_i^2\right) - \frac{(Y_1 + Y_2 + ... + Y_i)^2}{i}$

Total Variance

Total variance is:

S_T = Sum of square of all trial run results – $C.F.$

where $C.F. = T^2/N$ and $T = (Y_1 + ... + Y_N)$

10.2.4.10 Confidence Intervals

The calculations shown in the ANOVA table are only estimates of the population parameters. These statistics are dependent upon the size of the sample being investigated. As sample size increases, the precision of the estimate would be improved. For large samples, the estimates approach the true value of the parameter. In statistics, it is therefore customary to represent the values of a statistical parameter as a range within which it is likely to fall, for a given level of confidence. This range is termed as the confidence interval (*C.I.*). If the estimate of the mean value of a set of observations is denoted by *E(m)*, then the *C.I.* values for the mean and intervals are obtained according to the following procedure:

Upper confidence level = Mean + C.I.

Lower confidence level = Mean – C.I.

$C.I. = \sqrt{\frac{FxV_e}{N_e}}$ (Roy (1990))

where:

F = *F-value* from the F distribution Tables (F-ratio Tables) at a required confidence level and at *DOF 1* and error *DOF 8* (Roy (1990))

V_e = Variance of error term (from ANOVA)

N_e = Effective number of replications = {Total number of results (or number of S/N-ratios)}/{*DOF* of mean (=*1*, always) + *DOF* of all factors included in the estimate of the mean}or the total number of units in one level.

F-value is sometimes refereed to as F-ratio, used to test the significance of factor effects. It is statistically analogue to Taguchi's signal-to-noise ratio for control factor effect vs. the experimental error. The F-ratio uses information based on sample variances (mean squares) to define the relationship between the power of the control factor effects (a type of signal) and the power of the experimental error (a type of noise) (Fowlkes and Creveling (1995)).

10.2.4.11 Brainstorming

Brainstorming is an integral part of the Taguchi philosophy. Taguchi regards brainstorming as an essential step in the design of effective experiments. Through brainstorming sessions, clear statements of the problems are established, the objectives, the desired output characteristics, the methods of measurement and the appropriate experiments are designed. Taguchi does not prescribe a standard method of brainstorming as applicable to all situations. The nature and content of the brainstorming session will vary widely on the problem.

10.2.5 A Safety Optimisation Framework Using Taguchi Concepts

A safety optimisation framework using Taguchi concepts for maritime safety engineering applications presented in this Section has the following steps (Sii (2001), Sii et al. (2001)):

1. Define the problem.

   The first step is to describe the specific maritime safety problem in detail, either in qualitative or quantitative terms. Then define the objective parameter that is to be optimised.

2. Identify factors and their interactions.

   Brainstorming technique is normally used among a panel of experts to identify all the possible factors, levels, their interactions and other pertinent information about the optimisation problem. Sometimes factor screening may be required to provide a quick and simple way of ranking factors according to their importance in the optimisation. This will reduce the number of identified factors in order to perform the optimisation more efficiently.

3. Select an appropriate orthogonal array.

   In order to select the correct standard orthogonal array, it is necessary to determine the total degrees of freedom in order to find the minimum number of level combinations to be tested. The number of factors and their interactions as identified after the screening in Step 2 will determine the total degrees of freedom according to the equation given in the nest Section.

4. Conduct experiment.

This starts with the selection of a correct quality loss function to represent the description of loss attributed in the case. This is a purely mathematical analysis and *S/N-ratio* for each treatment is calculated according to the selected standard *S/N-ratio* expressions as described in Section 10.2.4.2. The calculated *S/N-ratios* are then normalised before proceeding to the next step.

5. Conduct analysis of variance (ANOVA) and other Taguchi-related analysis.

   This step mainly performs all the relevant operations in ANOVA. The main effects of each factor as well as interaction of factors are determined, then sum of squares for each main effect of factor is computed. The variance of each factor is calculated. The results are presented in a table.

6. Identify significant factors and their interactions.

   The contribution of each factor and their interactions are determined through division, i.e., the sum of square of each factor is divided by the total sum of squares of all the factors. Pooling is recommended when a factor is determined to be insignificant by performing a test of significance against the error term at a desired confidence level.

7. Find the optimal combination of factor levels to minimise system risk level.

   The non-linearity analysis is carried out to investigate the non-linearity of the *S/N*-ratio with respect to factor levels of each factor as well as their interactions to identify the optimal combination of factor levels. The non-linearity graphs are developed to demonstrate the outcomes of this investigation.

8. Recommend for implementation.

   Safety related recommendations pertaining to engineering design, operation and management are made based on the outcomes of the optimisation.

10.2.6 Application of Taguchi Concepts in Maritime Safety Studies

This example is designed for illustration purposes to demonstrate that the Taguchi method is a potential tool for maritime engineering safety studies (Sii et al. (2001)).

<u>Background information</u>

The ship's safety is substantially affected by many factors including ship owner management quality, crew operation quality, enhanced survey programme, degree of machinery redundancy, fire-fighting capability, navigation equipment level, corrosion control and preventive maintenance policy. In order to identify the salient factors and interactions that cause excessive variations, a trial application of Taguchi methods is performed here to optimise each factor to attain the optimal safety for the ship.

<u>Step 1: Define the problem</u>

Various factors such as design features, ship owner management quality, crew operation quality, etc. have different degrees of influence on ship's overall safety performance throughout its life cycle. This will further be complicated when all these factors are evaluated simultaneously to obtain the optimised solution. The prime objective of this study is to identify the factors and their associated reasons for high risks and to suggest measures that would reduce the overall risk level of the ship.

Step 2: Identify factors and their interactions

Brainstorming technique is used to gather relevant information to determine factors affecting ship's safety. The resulting list of significant factors affecting ship safety is given in Table 10.9. These factors are determined based on the information acquired. In the eight factors, seven have three levels, and one has two levels. There is a significant interaction between two factors, namely, the ship owner management quality and enhanced survey programme. The risk level values between 1 to 50 are assigned to each factor at each level by experts. The higher the risk level value, the more risky the system. These risk level values do not represent any absolute or exact degree of risk encountered by the ship and they are used only for relatively indicative purposes. To facilitate further discussion, the factors are assigned alphabet-identifiers.

Step 3: Select an appropriate orthogonal array

In order to choose an appropriate array, degrees of freedom must be computed first. Given seven factors with three-levels, one factor with two-levels, and one interaction of a two-level and a three-level factor, the number of degrees of freedom for the experiment is computed to be 7(3-1) + 1(2-1) + (3-1)(2-1) + 1 = 18. Table 10.10 shows the experimental design for an *L18* array. In all, *18* treatments must be used for the experiment with the factor levels as shown. For this study, however, three levels of risk are used in Table 10.11 representing judgements made by three experts.

The assignment of factors to columns is accomplished as follows:

- Since factor *E* is a two-level factor, it is assigned to column 1.
- As factors *E* and *D* are deemed to have significant interaction in brainstorming sessions, factor *D* is assigned to column 2.
- Other factors are thereafter assigned to columns 3 to 8 arbitrarily – factor *A* to column 3, factor *B* to column 4, and so on as clearly depicted in Table 10.11.

Interaction is not assigned to any column, since it can be computed without loss of any information or confounding.

Step 4: Conduct experiments

Three sets of experiments are conducted for each treatment as dictated by the *L18* array of Table 10.10 where the risk levels are assigned by three experts. The results are shown in Table 10.11. Then, for each treatment *S/N-ratio* is calculated using the following formula:

$$S/N\text{-}ratio \text{ for } i^{th} \text{ treatment} = -10\log_{10}\left(\frac{1}{n}\sigma^2\right) = -10\log_{10}\left(\frac{1}{n}\right)\left(Y_{i1}^2 + Y_{i2}^2 + Y_{i3}^2\right)$$

where Y_{ij}, j=1, 2 or 3 is the jth response of the ith treatment representing judgements made by three experts. These values are normalised by subtracting -27 (the average of the *S/N*-ratio) from each *S/N-ratio*. The *S/N*-ratios and their normalised values are also shown in Table 10.11.

Step 5: Conduct Analysis of Variance (ANOVA) and Step 6: Identify significant factors and their interactions

Based on the normalised *S/N-ratios* data in Table 10.11, analysis of variance is conducted. As a first step, for each level of each factor, the main effect is computed.

Example: for factor *A*, Level 1, *main effect* = 6.64 – 3.93 - 6.6 + 0.81 - 5.19 - 3.2 = -11.47

Factor A, Level 2, *main effect* = 10.04 + 10.4 + 0.32 + 9.29 –5.75 –6.14 = 18.16

Factor A, Level 3, *main effect* = -0.49 + 2.88 – 3.28 - 0.84 + 3.95 - 2.01 = 0.21

The main effects of other factors are computed likewise. For computing the effect of interaction of factors D and E, all possible combinations $(3\times2 = 6)$ of D and E are considered.

Level of factor D	1	2	3	1	2	3
Level of factor E	1	1	1	2	2	2
Interaction level assigned:	1	2	3	4	5	6

Thereafter, the effect of interaction is computed as an average of each level. Sum of squares for each main effect is computed using the standard methodology.

Calculation of $D\times E$ column:

D_1E_1 = 6.64 + 10.04 - 0.49 = 16.19

D_2E_1 = -3.93 + 10.4 + 2.88 = 9.35

D_3E_1 = -6.6 +0.32 –3.28 = -9.56

D_1E_2 = 0.81 + 9.29 - 0.84 = 9.26

D_2E_2 = -5.19 – 5.75 + 3.95 = -6.99

D_3E_2 = -3.2 –6.14 –2.01 = -11.35

Example: For the interaction of factors D and E, the formula used is slightly different since it has six levels. Specifically,

S_T = Sum of square of all trial run results – *C.F.*

where $C.F. = T^2/N$

$T = (Y_1 + Y_2 + Y_3 +\ldots Y_i)$

i = number of trials or treatments

$$\text{or } S_T = \left(Y_1^2 + Y_2^2 + \ldots Y_i^2\right) - \frac{(Y_1 + Y_2 + \ldots + Y_i)^2}{i}$$

S_T for $D\times E$ = $\{16.19^2 + 9.35^2 + (-9.56)^2 + 9.26^2 + (-6.39)^2 + (-11.35)^2\} - \{16.19 + 9.35 - 9.56 + 9.26 - 6.99 - 11.35\}^2 / 6$ = 703.91 – 7.94 = 696.43

S_T for A = $\{(-11.47)^2 + (18.16)^2 + (0.21)^2\} - (6.9)^2/3$ = 445.52

The main effects are shown in Table 10.12.

Then the table for ANOVA (analysis of variance) is ready to be developed. At the outset, a significance level of 0.05 or confidence level of 95% was set as the cut-off point for pooling an effect into error. The ANOVA table is developed as follows:

- The first column is simply the factor identifier.
- The second column is taken from Table 10.12, and is the sum of squares for each factor.
- The third column is developed by simply finding the percentage of each sum of square with respect to the total sum of squares of all factors and interaction.
- The fourth column lists the degree of freedom for each factor.

- The fifth column lists the variance for each factor. The variance values are computed by dividing sum of squares by degree of freedom of each factor.
- The sixth column tries to pool in error. In an attempt to find factors that can be pooled in the error, as a first step, all factors that contributed less than 2.5% to the overall sum of squares are pooled into the error term.
- The seventh column shows the *F*-values. Each *F*-value is the variance ratio which is computed by dividing the variance of the factor by the error variance.

  $F_A = V_A/V_e$

  where:

  F_A = *F*-value for factor *A*

  V_A = variance for factor *A*

  V_e = variance for error terms.

  In this case V_e = *sum of squares for* factors *B, C and G* = 2.27 + 1.49 + 0.72 = 4.48

Since factors *B*, *C* and *G* contribute less than 2.5% to the overall sum of squares, they are pooled into the error term.

F-value *for* factor *A* = 222.76 / 4.48 = 49.72

As shown in Table 10.13, this results in an *F*-value of 10.03 for factor *B*, 6.61 for factor *C* and 3.16 for factor *G*, which are not significant enough to be considered as independent main effects, based on our significance confidence level of 95%. Thus, factors *C, B* and *G* are pooled into the error term. They are found insignificant and the variations arising from these constituted the error variations. These *F*-values were compared against the *F*-values provided for 5% significance in appropriate *F* distribution tables (Roy (1990)). The effect of factors *A*, *B*, *D*, *E*, *F*, *H* and the interaction of factors of *D*×*E* is found significant at 99.5% confidence levels.

Non-linearity analysis

It is determined to investigate the non-linearity of these factors since most factors are at three levels. An investigation of the non-linearity of the *S/N*-ratio with respect to factor levels is carried out to identify the optimal combination of factor levels. Firstly, the average values of the main effects are computed for each factor as can be seen in Table 10.14. For factor *A*, for instance, the average value at three levels is computed as follows:

The total value of *S/N*-ratio when factor A at level 1 = -11.47 (refer to Table 10.11). Hence, the average value of *S/N*-ratio of factor *A* at level 1 = (-11.47 / 6 + 27) = 25.09, where 27 is added back which was originally subtracted in Table 10.11. Division by 6 is simply due to the fact that there are six terms containing factor *A* at level 1.

The other values for factor *A* at levels 2 and 3 are computed in a similar way. The same procedure yielded the rest of the main effects and interaction effects shown in Table 10.14. The upper and lower confidence levels of 1.99 are calculated in Step 7.

Step 7: Find the optimal combination of factor levels to minimise system risk level

Based on Tables 10.12 and 10.14, the non-linearity graphs for each of the factors (except factor E which has two levels and hence is not subject to non-linearity investigation) and interaction are developed. These are shown in Figures 10.7 to 10.15. The combination that yields the largest value of *S/N*-ratio is determined from these graphs to be as follows:

Factor	A	B	C	D	E	F	G	H	$D{\times}E$
Optimal level	2	1	2	1	1	3	2	1	1

Thus, A_2, B_1, C_2, D_1, E_1, F_3, G_2 and H_1 provide the best combination for the lowest possible risk level for the whole system. Keeping in mind that factors C and G are not significant, the management must keep factors A, B, D, E, F and H at the optimal levels in order to reduce risk level of the ship to the maximum extent.

<u>Confidence Intervals</u>

Finally, to get an idea about the current variability of each factor, confidence intervals are computed. These are also shown in Table 10.14. The following procedure is used to develop the intervals (Roy (1990)):

Upper confidence level = $Mean + CI$

Lower confidence level = $Mean - CI$

$$CI = \sqrt{\frac{FxV_e}{N_e}}$$

Thus, for factors A, B, C, D, E, F, G and H,

$$CI = \sqrt{\frac{5.32x4.48}{6}} = 1.99$$

The CI for interaction $D{\times}E$:

$$CI = \sqrt{\frac{5.32x4.48}{3}} = 2.82$$

These confidence intervals are reflected in Table 10.14.

<u>Step 8: Recommend for implementation</u>

As a result of the above study, the following changes are recommended in the design, operation and management system:

- Average level of preventative maintenance policy should be adopted. Factor A_2 is made operative.
- High degree of machinery redundancy is recommended. Factor B_1 is more preferable.
- Average fire-fighting capability is adequate for the system. Factor C_2 is selected.
- Ship owner management quality should be high. Factor D_1 is strongly urged.
- Enhanced survey programme should be adopted. Factor E_1 is strongly urged.
- Low navigation equipment is adequate. Factor F_3 is selected.
- Average corrosion control is recommended. Factor G_2 is selected.
- Competent crew operation quality is essential. Factor H_1 is strongly recommended.

10.2.7 Conclusion

This Section has introduced the Taguchi philosophy to maritime safety engineering. It provides a basic understanding and skill in utilising the Taguchi concepts and methodologies in safety related applications. A safety optimisation framework using Taguchi concepts is described and an application example is used to demonstrate how Taguchi concepts can be used to improve safety performance of a ship throughout its life-cycle via optimising its design features, operational characteristics and ship owner management quality. The results of this study show that the Taguchi methods, which have been employed for improving manufacturing processes, may provide an alternative tool for risk analysis in maritime safety engineering.

10.3 A Multiple Criteria Decision Making Approach

10.3.1 Introduction

Safety and cost are among the most important objectives that need to be considered in the design and operational processes of a large maritime engineering system. Formal multiple criteria decision making techniques including Multiple-Attribute Utility Analysis (MAUA) may be used to generate the best compromise compromise designs. Many multiple criteria decision making techniques need to be investigated in detail for their appropriate application in a practical environment. The theme of this section is to examine several multiple criteria decision analysis methods using examples for demonstrating their application in safety and cost synthesis.

10.3.2 Safety and Cost Modelling

10.3.2.1 Safety Modelling

Safety synthesis of an engineering system is usually conducted by aggregating safety assessments for its sub-systems, components and failure modes. A safety assessment framework may constitute a hierarchical structure with failure modes at the bottom level (Wang et al. (1996)). A failure mode could be described in several ways, for example in terms of failure likelihood, consequence severity and failure consequence probability using linguistic variables. This is a natural and sensible way for capturing ambiguity and uncertainty inherent in safety assessment.

Fuzzy sets are well suited to characterizing linguistic variables by fuzzy memberships to the defined categories for a particular situation. Failure likelihood, consequence severity and failure consequence probability could all be characterised using the same set of categories but different membership functions. In this way, the safety associated with a failure mode may also be modelled using fuzzy sets.

For example, the fuzzy safety description (*S*) associated with a failure mode can be defined as the following product of the fuzzy sets of the related failure likelihood (*L*), consequence severity (*C*) and failure consequence probability (*E*) (Wang et al. (1995, 1996)):

$$S = C \circ E \times L$$

where the symbol "$\circ$" represents the composition operation and "$\times$" the Cartesian product operation in fuzzy set theory. If seven categories are used to describe fuzzy sets, the above product could generate the safety description of the failure mode as a fuzzy set as follows:

$$S = [\mu_1/1,\ \mu_2/2,\ \mu_3/3,\ \mu_4/4,\ \mu_5/5,\ \mu_6/6,\ \mu_7/7]$$

where μ_i denotes the membership degree of the failure mode to the *i*th category.

Similar fuzzy sets could be generated for describing the safety of other failure modes, which could be aggregated using conventional fuzzy operations to generate safety descriptions for the components, the subsystems and the whole system of the assessment hierarchy. However, this process may lead to information loss.

Alternatively, safety could be more clearly expressed and communicated using linguistic variables (or assessment grades) such as "*Poor*", "*Average*", "*Good*" and "*Excellent*". Such assessment grades could be defined as distinctive safety standards on the basis of safety guidelines, regulations, laws and other situations specific to the engineering system in question. If the above four linguistic variables are used, then the safety of a failure mode could be described using the following expectation or distribution:

$$S = \{(\beta_1, Poor), (\beta_2, Average),\ (\beta_3, Good), (\beta_4, Excellent)\}$$

where β_j denotes the degree of belief that the safety of the failure mode should be assessed to the *j*th assessment grade. A safety distribution provides a panoramic view about the safety status of a failure mode, component, subsystem or the whole system. It can be used to identify areas for improvement and to simulate action plans to improve safety. β_j could be generated using various ways, for example by analysing historical data using statistical approaches if such data is available; otherwise expert judgements could be used to estimate β_j. If assessment grades are initially defined as fuzzy sets, then β_j could be generated from the fuzzy safety description using the best-fit method as described by (Wang et al. (1995)).

There are other ways to describe safety. The simplest approach would be to use a scale for scoring the safety of a failure mode. While this may be easy for safety aggregation to produce an average indicator about system safety, it could not capture uncertainty inherent in safety assessment and thereby the credibility of such assessment may become questionable. Unfortunately, several well known multiple criteria decision analysis methods, which could be used for safety synthesis, can only be implemented using certain types of scores. This will be discussed in detail in the next section.

10.3.2.2 Cost Modelling

Safety is closely related to cost. Although safety must have paramount importance over cost in most situations, there are cases where safety standards are already achieved and cost effectiveness needs to be given more attention. In such cases, cost should be analysed in conjunction with safety. Costs can be modelled using the methods discussed in Section 10.3.2.1. Costs related to safety improvement are usually affected by a number of factors (Wang et al. (1996)), including

a. Costs for the provision of redundancies of critical components, the provision of protection systems and alarm systems to reduce or eliminate the probabilities of occurrence of undesirable events, and the use of more reliable components.

b. Costs of labour incurred in redesign of the system.

c. Benefits resulting from the likelihood reduction of undesirable events and the improvement of system efficiency as a result of the improvement of system safety.

Ideally, costs could be estimated using precise numerical figures so that conventional methods could be applied to analyse costs together with safety. However, this is often not achievable due to the high uncertainty in estimation of safety related costs. Fuzzy sets provide an alternative way to model costs. For example, costs could be described using linguistic variables such as "*Very low*", "*Low*", "*Moderately low*", "*Average*", "*Moderately high*", "*High*" and "*Very high*". If the seven categories are used to describe fuzzy sets, a fuzzy cost description can be represented as $C = [\gamma_1/1,\ \gamma_2/2,\ \gamma_3/3,\ \gamma_4/4,\ \gamma_5/5,\ \gamma_6/6,\ \gamma_7/7]$ where γ_i is the membership degree of the cost to the ith category.

Alternatively, costs could be clearly described using expectations or distributions to indicate to what degrees costs are preferred, for example using linguistic variables (or assessment grades) such as "*Slightly preferred*", "*Moderately preferred*", "*Preferred*" and "*Greatly preferred*". Such an assessment grade could be defined as a clear cost threshold that an organisation determines for a specific situation. If the above four linguistic variables are used, then the cost of a design option could be described using the following expectation or distribution

$$C = \{(\beta_1, Slightly\ preferred), (\beta_2, Moderately\ preferred),$$
$$(\beta_3,\ preferred), (\beta_4, Greatly\ preferred)\}$$

where β_j denotes the degree of belief that the cost of the design option should be assessed to the jth assessment grade. A cost distribution provides a range of possible financial consequences with different probabilities, which may be incurred in order to develop and adopt the design option. As discussed before, β_j could be generated using various ways, either statistically or subjectively.

10.3.3.3 Safety and Cost Modelling – an Example

In safety modelling, the safety associated with a failure mode of a component may be judged by multiple designers. A diagram for synthesis of the safety for a failure mode is shown in Figure 10.16. Suppose there are *e* designers, each of whom is given a relative weight in the design selection process. The designers' judgements can be aggregated to generate assessments on the safety of failure modes, which can in turn be aggregated to produce assessments for component safety. Assessments for component safety can eventually be aggregated to generate an assessment for system safety using various methods such as those to be investigated in the next section.

In cost modelling, the cost incurred for each design option can also be judged by *e* designers. These judgements can be aggregated to generate an assessment for a design option using various methods such as those to be investigated in the next section. A diagram for synthesis of costs incurred for design options by multiple designers is shown in Figure 10.17.

In this section, safety and cost modelling is discussed for an engineering system in order to demonstrate the multiple criteria decision analysis methods in the next section. Consider a hydraulic hoist transmission system of a marine crane (Wang et al. (1995, 1996)), which is used to control the crane motions such as hoisting down loads as required by the operator. It consists of five subsystems: the hydraulic oil tank, the auxiliary system, the control system, the protection system and the hydraulic servo transmission system. Suppose there are four options for selection by four designers. The safety modelling and cost modelling of the four design options are described as follows using expectations or distributions. To simplify discussion and without loss of generality, the same set of evaluation grades are used to model both safety and cost, that is "*Slightly preferred*", "*Moderately preferred*", "*Preferred*" and "*Greatly preferred*". More detailed discussions about safety and cost modelling can be found in (Wang et al. (1996)).

Option 1: No failure mode is eliminated in the design review process.

For this first design option, suppose the safety assessments provided by the four designers are the same and are represented as the following expectation:

$$\begin{aligned} S_1^1 = S_1^2 = S_1^3 = S_1^4 \\ = \{&(0.122425,\ \textit{Slightly preferred}), \\ &(0.180205,\ \textit{Modertaely preferred}), \\ &(0.463370,\ \textit{Preferred}), \\ &(0.233999,\ \textit{Greatly preferred})\} \end{aligned}$$

For this option, there is no additional cost for eliminating failure modes. Suppose the four designers judge the cost incurred for this option as follows:

$$\begin{aligned} C_1^1 = C_1^2 = C_1^3 = C_1^4 \\ = \{&(0,\ \textit{Slightly preferred}), \\ &(0,\ \textit{Modertaely preferred}), \\ &(0,\ \textit{Preferred}), \\ &(1,\ \textit{Greatly preferred})\} \end{aligned}$$

Option 2: Eliminate "hoist up limit failure" and "hoist down limit failure" associated with the protection system.

For this second design option, suppose the safety assessments provided by the four designers are represented as follows:

$$\begin{aligned} S_2^1 = S_2^2 = S_2^3 = S_2^4 \\ = \{&(0.102676,\ \textit{Slightly preferred}), \\ &(0.156934,\ \textit{Modertaely preferred}), \\ &(0.38486,\ \textit{Preferred}), \\ &(0.355531,\ \textit{Greatly preferred})\} \end{aligned}$$

Suppose the four designers have different opinions about the costs incurred to eliminate the failure modes and their individual assessments are given as follows:

$$C_2^1 = \{(0.054309,\ \textit{Slightly preferred}),\ (0.066442,\ \textit{Modertaely preferred}),\ (0.821848,\ \textit{Preferred}),\ (0.057400,\ \textit{Greatly preferred})\}$$

$$C_2^2 = \{(0.102638,\ \textit{Slightly preferred}),\ (0.134831,\ \textit{Modertaely preferred}),\ (0.657202,\ \textit{Preferred}),\ (0.105330,\ \textit{Greatly preferred})\}$$

$$C_2^3 = \{(0,\ \textit{Slightly preferred}),\ (0,\ \textit{Modertaely preferred}),\ (1,\ \textit{Preferred}),\ (0,\ \textit{Greatly preferred})\}$$

$$C_2^4 = \{(0.067060,\ \textit{Slightly preferred}),\ (0.083011,\ \textit{Modertaely preferred}),\ (0.777240,\ \textit{Preferred}),\ (0.072689,\ \textit{Greatly preferred})\}$$

Option 3: Eliminate the failure modes involving "major leak" and "no output from the package motor" associated with the hydraulic servo transmission system.

For the third design option, suppose the safety assessments provided by the four designers are represented as follows:

$$S_3^1 = S_3^2 = S_3^3 = S_3^4 = \{(0.022722,\ \textit{Slightly preferred}),\ (0.033659,\ \textit{Modertaely preferred}),\ (0.073367,\ \textit{Preferred}),\ (0.870253,\ \textit{Greatly preferred})\}$$

Suppose the four designers' individual cost assessments are given as follows:

$$C_3^1 = \{(0.067604,\ \textit{Slightly preferred}),\ (0.084062,\ \textit{Modertaely preferred}),\ (0.777037,\ \textit{Preferred}),\ (0.071297,\ \textit{Greatly preferred})\}$$

$$C_3^2 = \{(0.102638, \textit{Slightly preferred}), (0.134831, \textit{Modertaely preferred}), (0.657202, \textit{Preferred}), (0.105330, \textit{Greatly preferred})\}$$

$$C_3^3 = \{(0.067060, \textit{Slightly preferred}), (0.083011, \textit{Modertaely preferred}), (0.777240, \textit{Preferred}), (0.072689, \textit{Greatly preferred})\}$$

$$C_3^4 = \{(0.067060, \textit{Slightly preferred}), (0.083011, \textit{Modertaely preferred}), (0.777240, \textit{Preferred}), (0.072689, \textit{Greatly preferred})\}$$

Option 4: Eliminate the two failure modes associated with the protection system in design option 2 and the two failure modes associated with the hydraulic servo transmission system in design option 3.

For the fourth design option, the safety assessments provided by the four designers are given by:

$$S_4^1 = S_4^2 = S_4^3 = S_4^4 = \{(0.013049, \textit{Slightly preferred}), (0.019045, \textit{Modertaely preferred}), (0.035897, \textit{Preferred}), (0.932027, \textit{Greatly preferred})\}$$

The four designers' individual cost assessments are given as follows:

$$C_4^1 = \{(0.059846, \textit{Slightly preferred}), (0.822751, \textit{Modertaely preferred}), (0.062553, \textit{Preferred}), (0.054850, \textit{Greatly preferred})\}$$

$$C_4^2 = \{(0.028571, \textit{Slightly preferred}), (0.912923, \textit{Modertaely preferred}), (0.031480, \textit{Preferred}), (0.027027, \textit{Greatly preferred})\}$$

$$C_4^3 = \{(0.057708, \textit{Slightly preferred}), (0.826250, \textit{Modertaely preferred}), (0.062819, \textit{Preferred}), (0.053223, \textit{Greatly preferred})\}$$

$$C_4^4 = \{(0, \textit{Slightly preferred}), (1, \textit{Modertaely preferred}), (0, \textit{Preferred}), (0, \textit{Greatly preferred})\}$$

10.3.3 Safety and Cost Synthesis Using Typical Multiple Criteria Decision Analysis (MCDA) Methods

Once safety and cost are assessed for a design option, there is a need to combine the assessments to provide an overall assessment for the option and eventually rank it against others. Several methods could be used in such a synthesis process. In this section, three methods are discussed and compared in dealing with the example presented in Section 10.3.2, including the additive utility function approach, the analytical hierarchy process (AHP) approach and the evidential reasoning approach. The assessment hierarchy for the example is shown in Figure 10.18.

Let ω_S and ω_C denote the relative weights of safety and cost, and ω_1, ω_2, ω_3, ω_4 the relative weights of the opinions of designers 1, 2, 3 and 4, respectively. For demonstration purpose, suppose safety is twice as important as cost and the opinions of designers 2 and 3 are twice as important as those given by designers 1 and 4 (i.e. $\omega_S = 2\omega_C$ and $\omega_2 = \omega_3 = 2\omega_1 = 2\omega_4$). Suppose the relative weights of the same group of criteria are normalised so that they are added to one. Then, we have $\omega_S = 0.6667$, $\omega_C = 0.3333$; and $\omega_2 = \omega_3 = 0.3333$, $\omega_1 = \omega_4 = 0.1667$. It should be noted that a range of weights could be assigned to test the robustness of the assessments generated.

10.3.3.1 Additive Utility Function Approach

Before this method can be applied, each assessment of a design option on either safety or cost given by a designer must be quantified using for example a score. Since an assessment in the example is represented as an expectation using the four evaluation grades, we need to quantify the grades first for example by using a scale or estimating the utilities of the grades (Winston (1994)). Suppose the utilities of the four evaluation grades are given as follows (Wang et al. (1996)):

$$u(\textit{Slightly preferred}) = 0.217$$

$$u(\textit{Moderately preferred}) = 0.478$$

$$u(\textit{Preferred}) = 0.739$$

$$u(\textit{Greatly preferred}) = 1$$

Then the scores of the four options on both safety and cost for each designer can be calculated as the following weighted average scores of the expectations with the degrees of belief used as weights:

Option 1

$u_1(safety/designer\ 1) = u_1(safety/designer\ 2)$

$= u_1(safety/designer\ 3) = u_1(safety/designer\ 4)$

$= 0.122425 \times 0.217 + 0.180205 \times 0.478 + 0.46337 \times 0.739 + 0.233999 \times 1$

$= 0.6891$

$u_1(cost/designer\ 1) = u_1(cost/designer\ 2) =$

$u_1(cost/designer\ 3) = u_1(cost/designer\ 4)$

$= 0 \times 0.217 + 0 \times 0.478 + 0 \times 0.739 + 1 \times 1 = 1$

Option 2

$u_2(safety/designer\ 1) = u_2(safety/designer\ 2)$

$= u_2(safety/designer\ 3) = u_2(safety/designer\ 4)$

$= 0.102676 \times 0.217 + 0.156934 \times 0.478 + 0.38486 \times 0.739 + 0.355531 \times 1$

$= 0.7372$

$u_2(cost/designer\ 1)$

$= 0.054309 \times 0.217 + 0.066442 \times 0.478 + 0.821848 \times 0.739 + 0.0574 \times 1$

$= 0.7083$

$u_2(cost/designer\ 2) = 0.6777$

$u_2(cost/designer\ 3) = 0.7390$

$u_2(cost/designer\ 4) = 0.7013$

Option 3

$u_3(safety/designer\ 1) = u_3(safety/designer\ 2)$

$= u_3(safety/designer\ 3) = u_3(safety/designer\ 4)$

$= 0.022722 \times 0.217 + 0.033659 \times 0.478 + 0.073367 \times 0.739 + 0.870253 \times 1$

$= 0.9455$

$u_3(cost/designer\ 1)$

$= 0.067604 \times 0.217 + 0.084062 \times 0.478 + 0.777037 \times 0.739 + 0.071297 \times 1$

$= 0.7004$

$u_3(cost/designer\ 2) = 0.6777$

$u_3(cost/designer\ 3) = 0.7013$

$u_3(cost/designer\ 4) = 0.7013$

Option 4

$u_4(safety/designer\ 1) = u_4(safety/designer\ 2)$

$= u_4(safety/designer\ 3) = u_4(safety/designer\ 4)$

$= 0.013049 \times 0.217 + 0.019045 \times 0.478 + 0.035897 \times 0.739 + 0.932027 \times 1$

$= 0.9705$

$u_4(cost/designer\ 1)$

$= 0.059846 \times 0.217 + 0.822751 \times 0.478 + 0.062553 \times 0.739 + 0.05485 \times 1$

$= 0.5073$

$u_3(cost/designer\ 2) = 0.4929$

$u_3(cost/designer\ 3) = 0.5071$

$u_3(cost/designer\ 4) = 0.4780$

Assessment of Design Options

The above scores show the average assessments of the four design options on both safety and cost provided by the four designers. Note that the four designers provided the same average assessment for each design option on safety. The additive utility function approach operates on average scores, as summarised in a decision matrix shown in Table 10.15.

One way to synthesize the assessments is to generate an overall weight for the cost provided by every designer. For example, the overall weight for the cost provided by designer 1 can be calculated as $0.3333 \times 0.1667 = 0.0556$. The overall weight multiplied by a score results in a weighted score. For example, the weighted score for the safety of design option 1 is given by $0.6667 \times 0.6891 = 0.4594$ and that for the cost of design option 1 provided by designer 1 is given by $0.3333 \times 0.1667 \times 1 = 0.0556$. All the other weighted scores are shown in Table 10.16.

In the additive utility (value in this case) function approach, the weighted scores on the safety and cost attributes are added up for an option, resulting in an overall score for the option. For example, the overall score for option 1 is given by:

$$u(\text{option 1}) = 0.4594 + 0.0556 + 0.1111 + 0.1111 + 0.0556 = 0.7928.$$

Similarly, the overall scores of the other three options are given by

$$u(\text{option 2}) = 0.7273,\ u(\text{option 3}) = 0.8615,\ u(\text{option 4}) = 0.8129.$$

The ranking of the four design options is then given on the basis of the magnitude of their overall scores as follows:

$$\text{option 3} \succ \text{option 4} \succ \text{option 1} \succ \text{option 2}$$

The additive utility (value) function approach provides a simple process for criteria aggregation. To use the method properly, however, one should be aware of its limits and drawbacks. Despite the loss of the original features and diversity of the distributed assessments given in Section 10.3.2.3, this approach assumes preference independence, a linear utility function for each criterion, and direct and proportional compensation among criteria. These assumptions are not always acceptable. For example, a linear utility function implies that the decision maker is neutral to risk. In many decision situations, however, decision makers are often averse to risk. This is particularly the case when safety is assessed. Preference independence means that tradeoffs between two criteria are independent of other criteria. While this is not easy to test, it is not appropriate to assume that this is always satisfied.

10.3.3.2 AHP

AHP is another method that can be used to deal with MCDA problems. AHP is based on the eigenvector method that is usually applied to estimating relative weights of criteria by means of pairwise comparisons. The basic theory on AHP has been described in Chapter 9. In this Chapter, some extra descriptions and discussions of this method are given in order to solve the above design selection problem.

Since it is already assumed that safety is twice as important as cost in selection of design options, a pairwise comparison matrix can be constructed as in Table 10.17, where the element "2" in the second row of the last column means that safety is twice as important as cost.

In AHP, the normalised right eigenvector of the pairwise comparison matrix with respect to its largest eigenvalue is employed as the weights of safety and cost. Suppose A represents the pairwise comparison matrix, or

$$A = \begin{bmatrix} 1 & 2 \\ \frac{1}{2} & 1 \end{bmatrix}$$

W a weight vector or $W = [\omega_s \;\; \omega_c]^{\mathrm{T}}$, and $\lambda_{\max}$ the maximum eigenvalue of the matrix A. Then, W is calculated using the following equation:

$$AW = W\lambda_{\max}$$

There are software packages that can be used to solve the above vector equation to find W (Saaty (1988)). An approximate solution procedure can be found in (Sen and Yang (1998)), as summarised below.

Step 1: Provide an initially normalised vector $W^0 = [1 \;\, 0 \;\cdots\; 0]^T$ and let $t = 0$.

Step 2: Calculate a new eigenvector as follows:

$$W^{t+1} = AW^t$$

Step 3: Calculate the maximum eigenvalue by:

$$\lambda_{\max} = \sum_{i=1}^{n} w^{t+1}$$

Step 4: Normalise and update the eigenvector as follows:

$$\overline{w}^{t+1} = \frac{w^{t+1}}{\lambda_{\max}}, \text{ and let } w^{t+1} = \overline{w}^{t+1} \text{ for all } i=1, \ldots, n$$

Step 5: Calculate the error between the old and new eigenvectors and then check if

$\left|w^{t+1} - w^{t}\right| \leq \delta$ for all i=1, …, n

where δ is a small non-negative real number (say $\delta = 1.0\times10^{-6}$). If the condition is satisfied, go to step 6. Otherwise, let t=t+1 and go to Step 2.

Step 6: Calculate the consistency index (CI) as follows:

$$CI = \frac{\lambda_{\max} - n}{n-1}$$

If $CI \leq 0.1$, the pairwise comparisons provided in the matrix A are satisfactorily consistent. Otherwise, the comparisons need to be revised.

Applying the above procedure to solve the eigenvector equation (n = 2) leads to the following results:

$$W = [\omega_s \ \omega_c] = [0.6667 \ \ 0.3333]$$

In the above pairwise comparison matrix, the weights between safety and cost are already made clear. Generally, if pairwise comparisons are provided for three or more criteria, they may not be completely consistent and as such it is not straightforward to obtain relative weights of criteria from the comparisons. The AHP method and several other methods can be used to generate weights using pairwise comparisons. For example, suppose the pairwise comparison matrix is provided for the importance of the opinions of the four designers, as shown in Table 10.18

A pairwise comparison matrix $A = (a_{ij})_{n\times n}$ is completely consistent if $a_{ij} = a_{ik}a_{kj}$ for all i, j, k =1, …, n where n is the dimension of the matrix. It is easy to show that the comparison matrix of Table 10.18 is completely consistent. Using the above procedure, the normalised eigenvector of the matrix is given by

$$W = [\omega_1 \ \ \omega_2 \ \ \omega_3 \ \ \omega_4] = [0.1667 \ \ 0.3333 \ \ 0.3333 \ \ 0.1667]$$

To use AHP for ranking design options, one also needs to compare them in a pairwise fashion with respect to each criterion. This may not be an easy task in general. Given the assessment data as in Table 10.15, however, a pairwise comparison matrix can be constructed for each criterion (Huang and Yoon (1981)). With respect to safety, for example, the four design options can be compared as in Table 10.19.

A number in Table 10.19 denotes the extent to which one option is more attractive than another. For example, the number "1.4085" in the second column of the last row means that option 4 is 1.4085 times as attractive as option 1 in terms of safety. It can be seen from Table 10.19 that the differences between the four design options are quite small and this would make it difficult to provide direct pairwise comparisons between the options without reference to the assessment data shown in Table 10.15.

In AHP, the scores of the four design options in terms of safety are generated, using the above solution procedure, as the eigenvector of the pairwise comparison matrix with respect to its largest eigenvalue, as shown in Table 10.20.

In a similar way, the pairwise comparison matrices of the cost criteria for the four designers can be generated, as shown in Tables 10.21 to 10.24.

The scores of the four design options in terms of cost for the four designers can also be generated by identifying the eigenvectors of the matrices in Tables 10.21 to 10.24 with respect to their respective largest eigenvalues, as shown in Table 10.25.

In AHP, different ways are suggested to aggregate the scores generated from the pairwise comparisons (Saaty (1988)). One way is to use the simple weighting approach for aggregation from one criteria level to another (Huang and Yoon (1981)). Firstly, aggregate the cost scores of the four designers by multiplying each score with the relevant weight and then adding up the weighted scores for each option, which leads to an aggregated cost score for each option, as shown in Table 10.26. Finally, the safety score and the cost score for an option are multiplied by their weights and then added up to generate an overall score for the option, as shown in Table 10.27.

Based on the overall scores of Table 10.27, the ranking of the four design options are given as follows:

$$\text{option } 3 \succ \text{option } 1 \succ \text{option } 4 \succ \text{option } 2$$

This ranking is different from that generated using the additive utility function approach in that the positions of option 1 and option 4 are swapped. In fact, the AHP method does not significantly differentiate the two options, as the difference between the overall scores of the two options is very small. AHP is usually applied to generating relative weights. However, the use of AHP to assess design options may lead to problems like rank reversal (Belton (1986), Islei and Lockett (1988), Stewart (1992), Barzilai (1997)), that is, the introduction of new options for assessment may cause the unexpected and irrational change of the ranking of the current options.

10.3.3.3 The Evidential Reasoning Approach

The evidential reasoning (ER) approach can be used to deal with multiple criteria decision analysis problems of both a quantitative and qualitative nature with uncertainty (Yang and Singh (1994), Yang and Sen (1994), Yang (2001)). It can process several types of information within an ER framework. The ER framework is different from most conventional MCDA modelling frameworks in that it employs a belief structure to represent an assessment as a distribution. In Section 10.3.2.3, four evaluation grades were defined as follows:

$$H = \{H_1, H_2, H_3, H_4\}$$
$$= \{Slightly\ preferred,\ Modertaely\ preferred,\ Preferred,\ Greatly\ preferred\}$$

Using the four evaluation grades, the assessment of an attribute A_1 on option O_1, denoted by $S(A_1(O_1))$, can be represented using the following belief structure:

$$S(A_1(O_1)) = \{(H_1, \beta_{1,1}), (H_2, \beta_{2,1}), (H_3, \beta_{3,1}), (H_4, \beta_{4,1})\} \text{ -- (}$$

where $1 \geq \beta_{n,1} \geq 0$ (n = 1, ..., 4) denotes the degree of belief that the attribute A_1 is assessed to the evaluation grade H_n. $S(A_1(O_1))$ reads that the attribute A_1 is assessed to the grade H_n to a degree of $\beta_{n,1} \times 100\%$ (n = 1, ..., 4) for option O_1.

There must not be $\sum_{n=1}^{4}\beta_{n,1} > 1$. $S(A_1(O_1))$ can be considered to be a complete distributed assessment if $\sum_{n=1}^{4}\beta_{n,1} = 1$ and an incomplete assessment if $\sum_{n=1}^{4}\beta_{n,1} < 1$. In the ER framework, both complete and incomplete assessments can be accommodated (Yang (2001)).

In the ER framework, a MCDA problem with M attributes A_i (i = 1, ..., M), K options O_j (j = 1, ..., K) and N evaluation grades H_n (n = 1, ..., N) for each attribute is represented using an extended decision matrix with $S(A_i(O_j))$ as its element at the ith row and jth column where $S(A_i(O_j))$ is given as follows:

$$S(A_i(O_j)) = \{(H_n, \beta_{n,i}(O_j)),\ n = 1,\cdots,N\} \quad i = 1, \ldots, M, \quad j=1, \ldots, K$$

It should be noted that an attribute can have its own set of evaluation grades that may be different from those of other attributes (Yang (2000)).

Instead of aggregating average scores, the ER approach employs an evidential reasoning algorithm developed on the basis of the evidence combination rule of the Dempster-Shafer theory to aggregate belief degrees (Yang and Singh (1994), Yang and Sen (1994), Yang (2001)). Thus, the ER approach is different from traditional MCDA approaches, most of which aggregate average scores.

Suppose ω_i is the relative weight of the attribute A_i and is normalised so that $1 \geq \omega_i \geq 0$ and $\sum_{i=1}^{L}\omega_i = 1$ where L is the total number of attributes in the same group for aggregation. To simplify the discussion, only the combination of complete assessments is examined. The description of the recursive ER algorithm capable of aggregating both complete and incomplete assessments is detailed in (Yang and Sen (1994), Yang (2001)). Without loss of generality and for illustration purpose, the ER algorithm is presented below for combining two assessments only.

Suppose the second assessment $S(A_2(O_1))$ is given by

$$S(A_2(O_1)) = \{(H_1, \beta_{1,2}), (H_2, \beta_{2,2}), (H_3, \beta_{3,2}), (H_4, \beta_{4,2})\}$$

The problem is to aggregate the two assessments $S(A_1(O_1))$ and $S(A_2(O_1))$ to generate a combined assessment $S(A_1(O_1)) \oplus S(A_2(O_1))$. Suppose $S(A_1(O_1))$ and $S(A_2(O_1))$ are both complete. Let

$$m_{n,1} = \omega_1\beta_{n,1} \ (n = 1, \ldots, 4) \text{ and } m_{H,1} = 1 - \omega_1\sum_{n=1}^{4}\beta_{n,1} = 1 - \omega_1$$

$$m_{n,2} = \omega_2\beta_{n,2} \ (n = 1, \ldots, 4) \text{ and } m_{H,2} = 1 - \omega_2\sum_{n=1}^{4}\beta_{n,2} = 1 - \omega_2$$

where each $m_{n,j}$ (j = 1, 2) is referred to as basic probability mass and each $m_{H,j}$ is the remaining belief unassigned to H_j (j = 1, 2, 3, 4).

The ER algorithm is used to aggregate the basic probability masses to generate combined probability masses, denoted by m_n (n=1, ..., 4) and m_H using the following equations:

$$m_n = k(m_{n,1}m_{n,2} + m_{H,1}m_{n,2} + m_{n,1}m_{H,2}), \qquad (n = 1, \ldots, 4)$$

$$m_H = k(m_{H,1}m_{H,2})$$

where

$$k = \left(1 - \sum_{t=1}^{4}\sum_{\substack{n=1 \\ n \neq t}}^{4} m_{t,1}m_{n,2}\right)^{-1}$$

The combined probability masses can then be aggregated with the third assessment in the same fashion. The process is repeated until all assessments are aggregated. The final combined probability masses are independent of the order in which individual assessments are aggregated.The combined degrees of belief β_n (n = 1, ..., 4) are generated by:

$$\beta_n = \frac{m_n}{1 - m_H} \qquad (n = 1, \ldots, 4)$$

The combined assessment for the option O_1 can then be represented as follows:

$$S(O_1) = \{(H_1, \beta_1), (H_2, \beta_2), (H_3, \beta_3), (H_4, \beta_4)\}$$

An average score for O_1, denoted by $u(O_1)$, can also be provided as the weighted average of the scores (utilities) of the evaluation grades with the belief degrees as weights, or

$$u(O_1) = \sum_{i=1}^{4} u(H_i)\beta_i$$

where $u(H_i)$ is the utility of the ith evaluation grade H_i. For i = 1, for example, we have $u(H_1) = u(\textit{Slightly preferred}) = 0.217$.

An intelligent decision system (IDS[1]) has been developed on the basis of the ER approach (Yang and Xu (2000)). The IDS software is designed to transform the lengthy and tedious model building and result analysis process into an easy window-based click and design activity. The rest of this sub-section is devoted to demonstrating the solution process of the above safety and cost-based design selection problem using the IDS software.

The main window of IDS for solving the design selection problem is shown in Figure 10.19, which has a menu bar, a tool bar and a model display window. The hierarchy of the assessment criteria can be readily constructed using the modelling menu or the related short cuts on the tool bar. IDS also provides an assistant model builder for building large-scale models that may have hundreds of criteria and options.

In the model display window, each criterion object is coloured in blue and has three boxes for displaying the criterion name, its weight and average score. For example, the criterion "1. Safety" has a weight of "0.6667" and its average score for "Design option 1" is "0.6956". Each alternative object is coloured in yellow and also has three boxes for displaying the alternative name, its ranking and overall average score. For example, "Design option 1" is ranked the third and has an overall average score of "0.776495". Apart from an average score, IDS is capable of generating a distributed assessment for each option on any criterion. Figure 10.20

[1] A free demo version of IDS can be obtained from Dr J B Yang via email: jian-bo.yang@umist.ac.uk

shows the overall distributed assessment of "Design option 1". In Figure 10.20, the degrees of belief to the evaluation grades clearly show the merits and drawbacks of the design option.

In IDS, a number of dialog windows are designed to support model building, data input, result analysis, reporting and sensitivity analysis. For example, Figure 10.20 is generated using an IDS dialog window for reporting results graphically. Figure 10.21 shows an IDS dialog window for data input for "Design option 1" on a cost criterion for "2.1 Designer 1". All data can be entered using similar dialog windows, whether they are precise numbers, random numbers with probabilities, or subjective assessments. Figure 10.22 shows the visual cross comparison of the four design options on both safety and cost generated using the IDS visual comparison dialog window.

In IDS, AHP and other methods are used for generating relative weights of criteria and the evidential reasoning approach is used to aggregate criteria from the bottom level of criteria to the top level criterion "Design selection". The overall assessment for each option can be characterised as shown for option 1 in Figure 10.20. In IDS, dialog window are designed to support visually scaling the evaluation grades or estimating the utilities of the grades. For example, Figure 10.23 shows a utility curve for the four evaluation grades. The curve can be changed onscreen to suit the requirements of individual designers. For the given utility curve, the average scores for the four design options are generated as shown in Table 10.28.

Based on the overall scores of Table 10.28, the ranking of the four design options are given as follows:

$$\text{option 3} \succ \text{option 4} \succ \text{option 1} \succ \text{option 2}$$

The above ranking is the same as that generated using the additive utility function approach. Apart from the average scores and the related ranking for the design options, however, the ER approach can provide much richer information for analysis. The distributed assessment at any attribute provides a panoramic view on each design option so that the benefits and risks involved in selecting an option are made clear to the designers.

10.3.4 Discussion of the Results

When designing a large maritime engineering product, especially at the initial design stages, there are usually several design options. It should be noted that such options are produced at the top level where only non-numerical data may be available. The information available for making decisions on which option to select at this stage may be incomplete. As a design proceeds to a more detailed stage, the selection of design options at lower levels is required and again a similar process for selecting a particular design option may be required. It should be noted that the decision making process at all levels needs to deal with multiple objectives and may involve uncertain or incomplete information. The MCDA methods described may prove useful to select the best design option by taking into account safety and other design objectives in a rational manner.

As the best design option is chosen, the design can further proceed. More and more information becomes available for more detailed safety analysis. Decision making may need to be carried out at the next level. At this stage, it may be the case that only part of the information is complete for quantitative safety estimate. This may also be true for modelling of other design objectives. In such cases, MCDA techniques may be required to combine safety estimate with other design objectives to arrive at the best designs within both technical and economic constraints.

As the design further proceeds, it reaches a stage where there is enough information for carrying out design optimisation based on quantitative safety assessment. At this stage, safety may be assessed using various safety assessment techniques in terms of likelihood of occurrence and magnitude of consequences. A mathematical model can be formulated and then again MCDA techniques can be used to process the model in order to optimise the design.

It is also worth mentioning that the MCDM techniques described can also be used to make decisions in maritime operations.

10.3.5 Conclusion

There is a great potential for MCDA methods to be applied in the design selection and optimisation processes. Appropriate application of MCDA tools can facilitate decision making in maritime engineering design and operations to improve efficiency.

10.4 References (Chapter 10)

1. ASII, (1989) "Taguchi Methods: Implementation Manual", American Supplier Institute Inc., Dearborn, MI, USA.
2. Barzilai J., (1997) "Deriving Weights from Pairwise Comparison Matrices", Journal of Operational Research Society, Vol. 48, pp.226-1232.
3. Belton V., (1896) "Comparison of the Analytic Hierarchy Process and a Simple Multi-Attribute Value Function ", European Journal of Operational Research, Vol. 26, pp.7-21.
4. Bendell A., (1988) "Introduction to Taguchi Methodology", Taguchi Methods: Proceedings of the 1988 European Conference, Elsevier Applied Science, London, England, pp. 1-14.
5. Bryne D.M. and Taguchi S., (1986) "The Taguchi Approach to Parameter Design", ASQC Quality Congress Transactions, Anaheim, CA, pp. 168.
6. Buxton I.L., Cuckson B.R. and Thanopoulos G.P., (1997) "The Assessment of Ship Susceptibility to Loss or Damage Using Classification Society Information", Conference Proceedings of Marine Risk Assessment: A Better Way to Manage Your Business, IMarE, London, 8-9, May.
7. Fowlkes W.Y. and Creveling C.M., (1995) "Engineering Methods for Robust Product Design: Using Taguchi Methods in Technology and Product Development", Addison-Wesley Publishing Company.
8. Gunter B., (1987) "A Perspective on the Taguchi Methods", Quality Progress, June, pp. 44-52.
9. Haykin S., (1999) "Neural Networks - a Comprehensive Foundation", Prentice-Hall, Inc.
10. Hecht-Nielsen R., (1990), "Neuralcomputing", Addison-Wesley Publishing Company.
11. Huang C.L. and Yoon K., (1981) "Multiple Attribute Decision Making Methods and Applications", A State-of-Art Survey. Springer-Verlag, Berlin.

12. Islei G. and Lockett A.G., (1988) "Judgmental Modeling Based on Geometric Least Square", European Journal of Operational Research, Vol. 36, pp.27-35.

13. Kackar R.N., (1985) "Off-line Quality Control, Parameter Design, and the Taguchi Methods - Response", Journal of Quality Technology, Vol. 17, No. 4, pp. 207-209.

14. Kuo C, (1998) "Managing Ship Safety", LLP, London.

15. Logothetis N. and Salmon J.P., (1988) "Tolerance Design and Analysis of Audio-Circuits", Taguchi Methods: Proceedings of the 1988 European Conference, Elsevier Applied Science, London, England, pp. 161-175.

16. Matlab, (1994) "Matlab Neural Network Toolbox User's Guide for use with MATLAB", The Math Work Inc.

17. Meisl C.J., (1990), "Parametric Cost Analysis in the TQM Environment", Paper presented at the 12th Annual Conference of International Society of Parametric Analysis, San Diego, CA, USA.

18. Nelson M. and Illingworth W.T., (1990) "A Practical Guide to Neural Nets", Addison-Wesley Publishing Company Inc.

19. Phadke S.M., (1989) "Quality Engineering Using Robust Design", Prentice Hall, Englewood Cliffs, NJ.

20. Roskilly A.P. and Mesbahi E., (1996) "Marine System Modelling Using Artificial Neural Networks: an Introduction to the Theory and Practice", Trans. IMarE, Vol. 108, Part 3, pp. 185-201.

21. Ross P.J., (1988) "Taguchi Techniques for Quality Engineering", NY, McGraw-Hill.

22. Roy R.K., (1990) "A Primer on the Taguchi Method," Van Nostrand & Reinhold, NY, USA.

23. Saaty T.L., (1988), The Analytic Hierarchy Process, University of Pittsburgh.

24. Sen P. and Yang J. B., (1998) "Multiple Criteria Decision Support in Engineering Design", Springer-Verlag, London.

25. Sii H.S., Ruxton T. and Wang J., (2000) "An Offshore and Maritime Safety-based Decision Support System: A Role for Artificial Neural Networking", 19th International Conference on Offshore Mechanics and Arctic Engineering, 14-17 February, New Orleans, USA.

26. Sii H.S., (2001) "Marine and Offshore Safety Assessment", PhD Thesis, Staffordshire University/Liverpool John Moores University, UK.

27. Sii H.S., Ruxton T. and Wang J., (2001) "Taguchi Concepts & and Their Applications in Marine and Offshore Safety Studies", Journal of Engineering Design, Vol. 12, No.4, pp. 331-358.

28. Stewart T. J., (1992) "A Critical Survey on the Status of Multiple Criteria Decision Making Theory and Practice", OMEGA International Journal of Management Science, Vol. 20, No. 5-6, pp.569-586.

29. Suh N.P., (1990), "The Principles of Design", Oxford University Press, UK.

30. Taguchi G., (1986) "Introduction to Quality Engineering", Asian Productivity Organisation, distributed by American Supplier Institute Inc., Dearborn, MI, USA.

31. Taguchi G. and Konishi S., (1987) "Orthogonal Arrays and Linear Graphs", American Supplier Institute Inc., Dearborn, MI, USA.

32. Taguchi G., Elsayed E. and Hsiang T., (1989) "Quality Engineering in Production Systems", McGraw Hill, NY, USA.

33. Wang J., Sii H.S. and Pillay A., (2001) "Some Novel Risk Modelling and Decision Making Approaches", Journal of UK Safety and Reliability Society, Vol. 20, No. 3, pp. 12-24.

34. Wang J., Yang J.B. and Sen P., (1995) "Safety Analysis and Synthesis Using Fuzzy Sets and Evidential Reasoning", Reliability Engineering & System Safety, Vol. 47, No. 2, pp.103-118.

35. Wang J., Yang J.B. and Sen P., (1996) "Multi-Person and Multi-Attribute Design Evaluations Using Evidential Reasoning Based on Subjective Safety and Cost Analysis", Reliability Engineering & System Safety, Vol. 52, pp.113-127.

36. Wille R., (1990) "Landing Gear Weight Optimisation Using Taguchi Analysis", Paper presented at the 49th Annual International Conference of Society of Allied Weight Engineers Inc., Chandler, AR, USA.

37. Yang J.B. and Singh M.G., (1994) "An Evidential Reasoning Approach for Multiple Attribute Decision Making with Uncertainty", IEEE Transactions on Systems, Man, and Cybernetics, Vol. 24, No. 1, pp.1-18.

38. Yang J.B. and Sen P., (1994) "A General Multi-Level Evaluation Process for Hybrid MADM with Uncertainty", IEEE Transactions on Systems, Man, and Cybernetics, Vol. 24, No. 10, pp.1458-1473.

39. Yang J.B. and Sen P., (1997) "Multiple Attribute Design Evaluation of Complex Engineering Products Using the Evidential Reasoning Approach", Journal of Engineering Design, Vol. 8, No. 3, pp.211-230.

40. Yang J.B. and Xu D.L., (2000) "Intelligent Decision System via Evidential Reasoning, Version 1.1, IDSL, Cheshire, England.

41. Yang J.B., (2001) "Rule and Utility Based Evidential Reasoning Approach for Complex Decision Analysis", European Journal of Operational Research, Vol. 131, No. 1, pp.31-61.

42. Yang J.B. and Xu D.L, (2002a) "On the Evidential Reasoning Algorithm for Multiattribute Decision Analysis under Uncertainty", IEEE Transactions on Systems, Man, and Cybernetics Part A: Systems and Humans, Vol. 32, No. 3, pp.289-304.

43. Yang J.B. and Xu D.L, (2002b) "Nonlinear Information Aggregation via Evidential Reasoning in Multiattribute Decision Analysis under Uncertainty", IEEE Transactions on Systems, Man, and Cybernetics Part A: Systems and Humans, Vol. 32, No. 3, pp.376-393.

Table 10.1 Artificial Neural Networks Characteristics

Number of layers	2
Number of input units	2
Number of output units	1
Learning rule	Back-propagation
Transfer function	Sigmoid: tan-sigmoid for hidden layer, purelin for output layer
Number of neurons (layer 1)	10
Number of neurons (layer 2)	1

Table 10.2 Training Pair Data from LR Defect Data for Bulk Carriers not Lost

	Training	Pairs	Output
Ship	dwt	Age	Hull Incidents Per Year
S1	72,000	16	1.25
S2	81,000	22	1.14
S3	26,500	12	1.58
S4	35,000	14	0.36
S5	26,500	11	1.91
S6	31,500	19	0.53
S7	20,000	13	0.92
S8	25,000	7	0.86
S9	25,000	18	0.50
S10	27,000	18	0.83

Table 10.3 Comparison of Predicted Results with Actual Data on Hull Incidents per year

Test Case	Ann Prediction [Failure Per Year]	Actual From LR Defect Data [failure per year]	Different
T1	1.04	1.18	11.9%
T2	1.53	1.33	15%
T3	1.95	1.64	19%
T4	0.93	0.85	9.4%
T5	0.61	0.61	0%

Table 10.4 ANN Characteristics

Number of layers	2
Number of input units	5
Number of output units	1
Learning rule	Fast back-propagation
Transfer function	Sigmoid: tan-sigmoig for hidden layer, purelin for output layer
Number of neurons (layer 1)	12
Number of neurons (layer 2)	1

Table 10.5 Hypothetical Input Data for Risk Prediction ANN

Ship Owner Management Quality	Operation Quality	Fire-Fighting Capability	Navigation Equipment Level	Machinery Redundancy	Possibility of Vessel Failure
Very low	Very high	Very high	Very high	Very high	Very high
Very high	Very low	Very high	Very high	Very high	Very high
Very high	Very high	Very low	Very high	Very high	Very high
Very high	Very high	Very high	Very low	Very high	Very high
Very high	Very high	Very high	Very high	Very low	Very high
Very high	Very high	Very high	Very high	Very high	Very low
High	High	High	High	High	Low
Average	Average	Average	Average	Average	Average
Very low	Very low	Very low	Very low	Very low	Very high
Very low	Low	Average	High	Very high	Very high
Low	Very low	Very high	Average	High	Very high
Average	High	Very low	Very high	Low	Very high
Very high	Low	High	Very low	Average	Very high
High	Very high	Low	Low	Very low	Very high
High	Very high	Low	Average	Low	High
Low	Average	Very high	Low	High	High
Average	Low	Low	Very high	Very high	High
Very high	High	Average	High	Low	High
Low	Average	High	Low	Average	High
Low	High	Low	Very high	Low	High
High	Low	Very high	Low	Low	High
Low	Low	High	Low	Low	Very high
Low	High	Low	Low	Low	Very high
Low	Low	Low	Low	Very high	Very high
Low	Low	Low	Low	Low	Very high

Table 10.6 Predicted Results by the ANN Model

Ship Owner Management Quality	Operation Quality	Fire-Fighting Capacity	Navigation Equipment Level	Machinery Redundancy	Possibility of Vessel Failure
Low	Very high	Very high	Average	Very high	High
Very high	High	Average	Low	Very low	Very high
Low to very low	High	Average	Low	Very high	Very high
High	Very low	High	Average	High to very High	Very high
High	High	Very low	High	Average	Very high
High	Average	High	High	Very high	Low
Average	High	Average	Average to high	Average	Average
Low to very low	Very high	Very high	Average	High	High to very high
Low	Low	Low	Low	Average	Very high
High	Average to low	Very high	Very high	Very high	High to very high

Table 10.7 A Typical Standardised L9 Orthogonal Array for up to Four Tree Level Control Factors

Combination	Control factor #1	Control factor #2	Control factor #3	Control factor #4
1	1	1	1	1
2	1	2	2	2
3	1	3	3	3
4	2	1	2	3
5	2	2	3	1
6	2	3	1	2
7	3	1	3	2
8	3	2	1	3
9	3	3	2	1

Table 10.8 Some Commonly Used Orthogonal Arrays

Orthogonal array	Number of factors	Number of levels per factor	Number of trial required by orthogonal array	Number of trials in a traditional full factorial experiment
$L_4 (2^3)$	3	2	4	8
$L_8 (2^7)$	7	2	8	128
$L_9 (3^4)$	4	3	9	81
$L_{12} (2^{11})$	11	2	12	2048
$L_{16} (2^{15})$	15	2	16	32768
$L_{16} (4^5)$	5	4	16	1024
$L_{18} (2^1) x (3^7)$	1	2	18	4374
	7	3		

Table 10.9 List of Factors Affecting Ship Safety

Factor Identifier	Factor	Level 1	Level 2	Level 3
A	Preventative maintenance policy	Adequate	Average	Sketchy (identify the malfunction parts)
B	Degree of machinery redundancy	75% High	50% Average	25% Low
C	Fire-fighting capability	High	Average	Low
D	Ship owner management quality	Good	Moderate	Poor (inadequate procedures)
E	Enhanced survey programme	Yes (adequate)	No	Nil
F	Navigation equipment level	High	Average	Low
G	Corrosion control	Good	Average	Poor
H	Crew operation quality	Competence (well-trained)	Average	Poor (inadequate knowledge)

Table 10.10 *L18* of Taguchi Experimental Design

Treatment	1	2	3	4	5	6	7	8
1	1	1	1	1	1	1	1	1
2	1	1	2	2	2	2	2	2
3	1	1	3	3	3	3	3	3
4	1	2	1	1	2	2	3	3
5	1	2	2	2	3	3	1	1
6	1	2	3	3	1	1	2	2
7	1	3	1	2	1	3	2	3
8	1	3	2	3	2	1	3	1
9	1	3	3	1	1	2	3	2
10	2	1	1	3	3	2	2	1
11	2	1	2	1	1	3	3	2
12	2	1	3	2	2	1	1	3
13	2	1	1	2	3	1	1	2
14	2	2	2	3	1	2	1	3
15	2	2	3	1	2	3	2	1
16	2	3	1	3	2	3	1	2
17	2	3	2	1	3	1	2	3
18	2	3	3	2	1	2	3	1

Table 10.11 Ship Safety in Terms of Risk Levels Under Various Treatments

Trial Number	Factor Identifier								Risk Level			S/N-Ratio	Normalised S/N-Ratio
Trial No.	*E*	*D*	*A*	*B*	*C*	*F*	*G*	*H*					-(-27) is added to each S/N-Ratio
1	1	1	1	1	1	1	1	1	6	13	11	-20.36	6.46
2	1	1	2	2	2	2	2	2	6	8	7	-16.96	10.04
3	1	1	3	3	3	3	3	3	24	25	22	-27.49	-0.49
4	1	2	1	1	2	2	3	3	25	28	25	-30.93	-3.93
5	1	2	2	2	3	3	1	1	7	6	7	-16.60	10.40
6	1	2	3	3	1	1	2	2	17	14	17	-24.12	2.88
7	1	3	1	2	1	3	2	3	46	47	50	-33.60	-6.60
8	1	3	2	3	2	1	3	1	22	17	25	-26.68	0.32
9	1	3	3	1	3	2	1	2	33	33	32	-30.28	-3.28
10	2	1	1	3	3	2	2	1	21	22	18	-26.19	0.81
11	2	1	2	1	1	3	3	2	8	8	7	-17.71	9.29
12	2	1	3	2	2	1	1	3	25	24	25	-27.84	-0.84
13	2	2	1	2	3	1	3	2	42	42	38	-32.19	-5.19
14	2	2	2	3	1	2	1	3	42	47	41	-32.75	-5.75
15	2	2	3	1	2	3	2	1	11	17	14	-23.05	3.95
16	2	3	1	3	2	3	1	2	33	33	31	-30.20	-3.20
17	2	3	2	1	3	1	2	3	47	47	42	-33.14	-6.14
18	2	3	3	2	1	2	3	1	25	33	26	-29.01	-2.01

Table 10.12 Main and Interaction Effects

Level	*A*	*B*	*C*	*D*	*E*	*F*	*G*	*H*	*D×E*
1	-11.47	6.53	4.45	25.45	15.98	-2.33	3.97	20.11	16.19
2	18.16	5.8	6.34	2.36	-9.08	-4.12	4.94	10.54	9.35
3	0.21	-5.43	-3.89	-20.9		13.35	-2.01	-23.75	-9.56
4									9.26
5									-6.99
6									-11.35
Total	6.9	6.9	6.9	6.9	6.9	6.9	6.9	6.9	6.9
Sum of Square	445.52	89.90	59.26	1074.2	314	184.76	28.33	1063.7	696.43

Table 10.13 The Final ANOVA Table After Pooling Insignificant Factors

Source/ Factors	Sum of squares	Sum of square (%)	Degree of freedom	Variance	Pooled in error	F-Value	Minimum confidence
A	445.52	11.26	2	222.76		49.72	>99.5
B	89.90	2.27	2	44.95	Pooled	10.03	>99.5
C	52.26	1.49	2	29.63	Pooled	6.61	>99.5
D	1074.2	27.15	2	537.1		119.9	>99.5
E	314	7.94	1	314		70.1	>99.5
F	184.76	4.67	2	92.38		20.62	>99.5
G	28.33	0.72	2	14.17	Pooled	3.16	>99.5
H	1063.7	26.89	2	531.85		118.72	>99.5
D×E	696.43	18.06	2	348.22		77.73	>99.5
Sum	3956.1						

Table 10.14 Confidence Interval and Optimal Setting Of Factors

Level	A	B	C	D	E	F	G	H	D×E
1	25.09	28.09	27.74	31.24	29.66	27.39	27.66	30.35	32.40
2	30.03	27.97	28.06	27.39	25.49	26.31	27.82	28.71	30.12
3	27.04	26.10	26.35	23.52		29.23	26.70	23.04	23.81
4									30.09
5									24.67
6									23.22
Upper confidence level	+1.99	+1.99	+1.99	+1.99	+1.99	+1.99	+1.99	+1.99	+2.82
Lower confidence level	-1.99	-1.99	-1.99	-1.99	-1.99	-1.99	-1.99	-1.99	-2.82
Optimal Level	2	1	2	1	1	3	2	1	1

Table 10.15 Decision Matrix for Design Selection

Attribute		Alternative design			
		Option 1	Option 2	Option 3	Option 4
Safety (0.6667)		0.6891	0.7372	0.9455	0.9705
Cost (0.3333)	Designer 1 (0.1667)	1	0.7083	0.7004	0.5073
	Designer 2 (0.3333)	1	0.6777	0.6777	0.4929
	Designer 3 (0.3333)	1	0.7390	0.7013	0.5071
	Designer 4 (0.1667)	1	0.7013	0.7013	0.4780

Table 10.16 Weighted Decision Matrix for Design Selection

	Option 1	Option 2	Option 3	Option 4
Safety	0.4594	0.4915	0.6304	0.6470
Cost by Designer 1	0.0556	0.0394	0.0389	0.0282
Cost by Designer 2	0.1111	0.0753	0.0753	0.0548
Cost by Designer 3	0.1111	0.0821	0.0779	0.0563
Cost by Designer 4	0.0556	0.0390	0.0390	0.0266

Table 10.17 Pairwise Comparison 1

	Safety	Cost
Safety	1	2
Cost	½	1

Table 10.18 Pairwise Comparisons between Designers

	Designer 1	Designer 2	Designer 3	Designer 4
Designer 1	1	½	½	1
Designer 2	2	1	1	2
Designer 3	2	1	1	2
Designer 4	1	½	½	1

Table 10.19 Pairwise Comparisons of Designs on Safety

	Option 1	Option 2	Option 3	Option 4
Option 1	1	0.9348	0.7288	0.7100
Option 2	1.0697	1	0.7797	0.7596
Option 3	1.3721	1.2826	1	0.9742
Option 4	1.4085	1.3165	1.0264	1

Table 10.20 Scores of Design Options on Safety

	Option 1	Option 2	Option 3	Option 4
Safety	0.2062	0.2206	0.2829	0.2904

Table 10.21 Pairwise Comparisons of Designs on Cost by Designer 1

	Option 1	Option 2	Option 3	Option 4
Option 1	1	1.4118	1.4278	1.9712
Option 2	0.7083	1	1.0113	1.3962
Option 3	0.7004	0.9888	1	1.3806
Option 4	0.5073	0.7162	0.7243	1

Table 10.22 Pairwise Comparisons of Designs on Cost by Designer 2

	Option 1	Option 2	Option 3	Option 4
Option 1	1	1.4756	1.4756	2.0288
Option 2	0.6777	1	1	1.3749
Option 3	0.6777	1	1	1.3749
Option 4	0.4929	0.7273	0.7273	1

Table 10.23 Pairwise Comparisons of Designs on Cost by Designer 3

	Option 1	Option 2	Option 3	Option 4
Option 1	1	1.3532	1.4259	1.9720
Option 2	0.7390	1	1.0538	1.4573
Option 3	0.7013	0.9489	1	1.3830
Option 4	0.5071	0.6862	0.7231	1

Table 10.24 Pairwise Comparisons of Designs on Cost by Designer 4

	Option 1	Option 2	Option 3	Option 4
Option 1	1	1.4259	1.4259	2.0921
Option 2	0.7013	1	1	1.4672
Option 3	0.7013	1	1	1.4672
Option 4	0.4780	0.6816	0.6816	1

Table 10.25 Scores of Options on Safety and Cost

		Option 1	Option 2	Option 3	Option 4
Safety (0.6667)		0.2062	0.2206	0.2829	0.2904
Cost (0.3333)	Designer 1 (0.1667)	0.3429	0.2429	0.2402	0.1740
	Designer 2 (0.3333)	0.3511	0.2379	0.2379	0.1731
	Designer 3 (0.3333)	0.3393	0.2507	0.2379	0.1720
	Designer 4 (0.1667)	0.3471	0.2435	0.2435	0.1659

Table 10.26 Aggregated Assessment of Options on Safety and Cost

	Option 1	Option 2	Option 3	Option 4
Safety (0.6667)	0.2062	0.2206	0.2829	0.2904
Cost (0.3333)	0.3451	0.2439	0.2392	0.1717

Table 10.27 Overall Assessment of Options on Safety and Cost (AHP Generated Results))

	Option 1	Option 2	Option 3	Option 4
Safety & cost	0.2525	0.2284	0.2683	0.2508

Table 10.28 Overall Assessment of Options on Safety and Cost (ER Generated Results)

	Option 1	Option 2	Option 3	Option 4
Safety & cost	0.7765	0.7407	0.9085	0.8818

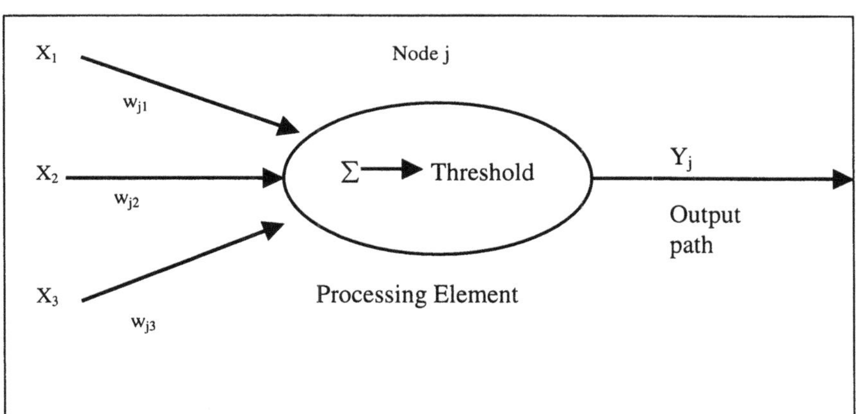

Figure 10.1 An ANN

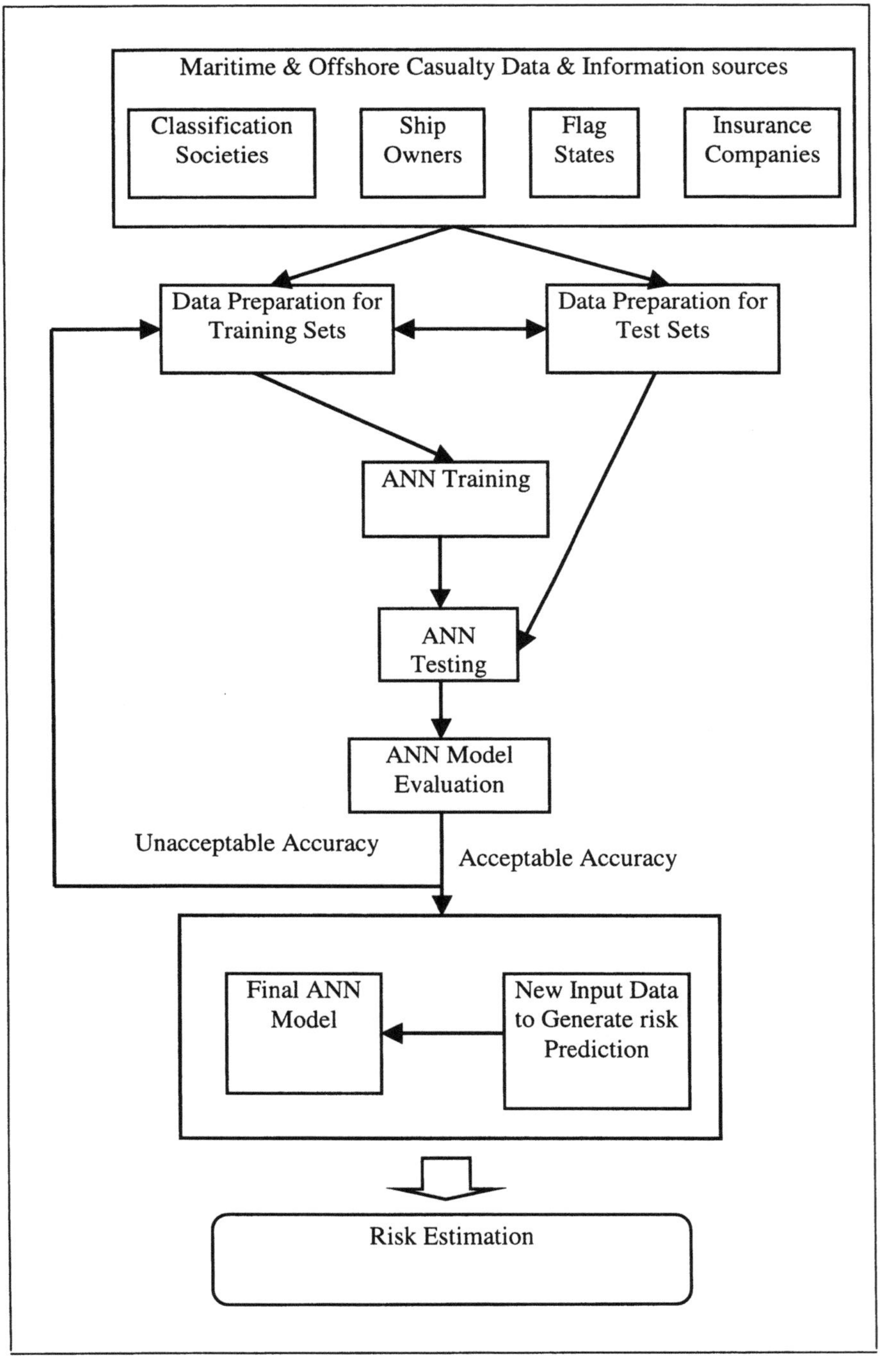

Figure 10.2 The risk estimation framework incorporating ANN

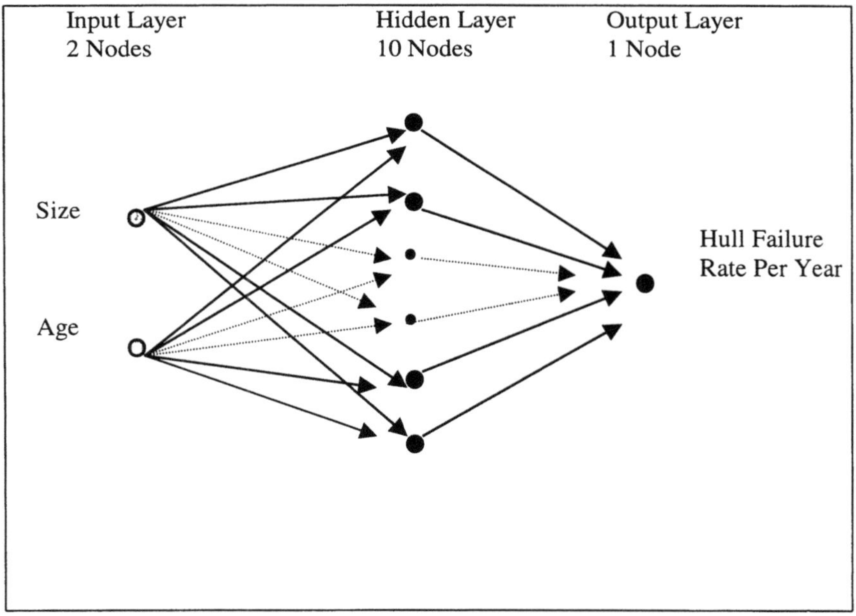

Figure 10.3 An ANN for bulk carrier hull failure prediction

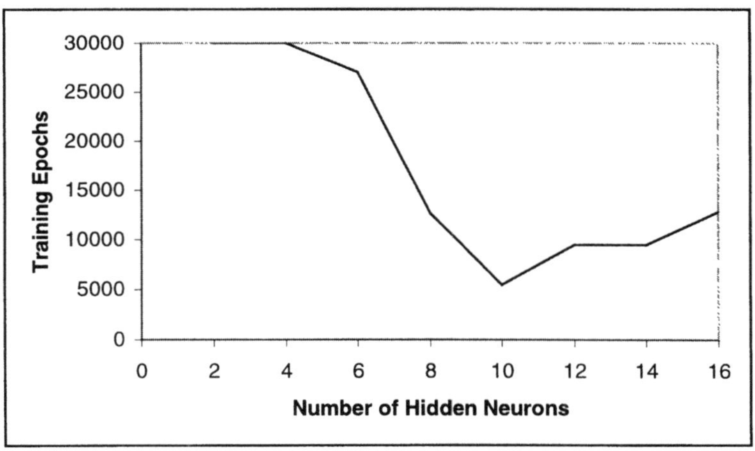

Figure 10.4 Effect of number of hidden neurons on training epoch for back-propagation learning algorithm

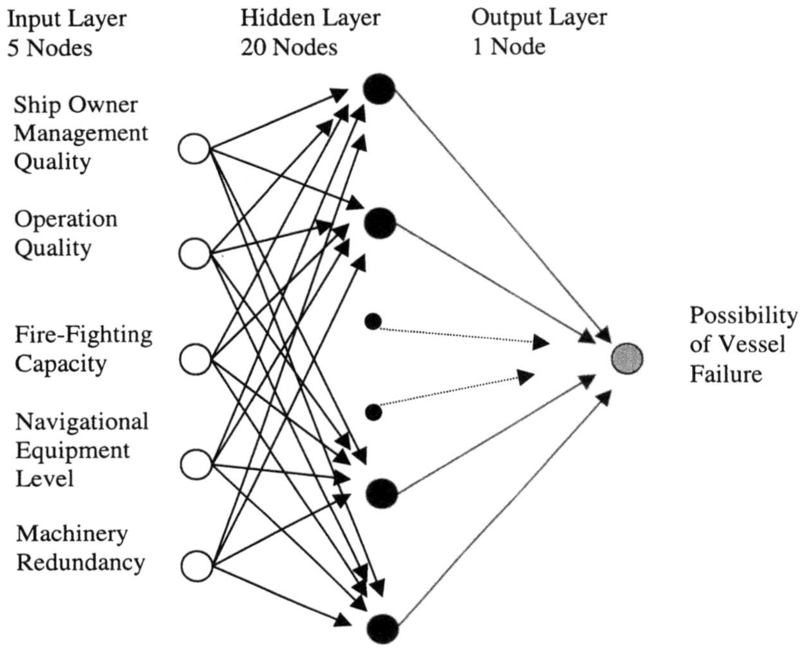

Figure 10.5 An ANN For vessel failure possibility prediction

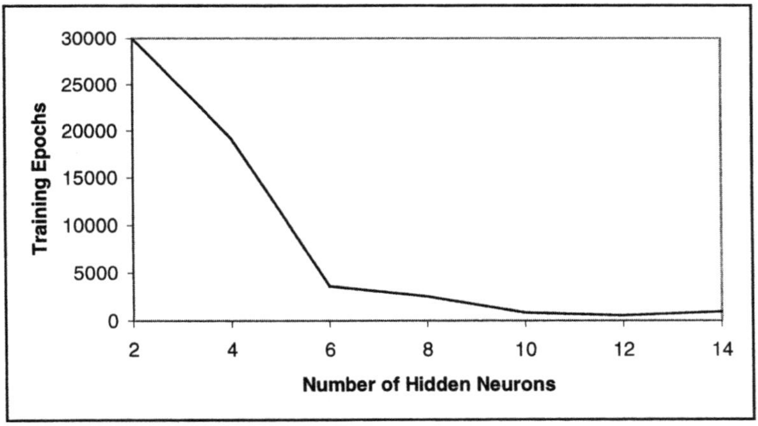

Figure 10.6 Effect of number of hidden neurons on training epochs for back-propagation learning algorithm with momentum and adaptive learning rate techniques

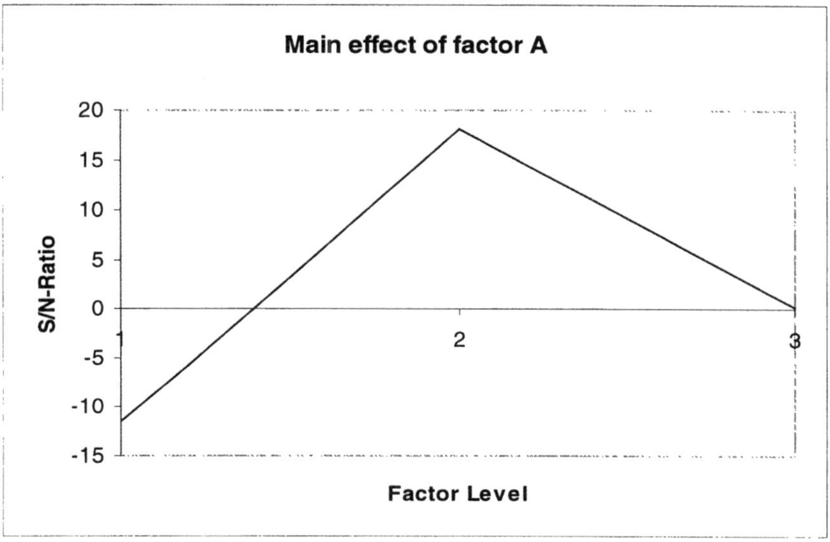

Figure 10.7 The non-linearity graph for factor A

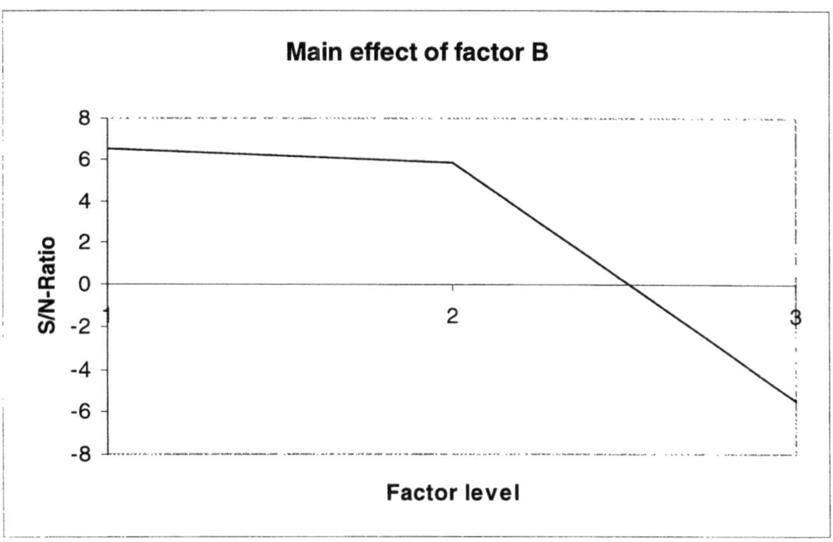

Figure 10.8 The non-linearity graph for factor B

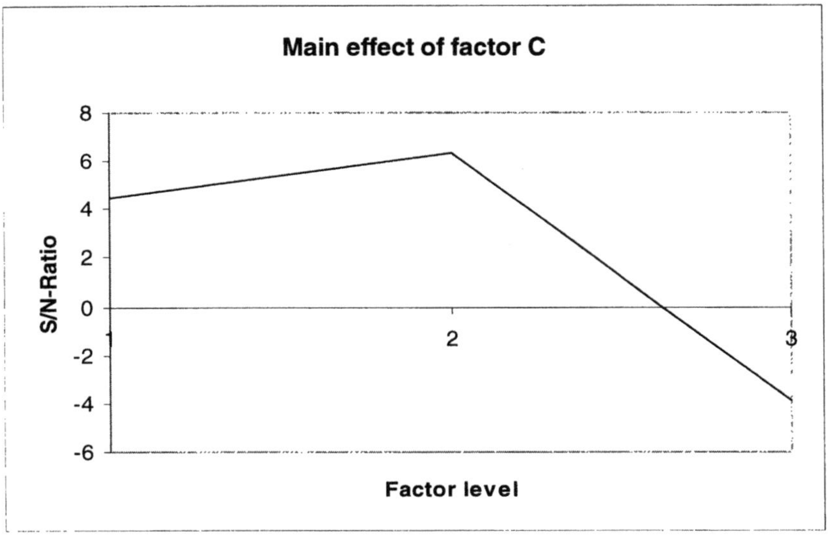

Figure 10.9 The non-linearity graph for factor C

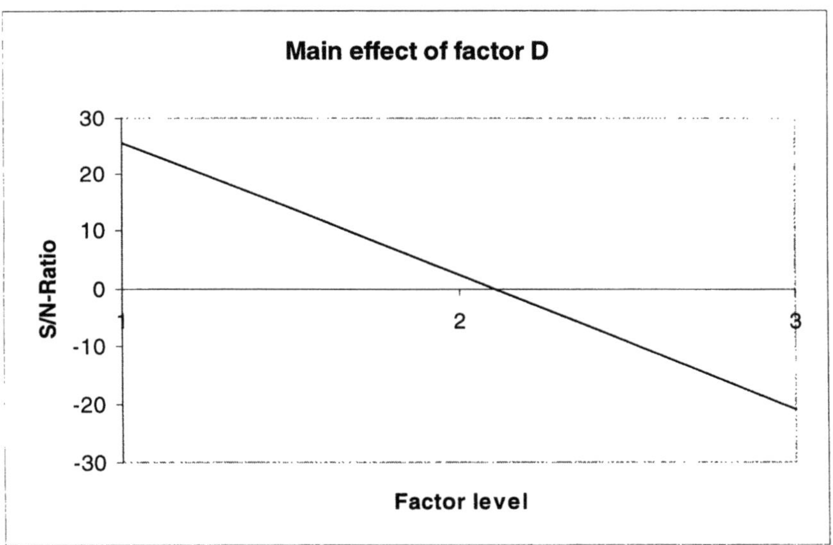

Figure 10.10 The non-linearity graph for Factor D

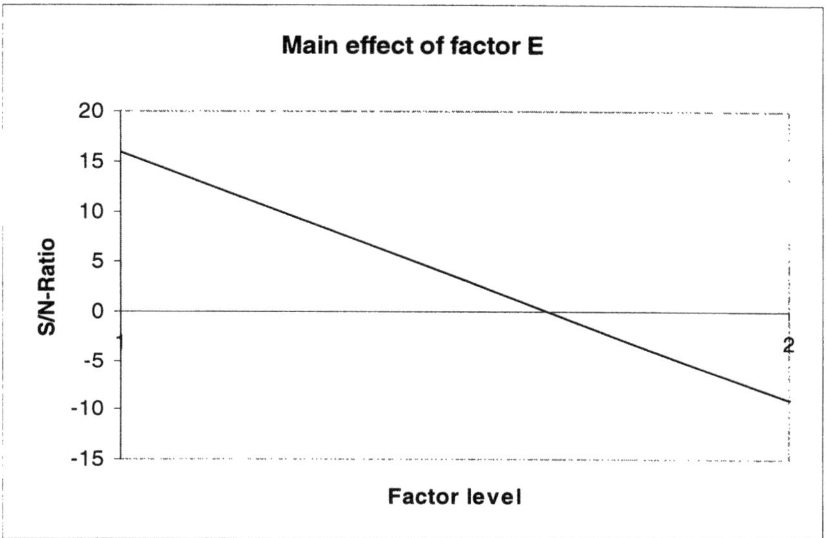

Figure 10.11 The non-linearity graph for factor E

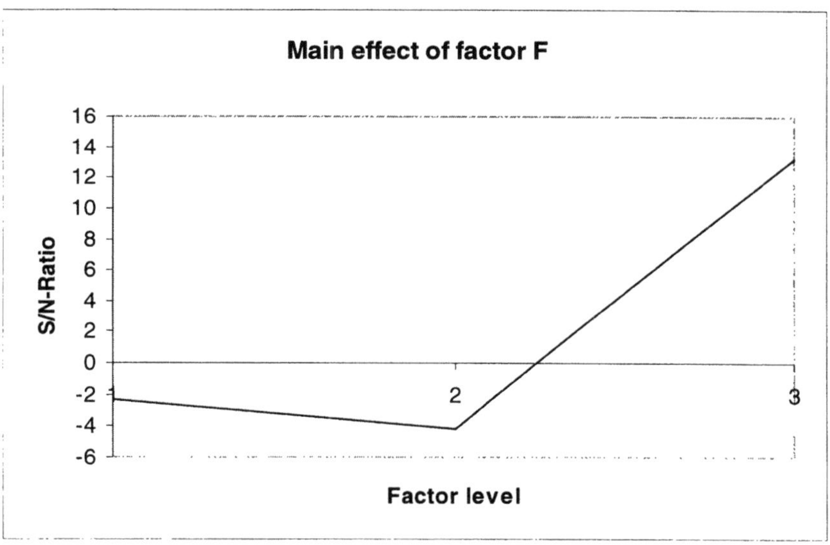

Figure 10.12 The non-linearity graph for factor F

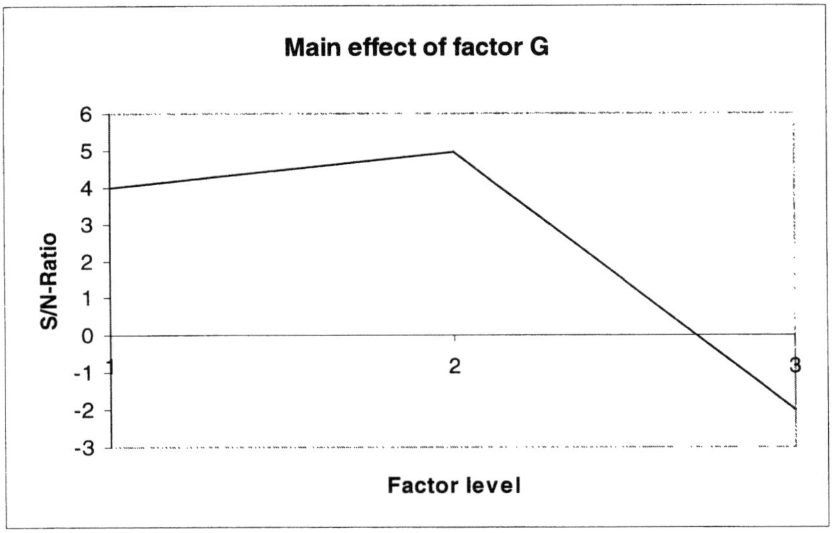

Figure 10.13 The non-linearity graph for factor G

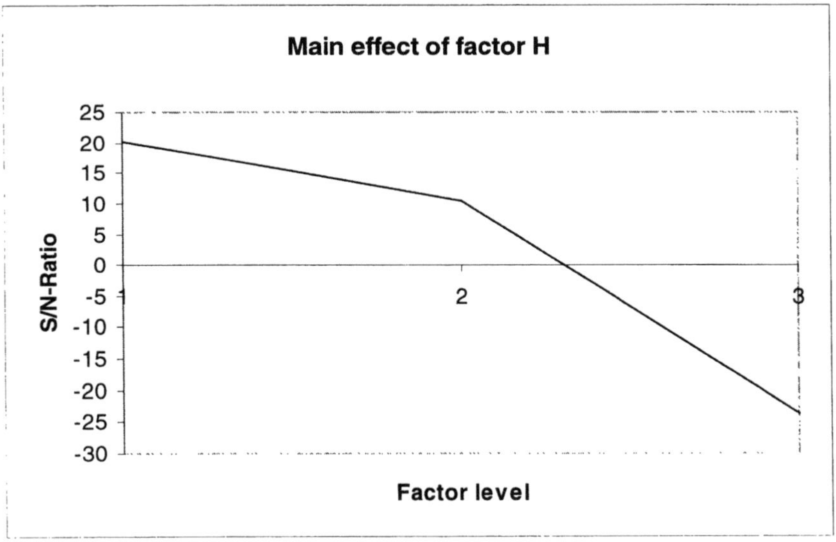

Figure 10.14 The non-linearity graph for factor H

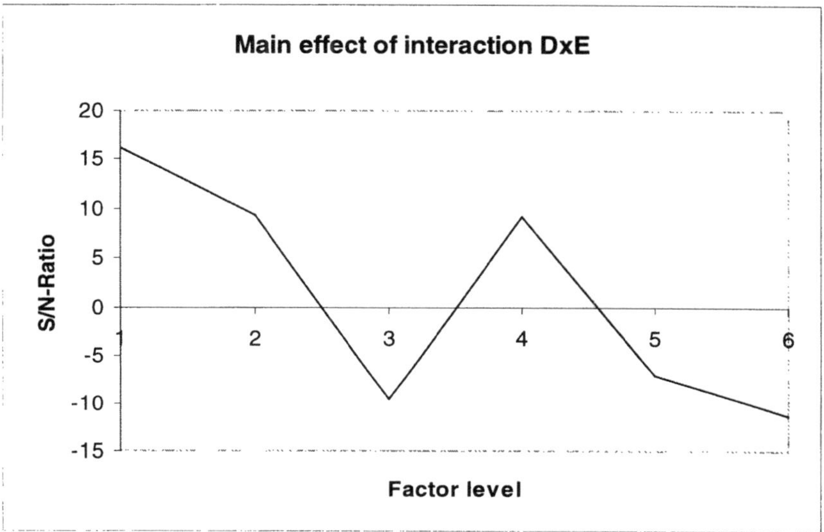

Figure 10.15 The non-linearity graph for factor D×E

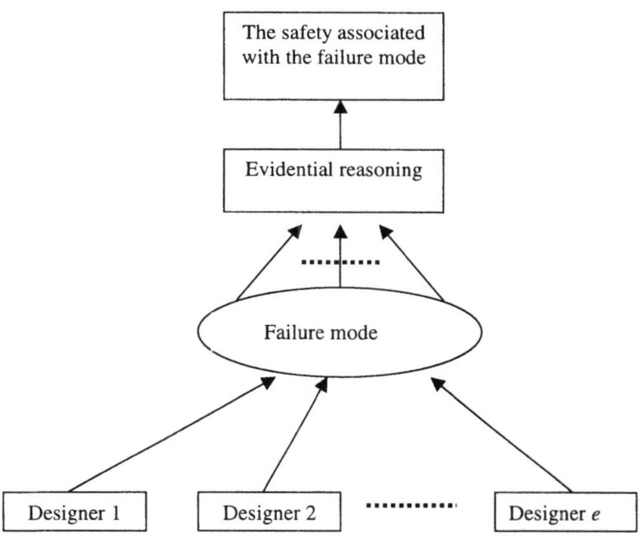

Figure 10.16 A diagram for synthesising the safety associated with a failure mode

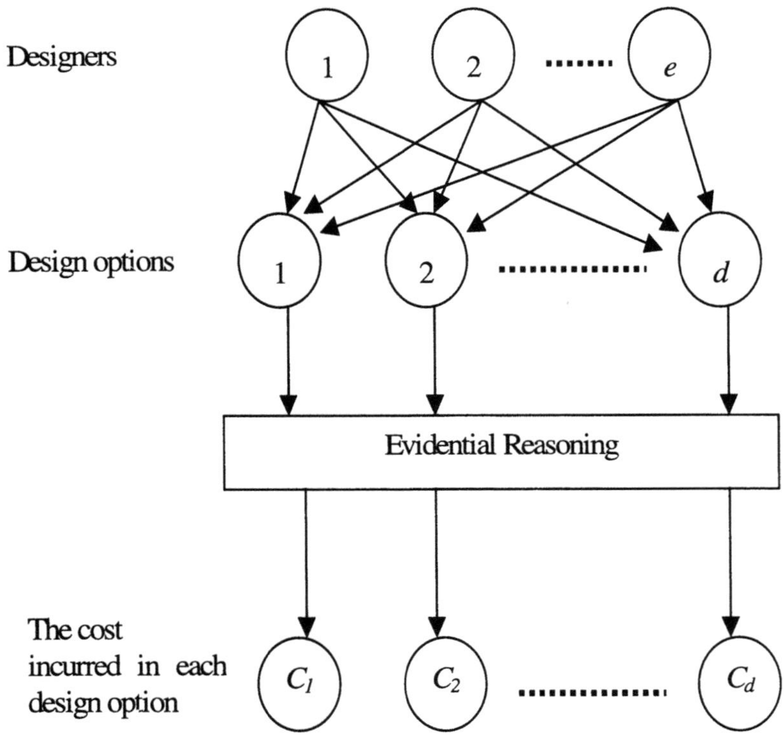

Figure 10.17 A hierarchical diagram of cost modelling

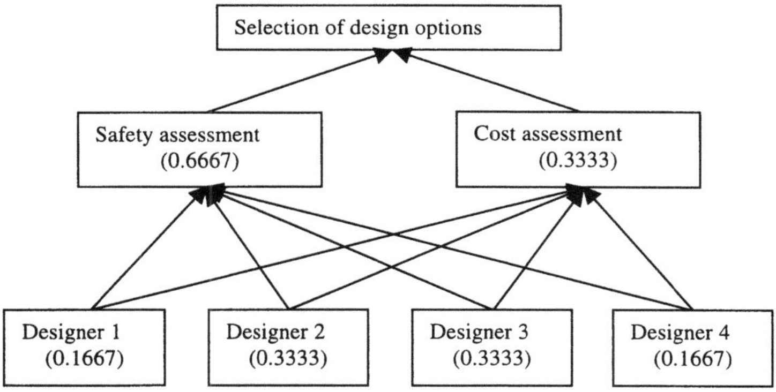

Figure 10.18 Safety and cost assessment hierarchy

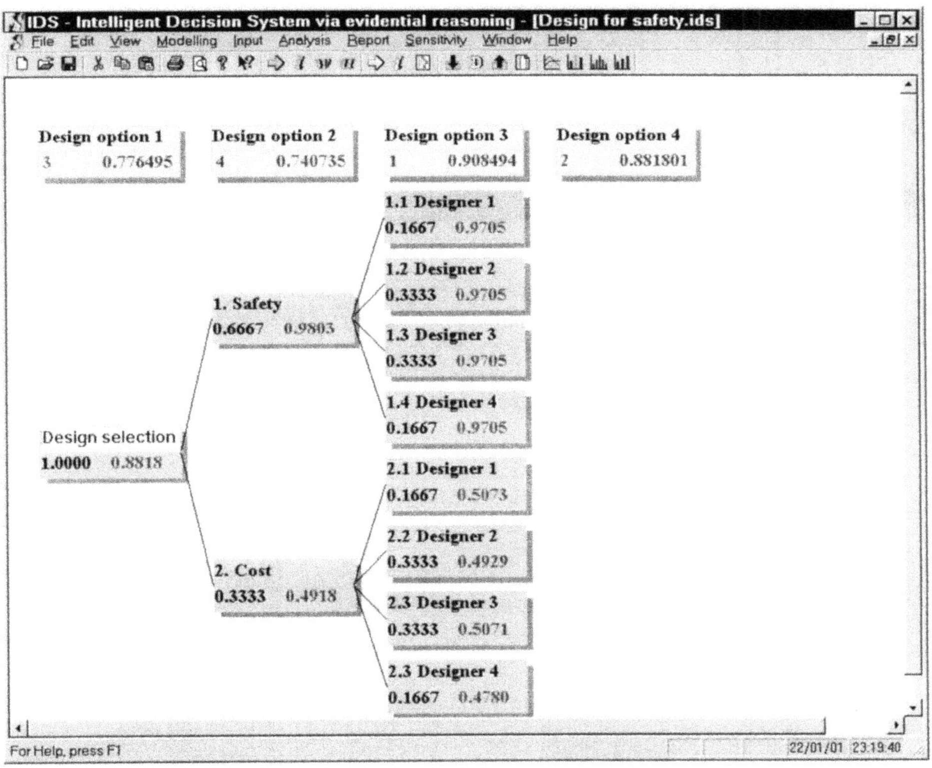

Figure 10.19 IDS Main Window for Safety & Cost Based Design Selection

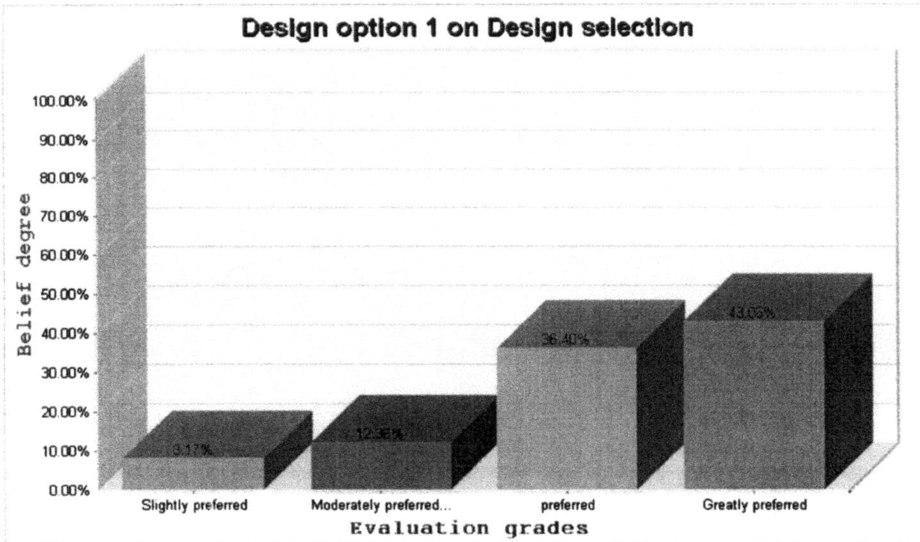

Figure 10.20 The overall distributed assessment of design option 1 generated by IDS

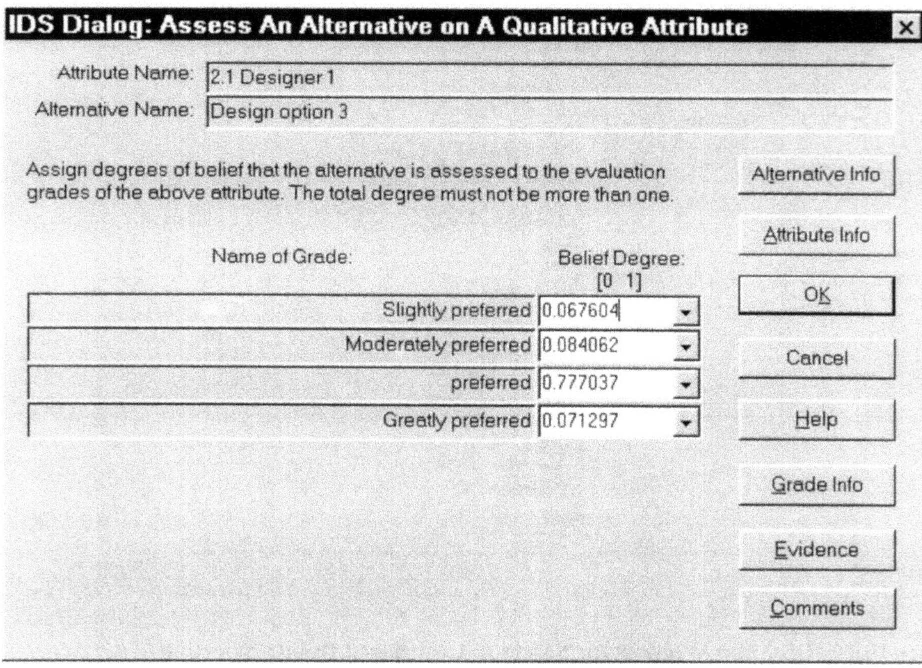

Figure 10.21 IDS Data input dialog window

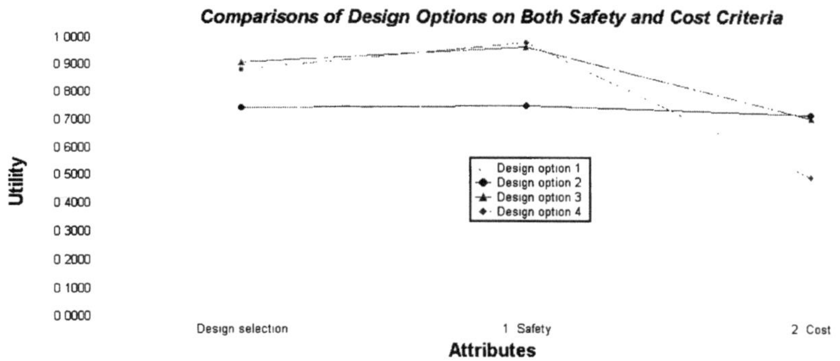

Figure 10.22 Comparison of design options on both safety and cost generated by IDS

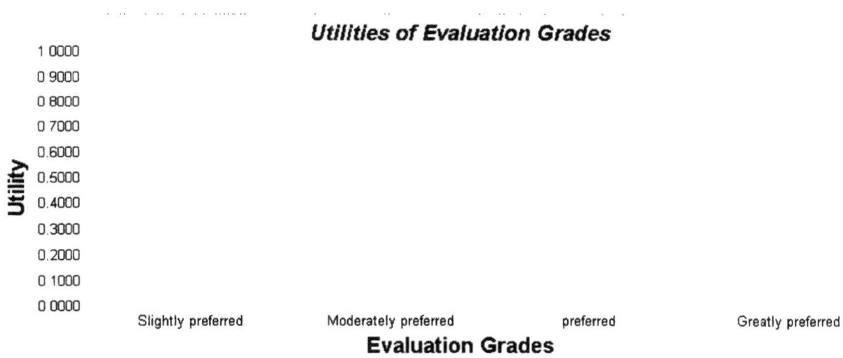

Figure 10.23 Utility curve of evaluation grades generated by IDS

Chapter 11

Conclusions

Summary

This Chapter briefly summarises that the risk assessment and decision making approaches described in previous Chapters would be of benefit in marine design and operations. The areas where further effort is required to improve the developed approaches are outlined.

The previous Chapters of this book have described the formal safety assessment framework in ship design and operations, and also a range of safety assessment and decision making approaches. The reasons behind the development of such approaches have been explained. Many valid reasons for using the developed safety analysis approaches have also been discussed.

The formal safety assessment framework has been described in a generic sense to be applicable to all design and operational problems of ships. It can be used as a basis for the development of various safety analysis methods and decision making procedures.

Obviously, in some cases, it could be time-consuming to conduct safety analysis of marine engineering systems using some of the described safety assessment and decision making approaches although more reasonable results would be obtained. It may take time to learn how to use such described approaches. It is believed that the described approaches possess enormous potential as valuable aids and effective alternatives in the areas of marine risk assessment and will gain increased usage in ship design and operations. It is also believed that practical applications of these approaches will result from utilisation by organisations that deal with safety problems, especially in situations where there are problems associated with a high level of uncertainty or insufficient safety data. The implementation of the described approaches could have a highly beneficial effect. In fact, it is widely accepted that any developed safety analysis approach should preferably be introduced into a commercially stable environment in order that the application has the chance to become established to prove feasible, otherwise it is more likely that its full potential will not be realised.

It would be useful if more test cases are applied to the described safety assessment and decision making approaches in order to further demonstrate their applicability. It would also be useful if more powerful and flexible risk modelling and decision making tools are developed to facilitate formal safety assessment of ships.

Appendix 1

Code of Practice for the Safety of Small Fishing Vessels

Check List of Requirements

Decked Vessels
10m and above Registered Length to less than 12m Registered Length

Lifejackets - 1 per person
Liferafts
2 Lifebuoys (1 with 18m buoyant line attached) or
1 Lifebuoy (fitted with 18m buoyant line) +1 buoyant rescue quoit
3 Parachute flares
2 Hand-held flares
1 Smoke signal (buoyant or handheld)
1 Fire bucket + lanyard
1 Multi-purpose fire extinguisher (fire rating 5A/34B)
1 Fire blanket (light duty) in galley or cooking area (if applicable)
1 Fire pump + Hose or
1 Fire bucket + 1 Multi-purpose fire extinguisher (fire rating 5A/34B) + 1 fixed fire extinguishing system for the machinery space
1 Multi-purpose fire extinguisher for oil fires (fire rating 13A/113B)
VHF Radio - fixed or hand held
Bilge pump
Bilge alarm
Navigation lights and sound signals
Compass
Waterproof torch
Medical kit

Notes:

I. Equipment need not be MCA approved provided it is fit for its intended purpose.
II. "Decked vessels" means a vessel with a continuous watertight weather deck that extends from stem to stern and has positive freeboard throughout, in any condition of loading the vessel.
III. VHF using DSC is highly recommended in view of cessation of the Coastguard's Channel 16 dedicated headset watch on 1st February 2005.

All Decked Vessels
Up to 10m Registered Length

Lifejackets - 1 per person
2 Lifebuoys (1 with 18m buoyant line attached) or
1 Lifebuoy (fitted with 18m buoyancy line) +1 buoyant rescue quoit
3 Parachute flares
2 Hand-held flares
1 Smoke signal (buoyant or hand held)
1 Fire bucket + lanyard
1 Multi-purpose fire extinguisher (fire rating 5A/34B)
1 Fire blanket (light duty) in galley or cooking area (if applicable)
1 Fire pump + hose or 1 fire bucket

1 Multi-purpose fire extinguisher for oil fires (fire rating 13A/113B)
VHF radio – fixed or hand held
Bilge pump
Bilge alarm
Navigation lights and sound signals
Compass
Waterproof torch
Medical kit

Notes:
I. Equipment need not be MCA approved provided it is fit for its intended purpose.
II. "Decked vessels" means a vessel with a continuous watertight weather deck that extends from stem to stern and has positive freeboard throughout, in any condition of loading the vessel.
III. VHF using Digital Selective Calling (DSC) is highly recommended in view of cessation of the Coastguard's Channel 16 dedicated headset watch on 1st February 2005.

Open Vessels
7m and above to less than 12m Registered Length

Lifejackets - 1 per person
2 Lifebuoys (1 with 18m buoyant line attached) or 1 lifebuoy (with 18m buoyant line) +1 buoyant rescue quoit
3 Parachute flares
2 Hand-held flares
1 Smoke signal (buoyant or hand held)
1 Fire bucket + lanyard
1 Multi-purpose fire extinguisher (fire rating 5A/34B)
1 Fire blanket (light duty) in galley or cooking area (if applicable)
1 Fire pump + hose or 1 fire bucket
1 Multi-purpose fire extinguisher for oil fires (fire rating 13A/113B)
VHF Radio – fixed or hand held
Bilge pump
Navigation lights and sound signals
Compass
Waterproof torch
Medical kit

Notes:
I. Equipment need not be MCA approved provided it is fit for its intended purpose.
II. VHF using Digital Selective Calling (DSC) is highly recommended in view of cessation of the Coastguard's Channel 16 dedicated headset watch on 1st February 2005.

OPEN Vessels
Less than 7m Registered Length

Lifejackets – 1 per person
1 Lifebuoy (with 18m buoyant line attached)
2 Parachute flares

2 Hand-held flares
1 Smoke signal (buoyant or hand held)
1 Fire bucket + lanyard
1 Multi-purpose fire extinguisher (fire rating 5A/34B) - if vessel has in- board engine
1 Fire blanket (light duty) if vessel has galley or cooking area
VHF Radio – fixed or hand held
Bailer
Navigation lights and sound signals
Compass
Waterproof torch
Medical kit

Notes:

I. Equipment need not be MCA approved provided it is fit for its intended purpose.

II. VHF using Digital Selective Calling (DSC) is highly recommended in view of cessation of the Coastguard's Channel 16 dedicated headset watch on 1st February 2005.

Appendix 2

Fishing Vessel (Safety Provisions) Safety Rules 1975

Arrangement of rules

PART I – GENERAL

Rule

27. Oil systems for lubricating, cooling and control – vessels of 24.4 meters in length and over.
28. Oil systems for lubricating, cooling and control – vessels of 12 meters in length and over but less than 24.4 meters in length.
29. Oil fuel installations (boilers and machinery) – general.
30. Oil fuel installations (boilers and machinery) – vessels of 24.4 meters in length and over.
31. Oil fuel installations (boilers and machinery) – vessels of 12 meters in length and over but less than 24.4 meters in length.
32. Oil fuel installations (cooking ranges and heating appliances).
33. Ventilation.
34. Liquefied petroleum gas installations (cooking ranges and heating appliances).
35. Storage of flammable liquids, toxic liquids, toxic gases and compressed gases.

E – BILGE PUMPING ARRANGEMENTS

36. Requirements for vessels of 24.4 meters in length and over.
37. Requirements for vessels of 12 meters in length and over but less than 24.4 meters in length.

F – ELECTRICAL EQUIPMENT AND INSTALLATIONS

38. General.
39. Distribution systems.
40. Electrical precautions.
41. Requirements for vessels of 24.4 meters in length and over.
42. Requirements for vessels of 12 meters in length and over but less than 24.4 meters in length.
43. Accumulator (storage) batteries and associated charging equipment.

G – MISCELLANEOUS PLANT AND EQUIPMENT

44. Watertight doors.
45. Steering gear – vessels of 24.4 meters in length and over fitted with rudders.
46. Steering gear – vessels of 12 meters in length and over but less than 24.4 meters in length fitted with rudders.
47. Steering gear – vessels of 12 meters in length and over fitted with steering devices other than rudders.
48. Electrical and electro-hydraulic steering gear.
49. Communication between wheelhouse and engine room – vessels of 24.4 meters in length and over.
50. Controllable pitch propellers.
51. Refrigerating plants.
52. Anchors and chain cables.
53. Spare gear.
54. Winches, tackles and lifting gear.

H – STRUCTURAL FIRE PROTECTION AND FIRE DETECTION

55. Structural fire protection – general.
56. Structural fire protection – vessels with hulls constructed of steel or other equivalent material.
57. Structural fire protection – vessels with hulls constructed of glass reinforced plastic.

92. Rations for lifeboats.
93. Security of equipment and rations in lifeboats, Class C boats and inflatable boats.
94. Equipment and rations for life rafts.
95. General provisions relating to the stowage and handling of life-saving appliances.
96. Stowage and handling of lifeboats and Class C boats.
97. Stowage and handling of inflatable boats.
98. Stowage and handling of life rafts, lifebuoys and lifejackets.
99. Embarkation into lifeboats, Class C boats, inflatable boats and life rafts.
100. Storage of pyrotechnic distress signals.

B – FIRE APPLIANCES

101. Requirements for vessels of 60 meters in length or over.
102. Requirements for vessels of 45 meters in length and over but less than 60 meters in length.
103. Requirements for vessels of 24.4 meters in length and over but less than 45 meters in length.
104. Requirements for vessels of 21 meters in length and over but less than 24.4 meters in length.
105. Requirements for vessels of 9 meters in length and over but less than 21 meters in length.
106. Requirements for vessels less than 9 meters in length.
107. Requirements for fire pumps.
108. Requirements for the fire main, water service pipes and hydrants.
109. Requirements for fire hoses, nozzles, etc.
110. Requirements for fire extinguishers.
111. Requirements for fire alarm and fire detection systems.
112. Requirements for fixed pressure water-spraying systems for machinery spaces.
113. Requirements for fixed fire smothering gas and steam installations.
114. Requirements for fixed foam fire extinguishing installations.
115. Requirements for fireman's outfits.
116. Means for stopping machinery, shutting off fuel oil suction pipes and closing openings.
117. Fire control plans.
118. Availability of fire-fighting appliances.

C – MUSTERS AND DRILLS

119. Muster list.
120. Training.
121. Inspections.

PART IV – EXCEPTIONAL PROVISIONS

122. Exceptional provisions.

PART V – SURVEYS AND CERTIFICATES

123. Surveys and periodical inspections.
124. Surveys.
125. Surveyor's report and declaration of survey.
126. Issue and form of fishing vessel certificates.
127. Duration of certificates.

Appendix 3

Influence Diagram

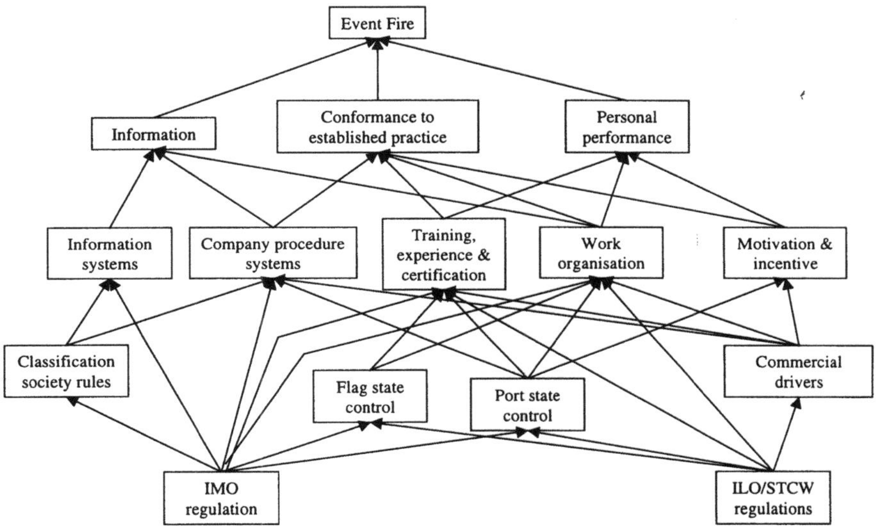

SUBJECT INDEX

www.ingramcontent.com/pod-product-compliance
Ingram Content Group UK Ltd.
Pitfield, Milton Keynes, MK11 3LW, UK
UKHW020238290926
15470UKWH00002B/614

9 780080 441481